CHROMIC MATERIALS

CHROMIC MATERIALS

Fundamentals, Measurements, and Applications

Michal Vik
Aravin Prince Periyasamy

Edited by
Martina Viková

Apple Academic Press Inc.
3333 Mistwell Crescent
Oakville, ON L6L 0A2 Canada

Apple Academic Press Inc.
9 Spinnaker Way
Waretown, NJ 08758 USA

Library and Archives Canada Cataloguing in Publication

Vik, Michal, author

Chromic materials : fundamentals, measurements, and applications / Michal Vik, Aravin Prince Periyasamy ; edited by Martina Viková.

Includes bibliographical references and index.

Issued in print and electronic formats.

ISBN 978-1-77188-680-2 (hardcover).--ISBN 978-1-351-17100-7 (PDF)

1. Chromic materials. 2. Color.   I. Viková, Martina, author, editor I. Periyasamy, Aravin Prince, author III. Title.

TA455.C45V55 2018   547'.135   C2018-903676-1   C2018-903677-X

Library of Congress Cataloging-in-Publication Data

Names: Vik, Michal, author.

Title: Chromic materials : fundamentals, measurements, and applications / Michal Vik, PhD, Aravin Prince Periyasamy ; edited by Martina Viková.

Description: Toronto ; New Jersey : Apple Academic Press, 2018. | Includes bibliographical refer-ences and index.

Identifiers: LCCN 2018027225 (print) | LCCN 2018030140 (ebook) | ISBN 9781351171007 (ebook) | ISBN 9781771886802 (hardcover : alk. paper)

Subjects: LCSH: Chromic materials.

Classification: LCC TA455.C45 (ebook) | LCC TA455.C45 V55 2018 (print) | DDC 620.1/12685f--dc23

LC record available at https://lccn.loc.gov/2018027225

Apple Academic Press also publishes its books in a variety of electronic formats. Some content that appears in print may not be available in electronic format. For information about Apple Academic Press products, visit our website at **www.appleacademicpress.com** and the CRC Press website at **www.crcpress.com**

CONTENTS

About the Authors...*vii*

About the Editor..*ix*

List of Abbreviations..*xi*

Preface..*xiii*

Introduction...*xv*

1. **Basic Terms**...1

    Michal Vik

2. **Type of Chromic Materials**...35

    Martina Viková

3. **Production of Chromic Materials**..109

    Aravin Prince Periyasamy and Martina Viková

4. **CIE Colorimetry**..155

    Michal Vik

5. **Instrumentation**...221

    Michal Vik

6. **Spectrophotometry of Color Change**......................................283

    Martina Viková

7. **Testing of Chromic Materials**...371

    Aravin Prince Periyasamy and Martina Viková

    Index...*407*

ABOUT THE AUTHORS

Michal Vik, PhD

*Associate Professor, Department of Material Science,
Technical University of Liberec, Czech Republic*

Michal Vik, PhD, is an Associate Professor in the Department of Material Science at the Technical University of Liberec, Czech Republic, and research consultant for the Centre for the Development of Engineering Research VÚTS, a.s. His scientific activities are in the areas of color science (color and appearance measurement, color difference formula development, quality control, and development and design of instruments), textile materials science (smart materials, advanced microscopy), and textile finishing (surface modification: plasma, photo-polymerization). Dr. Vik is a member of the Optical Society of America and has held positions with the Czech Society of Textile Chemists and Colorists and the ČNK CIE (Czech National Committee) (a national society of the International Commission on Illumination – CIE). He is a member of Division 1 of the International Commission on Illumination (CIE) and member of the CIE Technical Committees (TC1-55, TC1-63, TC1-72, TC1-95, TC2-61). He is an expert member of the European Technology Platform-Textile. Currently, he is head of the Colorimetry Group of the Czech Republic and Laboratory Color and Appearance Measurement of the Department of Material Engineering Technical University of Liberec. He is author or co-author of six books, about 20 scientific papers published in journals, more than 150 scientific contributions on the international conferences, and five patents.

Aravin Prince Periyasamy
Technical University of Liberec, Czech Republic

Aravin Prince Periyasamy is in the PhD degree program at the Technical University of Liberec, Czech Republic, under the supervision of Dr. Martina Viková and Dr. Michal Vik, on investigating the issues of kinetic measurement of photochromic textiles. His research interests include color chemistry; fiber production; kinetic measurement of photochromism and theoretical simulations of photochromic processes; production of photochromic materials with various techniques, namely mass coloration, screen-printing, and digital printing; and electrospinning. He is the author of three books and many articles, some of which were published in peer-reviewed journals. He has also presented several papers at professional conferences. He holds BTech and MTech degrees in textile engineering from the Anna University, Chennai, India.

ABOUT THE EDITOR

Martina Viková, PhD

Associate Professor, Material Science, Department of Material Engineering, Technical University of Liberec, Czech Republic

Martina Viková, PhD, is an Associate Professor of Material Science, Department of Material Engineering, Technical University of Liberec, Czech Republic. Her research interests include color chemistry, application of photochromic and thermochromic systems, synthesis of novel coloring matters, and dyeing of textiles. She has been the head of the research group IA3 of the project OP VaVpI "Innovative Products and Environmental Technologies." She is currently vice-head of the Colorimetry Group of the Czech Republic and Laboratory Color and Appearance Measurement of the Department of Material Engineering, Technical University of Liberec. She is author or co-author of three books, about 15 scientific papers published in journals, more than 80 scientific contributions at the international conferences, and five patents. She graduated with an MSc degree in textile engineering from the Technical University of Liberec and a PhD from Herriot-Watt University, School of Textile and Design, United Kingdom.

LIST OF ABBREVIATIONS

ANSI	American National Standards Institute
AOS	accessory optic system
BIS	Bureau of Indian Standards
BRDF	bidirectional reflectance distribution function
BSI	British Standards Institution
BSN	Badan Standard isasi Nasional
BT	black trap
C	chroma
CA	cellulose acetate
CCT	correlated color temperature T_{CP}
CEN	European Committee for Standardization
CES	Comisión Española de Solideces
CIP	Commission Internationale de Photometrie
CMOS	complementary metal-oxide semiconductor
CO	cotton
D	daylight
DEK	Deutsche Echtheits Kommission
DIN	Deutsches Institutfür Normung
DP	degree of polymerization
DR	drawing ratio
DSC	differential scanning calorimetry
EOS	Egyptian Organization for Standardization and Quality
FOV	field of view
FWHM	full width at half maximum
GUM	Guide to the Expression of Uncertainty in Measurement
HALS	hindered amine light stabilizers
I	isotropic
IPQ	Instituto Português da Qualidade
ISO	International Organization for Standardization
JISC	Japanese Industrial Standards Committee
K	kelvins
KATS	Korean Agency for Technology and Standards
KEBS	Kenya Bureau of Standards

LAV	large possible aperture
LC	liquid crystal
LCAM	Laboratory of Color and Appearance Measurement
LCST	lower critical solution temperature
LD	leuco dye
LGN	lateral geniculate nucleus
M	medium
MCDM	mean color difference from mean
ME	microencapsulation
MEMS	micro electromechanical systems
MFI	melt flow index
NBN	Bureau de Normalization
NEN	Nederland's Normalisatie-instituut
NMMO	N-methylomorpholine
PD	polymer dispersion
PDMS	polydimethylsiloxane
PKN	Polish Committee for Standardization
PLA	poly(lactic acid)
PMA	poly(methyl acrylate)
PMMA	poly(methyl meth-acrylate)
PMT	photomultiplier tube
PP	polypropylene
PS	photostationary state
PS	polystyrene
SABS	South African Bureau of Standards
SAC	Standardization Administration of China
SDC	Society of Dyers and Colorists
SFS	Finnish Standards Association SFS
SIS	Swedish Standards Institute
SNV	Schweizerische Normen-Vereinigung
SPD	spectral power distribution
SVHC	substance of very high concern
TISI	Thai Industrial Standards Institute
TLC	thermochromic liquid crystal
TSR	tele-spectroradiometer
UNI	EnteNazionaleItaliano di Unificazione
WS	white standard

PREFACE

We many times come across words like "smart materials" and "intelligent materials." Chromic materials, we can understand, are an important part of smart materials. The expression "chromic materials" refers to materials that show color change depending upon an external stimulus. The most important chromic phenomena—photochromism, thermochromism, ionochromism, and electrochromism—are dealt with in individual sections in this volume, each providing a description of the physicochemical principles underlying the color changes and a discussion of the molecular structures of the most important colorant classes.

A wide range of materials that exhibit color change effects have been investigated in recent decades, and numerous products have been introduced commercially. Currently, chromic materials are most commonly used in high technology. Some of the applications have been already developed to the technological level and have been commercialized successfully, whereas some are in the developmental stage. One of most common examples is color changeable glasses that are transparent in the shade or inside and become dark colored in the presence of sunlight. Due to this color change such adaptive glasses protect eyes from intense sunlight, UV radiation, respectively. Other examples of successful applications are temperature and humidity sensors that are broadly used in the food industry for evaluation of high quality and safe food products. Not least we can mention an electrochromic device (ECD) that controls optical properties such as optical transmission, absorption, reflectance, and/or emittance in a continual but reversible manner on application of a voltage (electrochromism).

The main goal of this book is to describe the phenomenon of color changeable materials from the point of application, spectrophotometry of color changeable materials, instrumentation, and their testing. We also discuss how to control quality of materials as well and also how to measure objectification of colorimetric parameters and absorption and remission spectra by using standardized colorimetric systems (CIE XYZ, CIE LAB, and CIE CAM).

Also in this book, we describe our experience with measurement of color changeable materials, their kinetic behavior in exposure, and decay phases from the point of colorimetric and spectrophotometric characteristics. And we try to show the way for quality control of these chromic materials.

We hope that this book will be helpful for researchers and innovators in this area and will help to facilitate good experiments, which are comparable, exact, adequate, and precise with a small error of measurement.

—Martina Viková
Editor

INTRODUCTION

Color changeable materials refer to stimuli-induced reversible change in color. Such stimuli can be light, temperature, chemicals, electricity, mechanical impact, etc. Based on that we can speak about photochromic, thermochromic, electrochromic, and piezochromic materials. Color changeable materials, we can call as molecular devices, as the name implies, are composed of molecules that are designed to accomplish a specific function. The simplest device that can be imagined is a switch. The defining characteristic of a switch is that of bistability, means it has an "ON" and "OFF" state. Thus, in its "ON" state, the switch must either perform some functions or allow another device to perform its function. In the "OFF" state the system must totally impede the function.

The bistability might be based on various properties of molecules like electron transfer, isomerization, differences in complexation behavior, and photocyclization; whereas light, heat magnetic or electric fields, chemical reactions, etc., can be used to achieve the change in bistable state. Molecules which experience a color change upon exposure to external light, heat or electrical stimuli are termed photochromic, thermochromic, and electrochromic compounds, respectively.

Fritzsche first discovered the phenomenon of photochromism when he observed the photo bleaching of tetracene. Photochromism is defined as a light-induced reversible transformation of a chemical species between two forms that have different absorption spectra. Molecules that are capable to interconvert reversibly interconverting between two isomers that differ in color are termed as photochromic compounds. In the above-mentioned example, the two isomers **A** and **B** can reversibly interconvert between the two forms by irradiating with appropriate wavelength of light. Irradiation of isomer **A** with one wavelength of light (hv_1) results in a photoisomerization reaction producing isomer **B**. The reverse reaction is carried out by irradiation of isomer **B** with a different wavelength of light (hv_2).

The term 'photochromism' was originally suggested by Hirshberg in 1950. The study of the excited states derived from the photochromic response and the transient species involved in the photoreactivity of photochromic molecules was facilitated by the development of techniques such as flash spectroscopy and laser photophysical methods. The interest in the applications of photochromic systems increased in the 1980s when the obstacle of the low fatigue resistance of photochromic compounds was overcome by synthesizing stable organic photochromic compounds, such as spirooxazine, naphthopyran, and mainly by diarylethene derivatives. Since then, commercial applications of photochromism, such as the plastic photochromic ophthalmic lenses, have become widespread.

Similarly, thermochromic pigments are materials which change color as a function of temperature. Types of pigments are known that change color either reversibly or irreversibly. The materials which change color reversibly and may be used in textiles are leuco dye based and cholesteric liquid crystal thermochromic pigments. The leuco dye based thermochromic pigments generally change from colored to colorless or to another color with an increase in temperature. The cholesteric liquid crystals exhibit 'color play' by passing through the whole spectrum with an increase in temperature. Thermochromic pigments have been commercialized since the late 1960s and used, for example, in thermographic recording materials. The thermochromic pigments are also used as temperature indicators such as measuring the body temperature, in food containers to determine the temperature or history of the food storage, in medical thermography for diagnosis purposes, in thermal mapping of engineering materials to diagnose faults in product design and in mechanical performance, in the cosmetic industry for moisturizing and as a carrier for vitamins, etc.

Coloration made by electroactive species that exhibits new optical absorption bands in accompaniment with an electron-transfer or redox reactions in which it either gains or loses an electron is termed as electrochromism. The most widely studied inorganic system is solid tungsten trioxide WO_3, also called as tungsten oxide or tungstic oxide, comprising WVI, in which the introduction of small amount of WV allows intense optical absorption or, with particular conditions, reflection. The adjective 'electrochromic' is often applied to a widely differing variety of fenestrate and device applications. For example, a routine web search using the phrase 'electrochromic window' yielded many pages describing a suspended-particle-device (SPD) window. Some SPD windows are also termed 'Smart Glass' – a term that, until now, has related to genuine electrochromic systems.

Chemochromic materials, frequently also called as halochromic, are found in a range of forms, resulting in a variety of applications – the most common form it is found in is a dye. The smart material is used in litmus paper, which detects the acidity and alkalinity of chemicals. The chemochromic chemical in the dye has a low acidity, and will change color depending on its pH level. For example, methyl red is yellow if pH > 6 and red if pH < 4. For example, in pregnancy tests, the chemicals used detect and respond to human chorionic gonadotropin (HCG) traces found in a pregnant woman's urine. Chemochromic dyes are also used to show the ripeness of fruit, as the chemical reacts with gases released by the fruit as it ripens.

The more sensitive the chemical, the better the product is for the user. In this case understanding how to measure dynamic colors correctly should be useful tool.

CHAPTER 1

BASIC TERMS

MICHAL VIK

CONTENTS

Abstract...1
1.1   Light...2
1.2   Blackbody Radiation..4
1.3   Interaction of Light with Matter 8
1.4   Lambert-Beer Law..10
1.5   Photometry...13
1.6   Color 15
1.7   Human Color Vision 19
Keywords...31
References..32

ABSTRACT

Measurement of color change can be covered by colorimetry. There is one fundamental, simple question at the heart of any discussion about color:

"How can a color be quantitatively measured?"

The answer and its ramifications comprise this book. This question is by no means as trivial as it seems, for color is not a tangible, objective entity. Rather, it is a sensation received by the brain. Indeed, it might seem that measuring color is essentially impossible; pain is a sensation, yet it certainly cannot be measured quantitatively. A restatement of the issue, which also points to an answer, is provided by the following sentence: The color of a physical sample depends primarily on the composition of the light reflected from it which enters the observer's eyes. This is governed by two factors: the reflectance characteristics of the sample and the composition of the light falling on it.

Colorimetry uses terms from three different scientific fields, namely Physics, Psychophysics, and Psychosensorics. From physics, we need to understand terms such as light, spectral power distribution, etc. From the psychophysics point of view, the important terms are chromaticity, color quality of light, and object color. And from the psychosensorial side, we will speak about the color itself. All these terms are important for understanding the connections between them and structure of the whole CIE colorimetry. The Commission International de l'Eclariage or CIE developed the most influential system for the description of color. This system is based on using a standard source of illumination and a standard observer. The system is used to obtain CIE standard-observer curves for the visible spectrum of the tristimulus values, which are converted to the unreal primaries X, Y, and Z [1]. Firstly, to define what is meant by color, we need to discuss what is meant by light, because color cannot be seen in the dark.

1.1 LIGHT

Light is radiant energy (electromagnetic radiation) that causes a visual sensation. Light is defined as electromagnetic radiation between 380 and 780 nm in wavelength. The present day description of light is "Light is a form of energy that displays dual nature as a particle or a wave, which is characterized by its spectrum of colors." Simple or monochromatic light (light of specific wavelength) can be considered as propagation of the electric and magnetic vectors at a mentioned speed [2]. Both vectors are in a rectangular position to each other and to the direction of motion as shown in Figure 1.1.

It is possible to describe monochromatic light by wavelength λ, frequency v, and/or by energy E. The following relationship connects these terms:

$$E = hv = h\frac{c}{\lambda} \qquad (1.1)$$

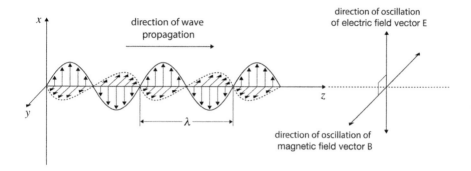

FIGURE 1.1   Representation of electric E and magnetic B vectors of light.

where E is the energy of a photon; $h = 6.626176 \times 10^{-34}$ J.Hz^{-1} is Planck's constant; $c = 2.9979245 \times 10^8$ m.s^{-1} is finite speed of light; and v and λ are frequency and wavelength of the radiation, respectively.

Many properties of light, particularly those related to absorption and emission, cannot be fully explained by the wave theory, and there is a commonly accepted concept that light really exist as a series of energy packets commonly known as photons. In modern physics, matter and energy are two sides of description of the same reality [3].

An example of this is the well-known Einstein equation:

$$E = mc^2 \tag{1.2}$$

where m is photon gravity, which relates to a specific frequency.

Based on substitution of Eq. (1.2) into Eq. (1.1), we obtain the following relationship:

$$m = \frac{hv}{c^2} = \frac{h}{\lambda c} \tag{1.3}$$

which indicates that light has higher energy when it has lower wavelength.

Brighter light means it contains more photons. When there are enough photons (high intensity), there exists a correspondence between the wavelength of light and its hue, as perceived by an observer. Light consisting of a single wavelength is monochromatic light, which looks to the eye as a pure color. The shortest perceptible wavelengths of the visible spectrum are violet (below which is the ultraviolet region). The longest visible wavelengths appear nearly pure red (after which it is the infrared region) (see Figure 1.2).

Similarly, in other cases like the electromagnetic spectrum, it is impossible to define exact borders. The entire range of wavelength is divided in Table 1.1, which represents spectral colors and approximate wavelength intervals associated with them.

1.2 BLACKBODY RADIATION

The temperature of a blackbody radiator can be used as a means to quantify the energy distribution of an illuminant. If a metallic object is heated, after

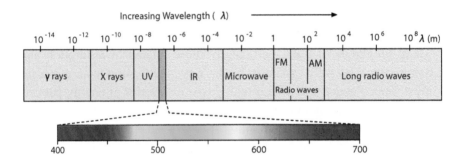

FIGURE 1.2   The electromagnetic spectrum.

TABLE 1.1   Color of Absorbed Light and Corresponding Complementary Colors

Wavelength (nm)	Energy (eV)	Color of absorbed light	Color seen
390 – 420	3.18 – 2.95	Violet	Green–yellow
420 – 450	2.95 – 2.76	Violet–blue	Yellow
450 – 490	2.76 – 2.53	Blue	Orange
490 – 510	2.53 – 2.43	Cyan	Red
510 – 530	2.43 – 2.34	Green	Magenta
530 – 545	2.34 – 2.28	Green–yellow	Violet
545 – 580	2.28 – 2.14	Yellow	Violet–blue
580 – 630	2.14 – 1.97	Orange	Blue
630 – 780	1.97 – 1.59	Red	Cyan

a short time, it becomes too hot to touch. This is due to emission of infra-red radiation. With more heating, the object begins to glow; first, a dull red color appears, followed by bright red, yellow, white, and even blue at higher temperatures. Blackbody radiation or cavity radiation refers to an object or system that absorbs all radiation incident upon it and re-radiates energy; this characteristic of the radiating system does not dependent upon the type of radiation which is incident upon it. The radiated energy can be considered to be produced by a standing wave or resonant modes of the cavity, which is radiating (Figure 1.3).

Classical description of blackbody radiation is based on Rayleigh-Jeans law:

$$E_v^0 = \frac{2\pi v^2}{c^2} kT \tag{1.4}$$

where $k = 1.380662 \times 10^{-23}$ J.K^{-1} is Boltzmann constant, and T is thermodynamic temperature in kelvins (K).

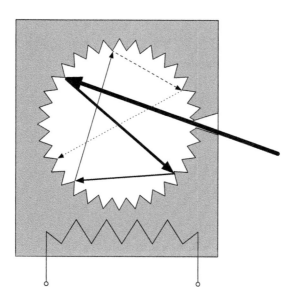

FIGURE 1.3 An approximate realization of a blackbody as a tiny hole in an insulated enclosure. Radiation entering the cavity has little chance of leaving the cavity before it is completely absorbed.

Careful analysis by Rayleigh and Jeans showed that the number of modes was proportional to the frequency squared. The problem is that the entire emitted energy is then infinitely high:

$$E^0 = \int\limits_0^\infty E_v^0 dv \tag{1.5}$$

Although the Rayleigh-Jeans law works for low frequencies, the predicted continual increase in radiated energy with frequency (dubbed the "ultraviolet catastrophe") did not occur [4].

Max Planck who developed a formula that agreed with the experimental data made a major breakthrough. His idea was that the oscillating electrons of the surface atoms of the blackbody emitted radiation according to Maxwell's laws of electromagnetism. Before Planck, it was assumed that these could have any value of energy, but Planck argued that the energy must go up in discrete amounts (quantized) because the frequencies of the oscillating electrons could only take certain values. As energy is proportional to the frequency – Eq. (1.3), if the frequency can only take discrete values, it means that energy is also quantized. The electrons have a fundamental frequency (like standing waves on a string) and the frequency can only go up in whole multiples of this frequency, called the quantum number. This assumption led Planck to correctly derive his formula [5]:

$$E_v^0 = \frac{2\pi v^2}{c^2} \frac{hv}{e^{hv/kT} - 1} \tag{1.6}$$

It is obvious that Planck's law at low frequencies ($hv \gg kT$) follows classical Raleigh-Jeans law. At high frequencies ($hv \ll kT$), the formula is changed as follows:

$$E_v^0 = \frac{2\pi hv}{c^2} e^{-hv/kT} \tag{1.7}$$

Based on Planck's law, it is possible to derive empirical Stefan-Boltzmann law for whole emitted energy of a blackbody radiator:

$$E^0 = \int_0^\infty E_v^0 dv = \sigma T^4 \qquad (1.8)$$

where $\sigma = 2\pi 5k4/15h3c^2 = 5.67 \times 10^{-8}$ W.m^{-2}.K^{-4}.

Consequently, it is possible to derive Wien's displacement law:

$$\lambda_m = \frac{b}{T} \qquad (1.9)$$

where $b = hc/4.965k = 2.896 \times 10^{-3}$ m.K and T is the absolute temperature in kelvin.

This rearranged equation shows why the peak wavelength decreases as temperature increases. This decrease in wavelength explains why objects glow first red, then orange-red, then yellow, and then even blue. These colors are successive decrease in wavelength.

The graph in Figure 1.4 shows that with increase in temperature, the peak wavelength emitted by the blackbody decreases (consequently below temperature of 600 K, the visible part of the spectrum is not connected, and the entire radiation is in the infrared part of the spectrum), This graph also shows

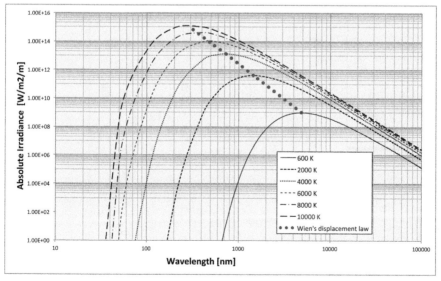

FIGURE 1.4 Plot of absolute energy against wavelength for a black radiator over a range of temperatures.

that as temperature increases, the total energy emitted increases, because of which the total area under the curve increases.

1.3 INTERACTION OF LIGHT WITH MATTER

When light strikes an object, it is reflected, absorbed, or transmitted. Because reflected light determines the color of a material, the appearance can change depending on the amount of light, the light source, the observer's angle of view, size, and background differences. Incident light on the object can be described as a specific stream of photons over time. These photons can migrate in the object. There are several essential interactions between the photons of light and the object. The photons of light targeted at an object can be reflected or transmitted into the object, and the transmitted photons may be scattered or absorbed in the object (Figure 1.5). The optical properties are given through the object nonuniformity (fibers in composite structure, resins, etc.), chromophores, and pigments. Each object type has its own optical

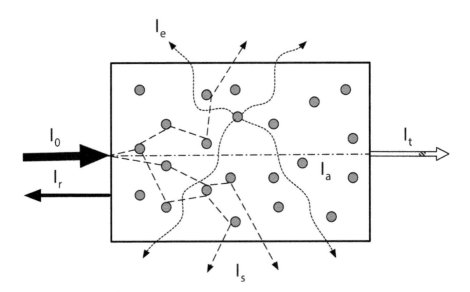

FIGURE 1.5 Diagram of interaction of a translucent textile substrate with light where I_o = intensity of incident light; I_r = reflected light intensity; I_e = intensity of the emitted light; I_s = scattered light; I_t = light intensity transmission; I_a = intensity of the absorbed light (attenuation of light flux is not illustrated due to clarity).

property and is described by the absorption, scattering, anisotropic factor, and refractive index.

1.3.1 SPECTRAL REFLECTANCE

The spectral reflectance $R(\lambda)$ is defined as the ratio of intensities of the reflected light to the incident light at an object boundary or an interface between two media under specified geometric conditions:

$$R(\lambda) \equiv \frac{I(\lambda)}{I_0(\lambda)} \qquad (1.10)$$

where $I(\lambda)$ and $I_0(\lambda)$ are the reflected and incident intensities, respectively.

From conservation of energy, it follows that reflectance may have values only in the interval 0 to 1, inclusively. Spectral reflectance applies to boundaries of all kinds of materials, including opaque, transparent, and translucent. There are exceptions such as adaptive materials that can change the reflectance by changing the relevant microphysical structure of its surface, but usually, a spectral reflectance is independent of the intensity of the incident light and is an intrinsic property of the material, which is called the spectral linearity. For given reflectance $R(\lambda)$ and the incident light intensity $I_0(\lambda)$, the material surface will alter the light intensity as follows:

$$I(\lambda) = I_0(\lambda) R(\lambda) \qquad (1.11)$$

Reflectance can be separated into two components: specular and diffuse reflectance. A more generalized dichromatic reflection model [6] states that light reflected from an object's surface is decomposed into two additive components: the body reflectance and the interface reflectance. Spectral reflectance of many natural materials is usually smooth, such as textiles, paper, paints, etc.

1.3.2 SPECTRAL TRANSMITTANCE

Spectral transmittance $T(\lambda)$ is the ratio of the transmitted and incident light intensities:

$$T(\lambda) \equiv \frac{I(\lambda)}{I_0(\lambda)} \tag{1.12}$$

where $I(\lambda)$ and $I_0(\lambda)$ are the intensities of the transmitted and incident lights, respectively. A filter with a specified spectral transmittance is particularly useful to change the appearance of a light source in order to obtain a special illumination. Such conversion filters then produce intensities according to the next equation:

$$I(\lambda) = I_0(\lambda) T(\lambda) \tag{1.13}$$

Quite a few types of optical filters are used commonly [7]: absorption filters, glass filters, gelatin filters, liquid filters, and liquid crystal tunable filters. In colorimetry, particularly in instrumentation blue glass filters are used as chromaticity conversion filters, allowing simulation of daylight sources.

1.4 LAMBERT-BEER LAW

The most widespread use of UV and VIS spectroscopy is in the quantitative determination of absorption species (chromophores), which is known as spectrophotometry. Absorption of electromagnetic radiation is quantified by the Lambert-Beer Law (sometimes called as Bouguer-Lambert-Beer law), which gives the fraction of monochromatic light transmitted through an absorbing system:

$$\frac{dI}{dx} = -\alpha I \tag{1.14}$$

After integration of this linear first-order differential equation (1.14), we get the well-known form of Lambert-Beer law:

$$I = I_0 e^{-\alpha(\lambda)d} \tag{1.15}$$

where I_0 is the incident light intensity, α is the absorption coefficient, and d is the thickness of the transparent sample.

Sometimes, it is possible to make corroborative measurements using a transmission spectrophotometer in order to check the data obtained in reflection spectroscopy. The connection between the absorption and extinction coefficient is [8]:

$$k(\lambda) = \frac{\lambda \alpha(\lambda)}{4\pi} \tag{1.16}$$

An important point is that transmittance and absorbance cannot normally be measured in the laboratory because the analyzed solution must be held in a transparent container or cell (cuvette). As shown in Figure 1.6, reflection occurs at the two air-wall interfaces as well as at the two wall-solution interfaces. Losses by reflection can occur at all the boundaries that separate the different materials, an interface of two media respectively. In addition, attenuation of a beam may occur as a result of scattering by particles in solution or sometimes by scratches, impurities, etc. on the wall of the cell.

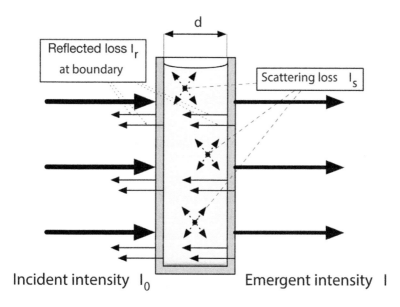

FIGURE 1.6 Reflection and scattering losses with a solution contained in a typical glass cell.

To compensate for these effects, the power of the beam transmitted by the analyzed solution is usually compared with that transmitted by an identical cell containing only solvent (blind sample). An experimental transmittance that closely approximates the true transmittance is then obtained with the equation:

$$T(\lambda) = \frac{P_{solution}(\lambda)}{P_{solvent}(\lambda)} \approx \frac{I(\lambda)}{I_0(\lambda)} \qquad (1.17)$$

The terms $P_{solution}$ and $P_{solvent}$ refer to the measured power of radiation, intensities of the transmitted and incident lights, respectively.

1.4.1 SCATTERING

Light scattering may be defined as the deflection of a ray of light due to the presence of a particle or optical discontinuity in the vicinity of the light ray. The mechanism by which light is scattered in case of solid objects may be considered reflection and refraction at the air-solid surface, refraction due to optical discontinuities, and diffraction of the light around the boundaries of solid object. Textile fabric, a sheet of paper, or other composite materials are a very complex material on a microscopic scale. There may be more or less linear structures, fiber fragments, inorganic or organic pigments and fillers, and various adhesives in such kind of objects, which are distributed in a nonuniform and mostly unknown manner. Based on this aspect, boundary conditions for the Maxwell equations, which are rigorously used for the description of propagation of electric and magnetic fields, are very complex and not very useful.

Scattering depends strongly on the size of the scattering particles. When the scattering particles are extremely small (to the order of 1000 nm), the light is scattered according to a simple law proposed by Rayleigh in which short wavelengths are scattered more than long wavelengths. For larger particles (to the order of 4000 nm and larger), the amount of scattering is according to Fresnel's equations: the amount of scattering depends upon the difference between the refractive index of the particle and of the medium in which it is dispersed and this difference is wavelength dependent [9]:

$$R_s = \frac{\left(n_2 - n_1\right)^2}{\left(n_2 + n_1\right)^2} \tag{1.18}$$

where R_s is specular reflection of light, and n_1 and n_2 are the refractive indices of the media

For these reasons, pigments are most efficient as light scatterers when their refractive index is quite different from that of resin with which they are to be used and their particle diameter is approximately equal to the wavelength of light. When pigments are of very small particle size and have approximately the same refractive index as resin with which they are used, they scatter so little light that they appear transparent. Scattering can, therefore, be controlled by selection of pigments with appropriate differences in refractive index or by control of their particle size [10].

Scattering is a dominant effect in numerous materials that affects light propagation. The effect can be divided to two main categories:

• Rayleigh scattering; and
• Mie scattering.

The different types are defined by the properties of the incident wave like wavelength and size. Rayleigh scattering describes the process of reflecting and refracting light rays at the obstacle boundaries. It is used when the wavelength of the incident light is much larger (approximately 10 times larger than the radius of the spherical particle) than the scattering structures.

In the intermediate range of the particle size to wavelength ratio, the Mie scattering is used. The Mie theory describes the scattering of incident light on a spherical scatterer for any size to wavelength ratio.

A number of light scattering theories that are also suitable for nonspherical particles have been developed, all having their pros and cons. Extensive overviews of available theories have been published by Wriedt [11], Kahnert [12], and more recently by Mishchenko [13].

1.5 PHOTOMETRY

Photometry is a part of radiometry, which is the science of measurement of electromagnetic radiation. To perform reflectance calculations in a correct way, it is important to take the energy distributions for incident and reflected light into account.

1.5.1 RADIANT AND LUMINOUS FLUX

Flux is the basic unit of optical power, expressed in watts (for radiant flux) or in lumens (for luminous flux). The term *luminous* is used when radiant flux is weighted with the spectral luminous function $V(\lambda)$, also named as the photopic efficiency function, defined by the CIE in 1924 [14]. One lumen (lm) is defined as the luminous flux emitted through unit solid angle (steradian) from a directional unit point source of one candela.

1.5.2 RADIANT AND LUMINOUS INTENSITY

Radiant and luminous intensity refers to the radiant and luminous flux emitted per unit solid angle, respectively. Candela (cd) is the luminous in a given direction, of a source emitting a monochromatic radiation at frequency of 540×10^{12} Hz (approximately a wavelength of 555 nm), the radiant intensity of which in that direction is 1/683 W per steradian.

1.5.3 IRRADIANCE AND ILLUMINANCE

Irradiance and illuminance refer to the radiant or luminous flux per unit area incident on a surface, respectively. Lux (lx) is the unit for illuminance, which is equal to the illuminance produced by a luminous flux of 1 lm (lumen) uniformly distributed over a surface area of 1 square meter.

1.5.4 RADIANCE AND LUMINANCE

The radiant flux emitted from a point source in a certain direction per unit solid angle per unit projected area perpendicular to the specified direction is called radiance. Radiance is denoted by $L_{e,\lambda}$ and it is equal to the double derivative of radiant flux with respect to projected surface area A_s and solid angle ω_s. In case of photometric quantity, the radiance is called as *luminance*. So, luminance is the luminous flux radiated from a point light source per unit solid angle and per unit projected area perpendicular to the specified direction. The basic unit for luminance is candela per square meter (cd.m^{-2}).

1.5.5 LUMINOUS EFFICACY OF LAMPS

Luminous efficacy has two meanings. For a light source, it means the ratio of luminous flux emitted to power consumed, an indicator for energy saving. For radiation, it means the ratio of luminous flux to radiant flux, which is expressed as K in Eq. (1.19).

$$K = \frac{\int_{380}^{780} P(\lambda) V(\lambda) d\lambda}{\int_{380}^{780} P(\lambda) d\lambda} \tag{1.19}$$

where $P(\lambda)$ is the radiant power, $V(\lambda)$ is the photopic efficiency function, and λ varies from 380 nm to 780 nm. The theoretical maximum value of luminous efficacy is 683 lm.W^{-1} when a light source emits monochromatic radiation at 555 nm. The values of luminous efficacy for all the commercial lamps range from 10 to 200 lm.W^{-1} [15]. For instance, tungsten lamps have low luminous efficacy with values normally under 20, while high pressure sodium lamps have luminous efficacy above 100, and top level of LEDs reaching more than 220 lm.W^{-1} [16].

1.6 COLOR

Color is a term that has many different meanings. Insufficiently precise meaning and unclear interpretation may cause misunderstanding in communication. Most frequently, color is associated with human eye perception. Color is, of course, related to light as its property and consequently with the objects. Color is the name given to visual sensations associated with the spectral content of the light entering our eyes.

When discussing about the color in general, we could be considering colored lights, colored solutions, or colored surfaces such as textiles, plastics, and paints. In almost all practical situations, we are concerned with colored surfaces, although, as we shall see, the properties of colored lights are used in the specification of the surface colors. It is important to realize that the color of an object depends on the light source used to illuminate its surface, the particular observer who views it, and the properties of the surface itself. Obviously, the nature of the surface is the most important factor. The term color is frequently related to chromatic colors rather than achromatic such as white, grey, and black [17].

It is commonly accepted that both specifications can be accurately defined in a three-dimensional space, that is, by specifying the following three attributes:

- Hue – λ
- Chroma (purity) – P
- Brightness (lightness) – L

Hue is the name we associate with a particular color sensation, in our own language. Our parents teach us this at a very early age. The first of these attributes specifies one of the colors of the spectral sequence or one of the nonspectral colors such as purple, magenta, or pink. This attribute is variously designated in different descriptive systems as HUE, dominant wavelength λ_D, chromatic color, or simply but quite imprecise as color. It should be noted that these terms do not have precisely the same meaning and therefore are not strictly interchangeable. The designation chromatic color is redundant in the strict sense, since chromatic already means colored, but it is still validly used as a distinction from achromatic colors, which refers to the sequence white, gray, and black. Verbal degrees of hue are *red, green, blue, yellow, purple,* etc.

Saturation defines the brilliance and intensity of a color. When a pigment hue is "toned," both white and black (grey) are added to the color to reduce the color's saturation. In terms of the "additive" light color model, saturation works on a scale based on how much or how little of achromatic color (white, grey or black) addition are represented in the color. It is possible to say that saturation is the closeness of the color to a monochromatic spectral color of the same hue or the difference from a grey of the same lightness. Other words that are legitimately used for surface colors are vividness and purity. This attribute gives a measure of the absence of white, gray, or black, which may also be present. Thus, the addition of white, gray, or black paint to a saturated red paint gives an unsaturated red or pink, which ultimately transforms into pure white, gray, or black as the pure additive is reached; with a beam of saturated colored light, white light may also be added, but the equivalent of adding black is merely a reduction of the intensity (see Figure 1.7). The important point is that saturation is colorfulness relative to the color's lightness, while chroma is colorfulness compared to white. When lightness changes, a change in saturation is perceived. Verbal degrees of saturation are *grayish, moderate, strong,* and *vivid*.

Lightness/Brightness specifies the achromatic (luminance) component, which is the amount of light emitted or reflected by the color. The term lightness refers to objects and is associated with reflected light. Verbal degrees of lightness are very light, light, pale, medium, dark, and very dark. The term brightness is used in light sources, typically for photometric or physical-optical energy measure. This statement is necessary because the human eye

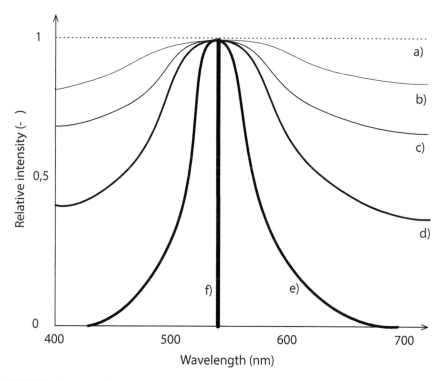

FIGURE 1.7 Simplified representation of green colors with different spectral purity. (a) Spectral curve of ideal white; (b)–(e) Decreasing of relative intensity in other colors than the dominant wavelength caused increasing of green color purity from greenness tint to green color; and (f) Spectral line of green color – purest color.

assesses the hue of color slightly differently with varying brightness. For a color having a given hue and saturation, there can be different levels variously designated as brightness, value, lightness, or luminance (once again, these terms are not strictly interchangeable), completing the three dimensions normally required to describe a specific color.

1.6.1 ADDITIVE AND SUBTRACTIVE MIXING

Most people's experience of color mixing is with paints. As children, we learned that the three primary colors are red, blue, and yellow. Why is green absent? When using little pans of watercolor, we mixed yellow and blue and made green. However, we were not mixing color, but rather colorants (pigments or dyestuffs). This type of color mixture is called subtractive colorant mixture because the individual colorants absorb – that is, subtract – portion of the incident light in certain spectral region (see Figure 1.8), leaving only the light from no absorbed spectral regions to be reflected or transmitted and observed. Thus, in the case of the mixture of yellow and blue watercolors, the yellow absorbs the violet and blue regions of the visual spectrum, and the blue watercolor absorbs the orange and red regions of the spectrum. As a

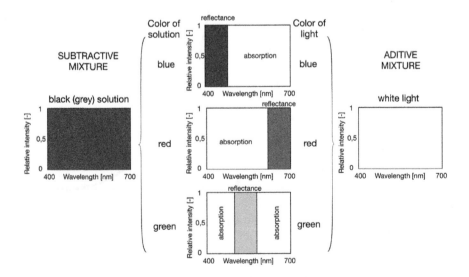

FIGURE 1.8   Simplified scheme of theoretical additive and subtractive primaries.

result, only green part of spectrum is left to be reflected. In this case, we can say that subtractive colorant mixture result is based on the addition of absorption spectra [15]. Subtractive mixture is of major concern when identifying colorants.

Contrary to subtractive mixture, the mixing of additive primaries (red, green, and blue) produces white. Yet how can red and blue be common to both mixing, additive and subtractive? How can nearly the same colors produce either black or white? The problem is that in additive color mixture, we are mixing light together. Thus, if we will have three surfaces reflecting red, green, and blue part of visual spectrum, as a result of combination of these three lights, white light is visible as shown in Figure 1.8.

Back to our question: How can nearly the same colors produce either black or white? We must separate the perception of color from the production of color. We recognize that color names such as red, green, yellow, and blue are vague. In fact, primaries in both mixing are not the same. The standard choice for additive primaries is red, green, and blue light, as this provides the maximum *gamut* (range of colors). A mixture of two color lights is additive in term of the colors produced, but the intensity of mixture color is not the sum of separate colors, but rather the average of them. A balanced mixture of these three primaries produces white, when mixed in pairs; the secondary colors are cyan, magenta, and yellow. It was mentioned that the standard choice of subtractive (colorants) primaries is cyan (turquoise blue), magenta, and yellow; these being the colors that absorb red, green, and blue light, respectively. When these are mixed in pairs, the secondary colors are red, green, and darkish blue/violet. The secondary additive colors are therefore the same as the subtractive primaries, and vice versa.

1.7   HUMAN COLOR VISION

Human beings interpret certain wavelengths of electromagnetic radiation as visible light. Light reflecting from an object and traveling through our visual system causes a color sensation (Figure 1.9). Also, the light illuminating the object can be interpreted to be of a certain color depending on its spectral power distribution. In order to have the color, we normally require an object to reflect or transmit the light from the source (see Figure 1.9A). Besides that, we can see also colors, which are directly emitted from the light source (see Figure 1.9B).

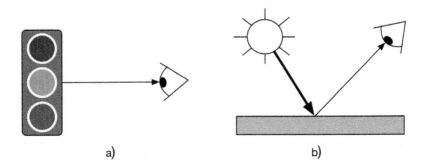

a)   b)

FIGURE 1.9   Schematic representation of the principles of viewing color. A) color lights, B) color surfaces

1.7.1   HUMAN EYE

A horizontal cross-section of the human eye is shown in Figure 1.10. The cornea is a thin transparent membrane through which light first enters the front of the eye. It is approximately a spherical section and takes up about

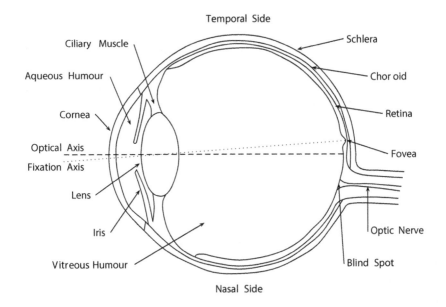

FIGURE 1.10   A horizontal section view of the human eye through the meridian.

one sixth of the eye's outer surface area. The remainder of the outer surface of the eye is the sclera or the white of the eye. The eye overall is approximately 25 mm in diameter with the volume primarily being taken up with the aqueous humor in the front of the eye and the vitreous humor at the rear, both of which have water as their main constituents.

The iris of the eye controls the quantity of light that passes to the sensory elements. The central aperture of the iris, through which light passes unhindered, is called the pupil. The diameter of the pupil is varied between 2 mm and 8 mm by the involuntary muscles.

The ability of the eye to focus on objects is due to curvature of the cornea, aqueous humor, lens, and vitreous humor and their respective refractive indices. The cornea of the eye provides most of the power of the eye due to its curvature and refractive index. The flexible lens provides the ability to vary focus by the action of ciliary muscles, which alter the curvature of the fore and aft surfaces of the lens. With good eyesight, objects at 15 cm to infinity can be focused on by changing the shape of the lens.

As previously mentioned, the human eye sensitivity covers a range of 380–780 nm. Light in the ultra-violet end of the spectrum tends to be absorbed by the cornea and the lens. Similarly, light at the infrared end of the spectrum tends to be absorbed by the aqueous humor and the vitreous humor [16].

1.7.2   RETINA

A cross-section of the retina reveals a series of layers, which, for simplicity, can be separated into two distinct areas, namely the receptors and the transducers. The receptor elements are rod and cone cells, which convert incident light into nerve impulses, whereas the transducer cell, such as the bipolar and ganglion cells, transcode the nerve impulses before they are conveyed to the brain via the optic nerve. The boundary of the retina is the pigment epithelium, which lies beyond the rod and cone cells. A high percentage of the light that reaches the retina is sensed by the receptors even though it must pass through several layers of transducer cells. In each eye, there are approximately 120 million rods, 7 million cones, and only about 1 million nerve fibers linking the eye to the brain. At the periphery of vision, several rods and cones share nerve fibers, whereas in the fovea, there is generally a one-to-one connection between nerve fibers and receptors. This, presumably, increases the visual acuity in the fovea.

In the center of the fovea is a small indentation, where the transducing cells that normally interfere with the light striking the sensory cells, are less inhibiting. This area, the *fovea centralis*, occupies a circular region of approximately 1 mm in diameter and is responsible for sensing the central 2 degrees of the field of vision. Hence, the ratio of rods to cones increases, as the angular displacement from the center of the fovea increases and the properties of vision in the central 2 degrees of vision is governed by the properties of the cones.

Of the rods and cones, much information is known about the rods because at the edges of vision, only rods are involved, and this means that the rod cells can be isolated in experiments on living or deceased eyes. The rods contain a pigment called rhodopsin, which seems to be a common chemical found in the eyes of several animals such as fish, cats, monkeys, and indeed humans. As such, much is known about this pigment and the properties it has when excited by light. When deprived of light, rhodopsin is a reddish-purple color. The action of light causes it to bleach to yellow, and, if the light is sufficiently intense or the exposure is long, the rhodopsin becomes colorless. Generally, only 60 percent of the light absorbed actually bleaches the rhodopsin, and thus, the remainder does not affect vision. Rhodopsin is most sensitive to wavelengths of light around 500 nm and tails off at wave lengths below 400 nm and above 600 nm. The regeneration of rhodopsin from fully bleached to the original reddish-purple state is an exponential process with 50% regeneration taking about 7 minutes. The reddish-purple state is commonly called the dark-adapted state.

Given higher sensitivity of the rods in the dark-adapted state and their higher concentrations at the periphery of vision, it is evident that they are primarily for night vision and provide a low fidelity peripheral vision to detect motion. The human visual system has three separate stages depending upon the intensity of light in the environment; these stages are known as scotopic, mesopic, and photopic. In scotopic vision, the light levels are too low for the cones to operate; hence, vision is exclusively governed by the action of the rods. Mesopic vision is the transitory state where both cones and rods can operate effectively. It is, however, a narrow region and not of much consequence. Photopic vision occurs at higher light intensities and is used for normal daytime vision. At these light levels, the effectiveness of the rods is greatly reduced as they are bleached by the light; hence, vision is primarily governed by the cones, although the rods still play a part, especially in the periphery of vision. The process of light adaptation reduces the sensitivity of the transduction cascade to an increase in illumination. This acts to reduce saturation, and although the sensitivity of the photoreceptors is decreased, discrimination of intensity differences over a range of light intensities improves [17].

A general approximation is that cones will only respond to light levels above 10^{-3} cd.m^{-2}; similarly, rods can respond to light levels down to 10^{-6} cd.m^{-2} [18]. Towards the center of vision, rods and cones are interspersed and cannot be isolated to discover their individual properties.

In the past 50 years, there have been several investigations on the cone receptors, which have shown that there are three types of cones that are distinguished from each other by the wavelength of light they absorb strongest. It is now accepted that cones should be referred to as long-, middle- and short- wavelength-sensitive (L-, M-, and S-), rather than red, green and blue, because the color descriptions correspond neither to the wavelengths of peak cone sensitivity nor to the color sensations elicited by the excitation of single cones [19], as shown in Figure 1.11.

The three types of cones are universally and most popularly known as R, G, and B cones. The R-G-B terminology is not very accurate, especially as the

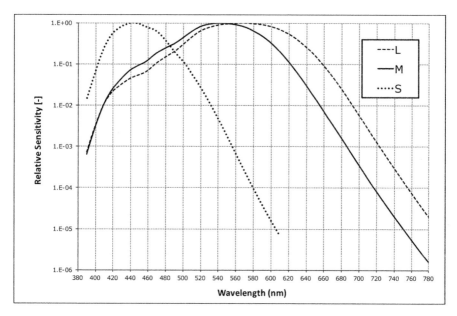

FIGURE 1.11   The 2-degree L-, M-, and S-cone fundamentals in terms of energy normalized to 1 at the maximum.

"red: receptor actually has its peak sensitivity in the yellow region. Though the three types of cones are red-sensitive, green-sensitive and blue-sensitive, the peaks of the sensitivity curves of the three types of cones do not really correspond to the red, green, and blue spectral colors. It is, therefore, more accurate to describe the three types of cones as long, medium, and short wavelength (L, M, S) photoreceptors. Stiles named them as π_5, π_2, and π_1 [23], while Hunt [24] named them ρ, γ, and β (Greek symbols for the three colors), respectively. The three types of cones are distributed more or less randomly in the retinal mosaic of receptor, but they are not equal in number. The numerical distribution of the cones (L:M:S of about 40:20:1, but widely varying across observers) is important in constructing the opponent signals present in the visual system [25].

1.7.3   VISUAL PATHWAYS AND VISUAL CORTEX

The optic nerve, on coming out of the optic disc, forms what is known as optic chiasm, as shown in Figure 1.12. This results in a cross-mapping of the visual field: the left part of the visual field goes to the right half of the cortex, and vice versa. In each cerebral hemisphere, two pathways emerge from the optic chiasm. The smaller pathway ends in a visual center located outside the cerebral hemisphere called superior colliculus and is thought to be responsible for eye movement. After the chiasm, about 80% of the optic nerve connects to the lateral geniculate nucleus (LGN) of the thalamus. The human LGN consists of six main layers and numbered from bottom (1) to top (6). Although the LGN receives inputs from both eyes, the visual signal from the two eyes remain separate in the LGN, meaning that each of the six layers receives inputs from only one eye [26, 27]. Individual neurons (nerve cells) in the LGN can be activated by any change in brightness or color within the area of view (receptive field) of any one eye [28]. Neurons in V1 transmit visual information to various distinct cortical regions located in the posterior temporal and parietal cortex. Almost half of the cortex is involved in visual function [29].

Two systems have been widely studied: the retino-occipital or retino-cortical visual pathway (see Figure 1.12A) and the retino-collicular or retino-tectal visual pathway (see Figure 1.12B). But there also exist three other systems: the retino-pretectal (see Figure. 1.12C) and the retino-hypothalamic (see Figure 1.12D) visual pathways as well as the accessory optic system (AOS) (not illustrated), which play a crucial role in vision though they are less known [30].

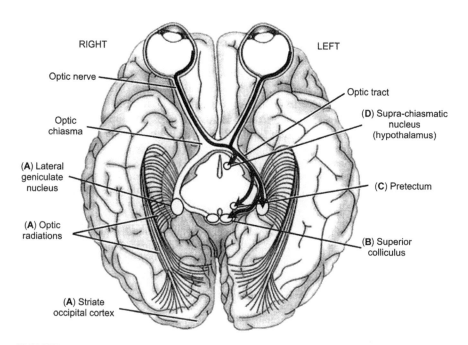

FIGURE 1.12 Inferior view of the brain showing the five main visual pathways, in which left is right and vice-versa. (A) Retino-occipital or retino-cortical visual pathway. (B) Retino-collicular or retinotectal visual pathway. (C) Retino-pretectal visual pathway. (D) Retino-hypothalamic visual pathway. The accessory optic visual pathway is not shown. Thick arrows show the different trajectories of ganglion neurons originating in leftward hemi-retinas (temporal hemi-retina of the left eye and nasal hemi-retina of the right eye). Ganglion neurons originating in rightward hemi-retinas exhibit mirror trajectories but are not illustrated here for clarity of display. (Reprinted from Coubard OA, Urbanski M, Bourlon C and Gaumet M (2014), Educating the blind brain: a panorama of neural bases of vision and of training programs in organic neurovisual deficits. Front. Integr. Neurosci. 8:89. doi: 10.3389/fnint.2014.00089. © 2014 the authors. https://creativecommons.org/licenses/by/4.0/).

1.7.4 THEORIES OF COLOR VISION

Many theories have been proposed over the centuries on how human color vision operates, and it has only been through recent experiments that confidence in these theories has been gained. The oldest and still the most accepted theory of color vision was first proposed by Thomas Young in 1802 [11]. His reasoning was that as the number of nerve pathways from the eye to the brain was limited, and that the concept of three color channels had no foundation in the theory of electromagnetic radiation, the eye therefore has three separate signaling mechanisms, which respond to red, green, and blue. Herman von Helmholtz showed that more than three primary colors were required to match some colors exactly, which was contrary to Young's theory and led to its rejection for many years.

Subsequently, Helmholtz showed that if the three primaries were not pure, then this was sufficient to match any color. Hence, while a red receptor will respond more vigorously to light in the red wavelengths, it will still respond in less vigorously to green and blue wavelengths. This notion of the eye having three receptors with overlapping response curves became known as the Young-Helmholtz theory and has been widely accepted since its inception. It has only been this century that direct evidence for the existence of three cone receptors has been found.

Ewald Hering, at the beginning of the 20th century, discovered that some colors tended to cancel each other out, for example, red cancels green and blue cancels yellow. Hering subsequently postulated that there were indeed three different receptors in the eye; however, rather than responding to red, green and blue, they respond to red versus green, yellow versus blue, and white versus black. This theory became known as the Hering Opponent Color theory and was the most accepted theory in opposition to that of the Young-Helmholtz theory.

The ability to perform electrical analysis of the process of seeing has provided great insight into the operation of the eye, and in particular, communication between the eye and brain. These analyses have provided much understanding of how the human color vision system works and its limitations.

Whilst there is evidence that there are indeed three types of cones as predicted by the Young-Helmholtz theory, arguably the most interesting result from electrical analysis is that early analysis of the retina of fish showed that certain cells gave potentials that became more negative for increasingly blue light, and more positive for increasingly yellow light. Other cells were also found to respond in a similar manner for red versus green light. This experiment was repeated on the macaque monkey, which has a color vision system that more closely resembles that of humans. Similarly, cells were found that produced spike potentials that increase the rate of firing depending on whether red versus green or yellow versus blue light is incident upon them. Such cells are called opponent cells, and conversely, there exist nonopponent cells that respond proportionally with luminosity.

Current understanding of the human color system can be summarized as follows: rods are receptors that detect luminosity levels, primarily at low light levels and at the periphery of vision. There are three different types of cone receptors responding to blue, green, and red light (the red cone actually responds more strongly to yellow), as predicted by the Young-Helmholtz theory. The outputs of the cone and rod cells are coded into spike potentials with the frequency being the measure of intensity. There are four distinct type of signals sent to the brain, the luminosity from the rods, the luminosity from the sum of the cones, a red versus green signal derived from the red and green cones, and a yellow versus blue signal derived from a compounding of red and green cones and the blue cone. This signaling method is close in concept to the Hering Opponent Color theory [31].

1.7.5 COLOR DEFICIENCY

Color deficiency is defined in terms of color discrimination tasks. A color-deficient person will perceive certain colors as being identical, which a color-normal person will be able to distinguish as different. With decreased red-green sensitivity (the most common), colors that are primarily defined by their red or green components, such as rose, beige, and moss green, may appear identical, as the color-deficient observer is less sensitive to the red and green that distinguish to them. Moreover, many color deficient people are not even aware of their deficiency.

Approximately 8% of the male population (less for non-Caucasians than for Caucasians) and slightly less than 1% of the female population suffer from some genetic color deficiency. The most common deficiency (5% of males, 0.5% of females) is deuteranomaly. Deuteranomaly is an anomalous trichromacy caused by an abnormal M-type cone, resulting in abnormal matches and poor discrimination between colors in the medium (M) and long (L) range of wavelengths. The typical result of deuteranomaly is what is referred to as a red-green deficiency: the inability, or at least a great difficulty, to discriminate reds and greens. The abnormal M-type cone has its peak sensitivity much closer to the peak of the L-type cone; thus, the two overlap more than in normal population, causing significantly reduced discrimination [32].

A similar deficiency can cause by an abnormal L-type cone. In this case, the peak of the L-type cone is shifted closer to the M-type cone, causing again reduced red-green discrimination.

A more severe case that causes red-green deficiency is *deuteranopy* (deutan defect) or a complete lack of the M-type cone. Similar deficiency can be caused by a complete lack of the L-type cone and we call it as *protanopy* (protan defect).

These, as well as other deficiencies, which are caused by a missing or an abnormal cone type S are much less common. Such defects are called as *tritanopy* (tritan defect) or *tritanomaly*.

However, color naming may not be impaired in color-deficient observers as it depends on learning. Also, luminance, and other cues can contribute to the ability of color-deficient observer to call colors by their correct names (for example, red is dark, yellow is bright). It is well known that red-green deficient drivers use the physical position of the red light at the top and the green light at the bottom of traffic lights to distinguish them. The (rare) practice of placing traffic lights horizontally removes this additional physical clue (for example, in Japan, traffic lights are placed horizontally and nevertheless green is replaced by blue signal light).

Beside such congenital color deficiencies, which have been found in people without previous history, we can meet with acquired color blindness. A defect in color perception in the center or the periphery of the visual field may occur in any disease affecting the retina, the optic nerve, or the optic cortex in the occipital lobe of the brain. The most common disease of the central nervous system that causes central color defects is multiple sclerosis. The most common occurrences of a defect of color perception in the center of retina are connection with toxic amblyopia, and the poisons that commonly depress the color sensitivity such as carbon disulfide, lead and thallium poisoning, narcotics, tobacco, and alcohol.

The clinical assessment of color vision is not difficult, and it may be worth giving thought on why it is not always done in routine clinical practice. There are a number of possible reasons. The first is that there is no single test of color vision that provides the clinician with all the information needed to advise tested persons. Many practitioners use only the Ishihara test to assess color vision. This test is insufficient for industry in many reasons; although the Ishihara test is very good at detecting red-green abnormal color vision, it provides no useful information about its severity, it does not always classify protan/deutan, and it does not include a test for the tritan color vision deficiencies [15].

The Ishihara test has to be included in the basic battery of tests for the assessment of color vision in clinical practice because it is widely accepted, readily available, and not unduly expensive. It is a splendid test for the detection of red-green color vision deficiency: it has high sensitivity and specificity and can perform acceptably even when it is given under the wrong illumination (test is designed for daylight). Its only problem is that it is so readily available that those who wish to pass, so that they can pass the color vision standard for an occupation, can obtain a copy and learn the correct answers. Generally, we put the Ishihara test, which is famous, into pseudo-isochromatic plates. These plates were designed based on the results of the color matching experiment and can be used to identify different types of dichromats based on a few simple judgments.

Each plate consists of a colored test pattern drawn against a colored background. The test and background are composed of circles of random sizes; they are distinguished only by their colors. The color difference on each plate is invisible to one of the three classes of dichromats. Hence, when a subject fails to see the test pattern, we conclude that the subject is missing the particular cone class.

The Farnsworth-Munsell 100 Hue test is also commonly used to test for dichromatic deficiency. In this test, which is much more challenging than the Ishihara plates, the observer is presented with a collection of cylindrical objects, roughly the size of bottle caps and often called caps. The colors of the caps can be organized into a hue circle, from red to orange, yellow, green, blue-green, blue, purple and back to red. Despite the name of the test, there are a total of 85 caps, each numbered according to its position around the hue circle. The color of the caps differs by roughly equal perceptual steps.

The observer's task is to take a random arrangement of the caps and to place them in order around the color circle. At the beginning of the task, four of the caps (1, 23, 43, and 64) are used to establish anchor points for the color circle. The subject is asked to arrange the remaining color caps to form a continuous series of colors.

The hue steps separating the colors of the caps are fairly small; subjects with normal color vision often make mistakes. After the subject finishes sorting the caps, the experimenter computes an error for each of the 85 positions along the hue circle. The error is equal to the sum of the differences between the number on the cap and its neighbors. For example, in a correctly ordered series, the caps are ordered continuously, say 1-2-3. In that case, the difference between the cap in the middle and the one on the left is −1, and the one on the right is +1. The error score is 0 in this case.

If the caps are ordered 1-3-2, the two differences are +2 and +1 and the error is 3. Normal observers do not produce an error greater than 2 or 3 at any location.

The subject's error scores are plotted at 85 positions on a circular chart as shown in Figure 1.13. An error score of zero plots at the innermost circle and increasing error scores plot further away from the center. Subjects missing the L cones (protanopy), M cones (deuteranopy), and S cones (tritanopy) show characteristically different error patterns that cluster along the different portions of the hue circle (see Figure 1.14).

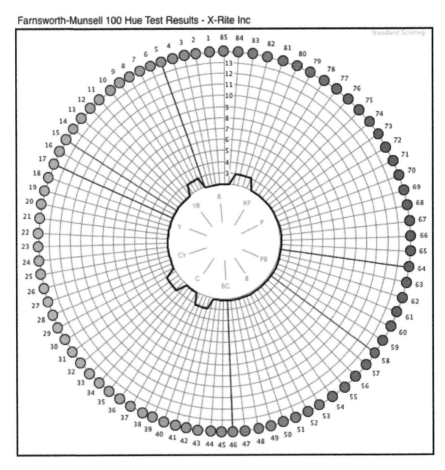

FIGURE 1.13 FM 100 Hue test – color superior discrimination.

Farnsworth-Munsell 100 Hue Test Results - X-Rite Inc

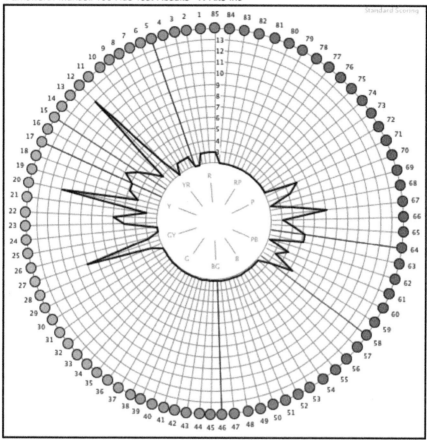

FIGURE 1.14   FM 100 Hue test – protan color deficiency.

KEYWORDS

- **color**
- **colorimetry**
- **human color vision**
- **optical properties of solids**
- **photometry**

REFERENCES

1. Hunt, R. W. G., & Pointer, R. M., (2011). *Measuring Color*, 4th Edition; Wiley, Chichester, (The Wiley-IS&T Series in Imaging Science and Technology), pp. 26–38.
2. Feynman, R. P., Leighton, R. B., & Sands, M., (1970). *The Feynman Lectures on Physics*, Addison-Wesley Longman: Boston, vol. 1, pp. 28(1)–28(4).
3. Parnov, I. E., (1974). *At the Crossroads of Infinity* (translation from the Russian into Czech), Orbis Praha, pp. 35.
4. Moore, W. J., (1963). *Physical Chemistry* (4th edn.), Longmans Green & Co., London, pp. 467.
5. Tipler, P. A., & Mosca, G., (2008). *Physics for Scientists and Engineers*, W. H. Freeman and Company, New York, pp. 683–685.
6. Tominaga, S., (1994). Dichromatic reflection models for a variety of materials, *Color Res. Appl., 19*(4), 277–285.
7. Wyszecki, G., & Stiles, W. S., (1982). *Color Science: Concepts and Methods, Quantitative Data and Formulae*, 2nd Edition, Wiley New York, pp. 30–48.
8. Räty, J., Peiponen, K. E., & Asakura, T., (2004). *UV-Visible Reflection Spectroscopy of Liquids*, Springer, Berlinpp, p. 19.
9. Bukshtab, M., (2012). *Applied Photometry, Radiometry, and Measurements of Optical Losses*, Springer, Dordrecht, pp. 18–32.
10. Donald, B., & Mathew, R., (1988). Refractive index – A key to understanding color differences, *J. Vinyl. Addit. Technol., 10*(4), 205–209.
11. Wriedt, T., (1998). A review of elastic light scattering theories, *Part. & Part. Syst. Charact., 15*(2), 67–74.
12. Kahnert, F. M., (2003). Numerical methods in electromagnetic scattering theory, *J. Quant. Spectrosc. Radiat. Transf., 79–80*, 775–824.
13. Michael, I. M., (2009). Electromagnetic scattering by non-spherical particles: A tutorial review, *J. Quant. Spectrosc. Radiat. Transf., 110*(11), 808–832.
14. Luo, R., (2016). *Encyclopedia of Color Science and Technology*, Springer Reference, New York, pp. 1152–1154.
15. DiLaura, D. L., Houser, K. W., Mistrick, R. G., & Steffy, G. R., (2011). *The Lighting Handbook, Tenth Ed. Reference and Application*, Illuminating Engineering Society of North America, New York, pp. 7, 22–71.
16. van Driel, W. D., & Fan, X. J., (2013). *Solid State Lighting Reliability, Components to Systems,* Springer, New York, pp. 15.
17. McLaren, K., (1983). *The Color Science of Dyes and Pigments*, Adam Hilger Ltd. Bristol, pp. 10–152.
18. Vik, M., (2017). *Colorimetry in Textile Industry*, VÚTS, a.s., Liberec, pp. 11.
19. Overington, I., (1976). *Vision and Acquisition*, Pentech Press London, pp. 7.
20. Fain, G. L., Matthews, H. R., Cornwall, M. C., & Koutalos, Y., (2001). Adaptation in vertebrate photoreceptors, *Physiological Reviews, 81*(1), 117–151.
21. Wördenweber, B., Wallaschek, J., Boyce, P., & Hoffman, D., (2007). *Automotive Lighting and Human Vision*, Springer Berlin.
22. Stockman, A., & Sharpe, L. T., (1999). Cone spectral sensitivities and color matching, in: *Color Vision, From Genes to Perception*, Gegenfurtner, K. H., & Sharpe, L. T., (eds.), Cambridge University Press.

23. Stiles, W. S., (1978). *Mechanisms of Color Vision.* Selected papers of W. S. Stiles, with a new introductory essay, Academic Press.

24. Hunt, R. W. G., (2004). *The Reproduction of Color*, John Wiley & Sons Inc., Hoboken, pp. 10–11.

25. Fairchild, M. D., (2013). *Color Appearance Models*, 3rd edn., John Wiley & Sons, Ltd. Chichester, pp. 16–18.

26. Kaplan, E., (2014). The M. P., and K. pathways of the primate visual system. In: Werner, J. S., & Chalupa, L. M., (edited): *The New Visual Neurosciences.* The MIT Press. Cambridge. Massachusetts, pp. 215–226.

27. Casagrande, V., & Norton, T., (1991). Lateral geniculate nucleus: A review of its physiology and function. In: Cronly-Dillon, J., & Leventha, A. G., editors, *Vision and Visual Dysfunction: The Neural Basis of Visual Function, 4,* Macmillan, London, pp. 41–84.

28. Deeb, S. S., & Motulsky, A. G., (1996). Molecular genetics of human color vision. *Behav. Genet., 26,* pp. 195–207.

29. Palmer, S. E., (1999). *Vision Science: Photons to Phenomenology*, MIT Press, Cambridge.

30. Coubard, O. A., Urbanski, M., Bourlon, C., & Gaumet, M., (2014). Educating the blind brain: a panorama of neural bases of vision and of training programs in organic neurovisual deficits. *Front Integr. Neurosci., 8,* 89. doi: 10.3389/fnint.2014.00089.eCollection.

31. Ebner, M., (2007). *Color Constancy*, John Wiley & Sons, Ltd. Chichester, pp. 12–16.

32. Voke, J., (1983). Significance of defective color vision, *Rev. of Prog. in Coloration, 13,* 1–9.

CHAPTER 2

TYPE OF CHROMIC MATERIALS

MARTINA VIKOVÁ

CONTENTS

Abstract...35
2.1   Definition of Chromic Phenomena...36
2.2   Materials with Photochromic Colorants 37
2.3   Materials with Thermochromic Colorants.....................................55
2.4   Materials with Chemochromic Colorants.......................................71
2.5   Materials with Electrochromic Colorants......................................87
2.6   Materials with Luminescent Colorants..95
Keywords..100
References...100

ABSTRACT

This chapter contains an overview about chromic phenomena. Many books have been written on this topic. However, these books are conceived to explain about basic classes of chromism and chromic phenomena, types of chromic materials, etc. [1–5]. The basic class of chromic dyestuff and pigments are suitable for application in many industries such as textile, paint, and paper industries. There are many variations of chromic colorant that can be applied in many areas of human living and that can help in improving the well-being in our lives. Some examples are textile sensors with chromic dyestuff monitoring external stimulus like the amount of UV light and temperature and also the amount of dangerous chemical substances.

2.1   DEFINITION OF CHROMIC PHENOMENA

There are many magic color changes reacting on the different external stimulus. The color changes can be from white to black, colorless to colored, or from one color to another. These chromic phenomena represent changes in absorption, reflection, or refraction spectrum. These color change phenomena are classified and named after the stimulus that causes the change. The external stimulus is usually on chemical, physical, or mechanical basis. These external stimuli can be light, temperature, chemicals, magnetism, moisture, electrical current, potential, etc. This color change is obviously reversible in the colors of compounds.

Major types of chromism may be divided into five groups based on the causes of change in color. The first group is vibration and simple excitations – external heat or energy transfer within molecules. The second group is ligand field effects, and the third group is transition between molecular orbitals against the fourth group, where transitions is realized between energy bands. The fifth group, the last one, is geometrical and physical optics. This involves the classification of the various chromic phenomena and involving the following process, mainly:

- the reversible color change;
- the absorption and reflection of light;
- the absorption of energy followed by the emission of light;
- the absorption of light and energy transfer; and
- the manipulation of light.

The phenomenon of color change (reversible or irreversible) is caused through molecular conversion under the influence of different stimulus mentioned above. According of the stimulus of color change, we can classify the changes as:

- photochromism (light);
- thermochromism (temperature);
- chemochromism (chemicals);
- electrochromism (electrical current or potential); and
- ionochromism (ionic species).

Many authors have mentioned about these types of smart materials suitable for use in design [5–7]. But, these authors have presented them as potential materials for sensors that indicate intensity of external stimulus (e.g., amount of UV light, temperature, chemicals electrical current, potential, etc.). Examples of textile sensors are presented in Figure 2.1. There is visible color changeable part of textile UV sensor and stable scale that correspond in coloration with the value of simple UV index.

2.2 MATERIALS WITH PHOTOCHROMIC COLORANTS

Solar light is the key factor for the growth and development of living organisms. Incoming solar energy is transformed into chemical energy or is the signal mediator for sensory processes. One of the most significant processes is the process of vision. On the Earth, the ambient light provided by the Sun is crucial for competition and surviving not only for plants and phototrophic (*see* Phototropism in Section 2.2.1) organisms, but even for insects and mammals (e.g., circadian clock circuits).

Photochromic materials may be based on organic, inorganic, and biological field based on their application [8]. In general, organic photochromic materials are widely available, and the major compounds are few of anilines, disulfoxides, hydrazones, oxazones, semi-carbozones, stilbene derivatives, succinic anhydride, camphor derivatives, o-nitrobenzyl derivatives, and spiro compounds [9–11]. These compounds are excited by the UV range from 200 to 400 nm, but some compounds were extended till 430–435 nm, and very few are excited at the visible region. Silver halide in borosilicate or alumino borosilicate is an example for inorganic photochromic

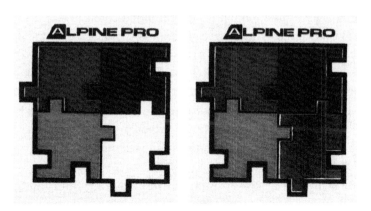

FIGURE 2.1 Textile sensors and after UV irradiation.

material, and it is excited in the range of 200–400 nm. Even though some compounds were extended till 430–435 nm, very few are excited at the visible region [12].

2.2.1 PHOTOCHROMISM OF ORGANIC COMPOUNDS

There are many conformational changes that can take place in the excitation process, which lead to changes in electronic absorption spectra, resulting in a visible color change. If the changes are thermally reversible, after removal of the irradiation that activates the changes, the system returns to the state before irradiation and the induced absorption or color spontaneously disappears. This was previously referred to as phototropism [13], and now more correctly as photochromism [14–22]. If the color changes are caused by heat, then this is termed thermochromism. It is possible to distinguish between photochromism, a reversible coloration process, and irreversible photo-induced chemical reaction.

Before 1950s, many of the photochromic studies were based on the sunlight; it means that they observed color changes at day and reversible at night. But in the second half of century, many of the artificial light sources (almost equal to sunlight) were found and that can be utilized for irradiation with respect to the specific wavelength and intensities of the whole spectrum (mainly from infrared to ultraviolet). The case of so-called positive photochromism is present, i.e., the distance absorption maximum of form **A** is located in the shorter wavelength range than that of form **B**, as shown in Figure 2.2. The thermodynamically more stable, colorless, or slightly yellow form **A** is by irradiation with UV light in the color; in this case, blue in the form **B** transferred. The reverse reaction is spontaneous in the dark and/or photochemically by irradiation with the light of wavelength in the range of the absorption maximum form **B**. In case of negative photochromism, discoloration according to the photo chrome system with UV light activation is colored in the dark and uncolored according to UV radiation [23, 24]. Simply, the photochromic processes may be described as follows [3]; the reaction scheme and UV–VIS spectra of photochromic compounds are shown in Figure 2.2.

The formation of a new absorption band resulting from the transition $S_0B \rightarrow S_1B$ from various vibrational levels in the excitation of a colorless

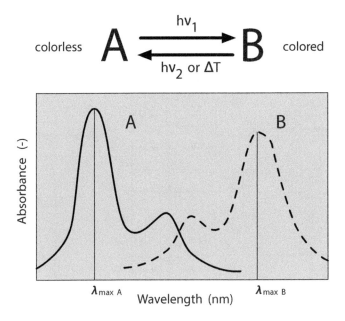

FIGURE 2.2 Graphic representation of a photochromic system.

molecule S_1A is schematically described in Figure 2.3. After the molecule is in the excited state S_1B, the colored molecule is deactivated to the ground state S_0B. Then, there follows an exergonic process in which the regeneration of the colorless form occurs via a radiation-less transition with the creation of the original form in its ground state S_0A. The system **B** is not thermodynamically stable and therefore spontaneously returns to the state **A**. Frequently, the backward reaction **B→A**, the so-called thermal conversion from S_0B to S_0A, proceeds via the transition state **X**, whose energy is higher than the singlet state of colored form S_0B. The process is thermally activated. The reaction **B→A** may also be induced with long wavelength light (infrared irradiation) or by light near to the new absorption band.

With regard to the energy difference $E(S_0B) - E(S_0A)$, the photochromic change can be used to provide energy accumulation.

As with the yield of the light energy conversion to the thermal energy, the accumulative capacity (difference of heats of colored **B** and colorless **A** forms) depends on the chemical structure of the meta-stable photoproduct, which has nonconventional bond lengths and angles and on the dissipation

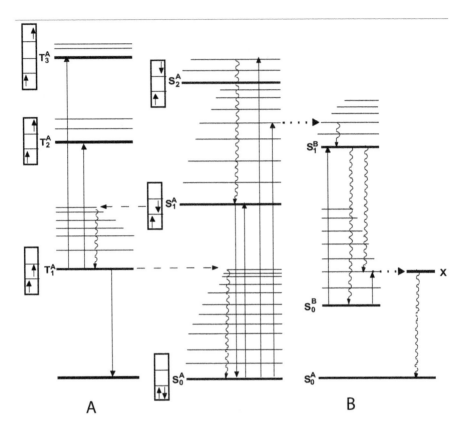

FIGURE 2.3 The Jablonski diagram: Representation of electronic, radiative, and nonradiative transitions (reproduction limited to the first excited states).

of resonance energy and involves increased stability caused by delocalization of π electrons.

The limitation of the effect results from the second theory of thermodynamics. Because it may be described as a thermal machine that is working between temperatures T and T' (photochemical reaction **A** (S_0A) →**A** (S_1A) →**B** (S_0B) at temperature T and exothermal reaction **B** (S_0B) →**X**→ **A** (S_0A) at temperature T'), the final yield of accumulative capacity increases at temperature T', as does the energy of the ground state of the "colored" molecule S_0B.

The highest specific accumulative capacity of a well-known system is around 2 kJ.kg^{-1} (Norbornadiene → Quadricyclane, as shown in Figure 2.4).

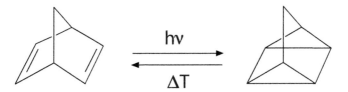

FIGURE 2.4   Reaction of Norbornadiene → Quadricyclane.

2.2.2 CLASSIFICATION OF PHOTOCHROMIC MATERIALS

The photochromic process may be classified into several main groups according to the mechanism of conformation changes [5, 25] as given below:

2.2.2.1 Triplet–Triplet Photochromism

This system has its own absorption band in the long wavelength region of the spectrum characterized as the colored form **B**, which include photochromic species whose triplet state has a sufficiently long lifetime (approximately 1s) and show strong triplet-triplet excitation. Typical examples are penthacene (Figure 2.5a) or dibenzanthracene (Figure 2.5b)

These materials provide very quick response photochromic media with a bleaching time shorter than 1 second. The bleaching time refers to term for the reversion of photochromic color change, in the case of positive photochromism, proceeding from colored to colorless forms. The triplet state of aromatic hydrocarbons (anthracene, dibenzanthracene, etc.) is quenched by

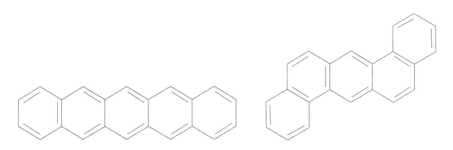

FIGURE 2.5 A   Penthacene.   **FIGURE 2.5 B**   Dibenzanthracene.

nitrobenzenes through outer sphere electron transfer reactions. This triplet state has been referred to as a contrary type [26]. This is defined in photochromic effect of the triplet type, where three independent steps are required. A chromophore quickly absorbs activating radiation to raise ground state molecules to an excited singlet state; as the second step, the excited singlet state proceeds by a process known as intersystem crossing to convert to a triplet state; thirdly, triplet state molecules absorb incident radiation to convert from the first triplet state level to a higher level.

The quantum yield of triplet-triplet photochromism depends on the concentration of oxygen in the system, because oxygen creates triplet excitations. Then quantum yield depends on the relative contributions from the processes, which have an influence on the population of the triplet state such as phosphorescence, radiation-less transitions, intersystem conversion, etc.

2.2.2.2   Heterolytic Cleavage

In this type of system, a covalent bond is broken in the excited molecule, and new conformations are created with zwitterionic (dipolar) structures [27, 28]. The ionic species either continue to react or they are stable, and they can exist separately [29]. Typical structures within this photochromic group are pyrans (Figure 2.6):

Under the influence of UV irradiation, the bond is broken between carbon and oxygen, and the pyran ring is opened. Ionic structure (**M1**) is formed, which is similar to merocyanine colorants and which provides intense coloration. There is also a resonance contribution from neutral form (**M2**). In the photochromic process involving pyrans, cis-trans conversion (isomer **M3**) and triplet–triplet absorption also play an important role. The balance between the contributing forms (**M1, M2,** and **M3**) determines the resulting color after irradiation.

The rate of thermal bleaching depends on substituents R3 and R4. Electron donor substituents accelerate the reverse ring closure to the pyran ring. Similarly, steric influence, for example with a bulky substituent R2 decreases the probability of the formation of colored planar conformers. The reason is the fact that the sp3 carbon creates a point of contact for the two planes of the molecules (see Figure 2.7), which have to rotate about 90° to create the new colored isomers.

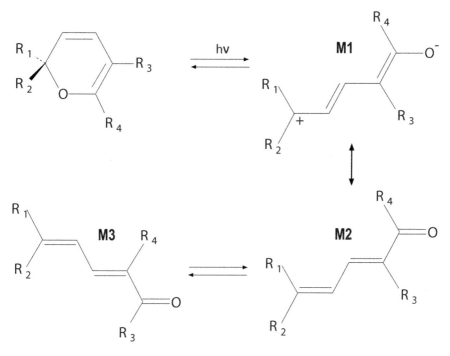

FIGURE 2.6 Photochromic reaction of pyrans.

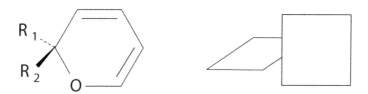

FIGURE 2.7 Creation of contact point for planes with rotation of about 90°.

2.2.2.3 Homolytic Cleavage

In this system, with the absorption of radiation, a bond is broken, and the dissociation of the molecule takes place to give two or more parts, which contain azygous electrons. In comparison with the mechanism of heterolytic cleavage, where the products have some ionic characteristic, during homolytic cleavage, paramagnetic radical products are formed. The characteristic of the cleavage strongly depends on the bonding configuration to accommodate the dissociation of electrons. An example is the dissociation of hexaphenyl bisimidazole (Figure 2.8) to the two 2,4,5-triphenylimidazole radicals [1, 5].

Another example is the cleavage of a C-Cl bond in 2,3,4,4-tetrachloro-1-oxo-1,4-dihydronaphthalene (Figure 2.9) with the formation of an additional chlorine radical fragment and an optical absorption around 530 nm [1].

2.2.2.4   Trans-Cis Isomeration

Thermal trans-cis isomeration is common in many chemical reactions. Thermal energy can cause a change from the trans state to the cis state, in most cases around a central unsaturated bond. This type of isomeration is also involved as part of the process of vision, in which the photon absorbed by the molecule of rhodopsin changes the membrane permeability of cones so that sodium ions can pass through. The chromophoric part of the molecule of rhodopsin is created by the (nonatetraenyliden) alkylamine radical in the 11-cis, 12-cis configuration [29–38]. After absorption of a photon, part of

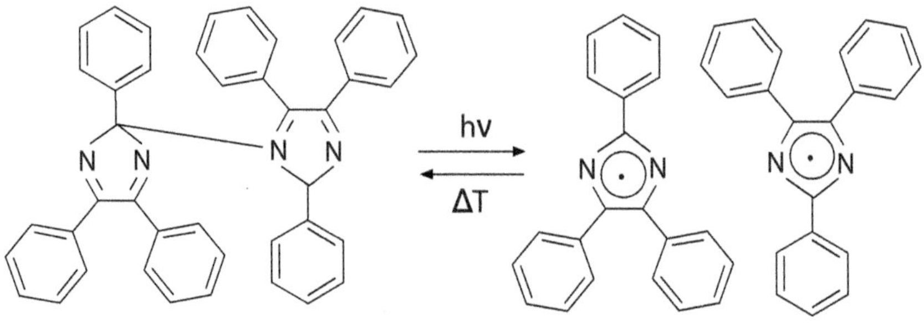

FIGURE 2.8   Photochromism of hexaphenyl bisimidazole.

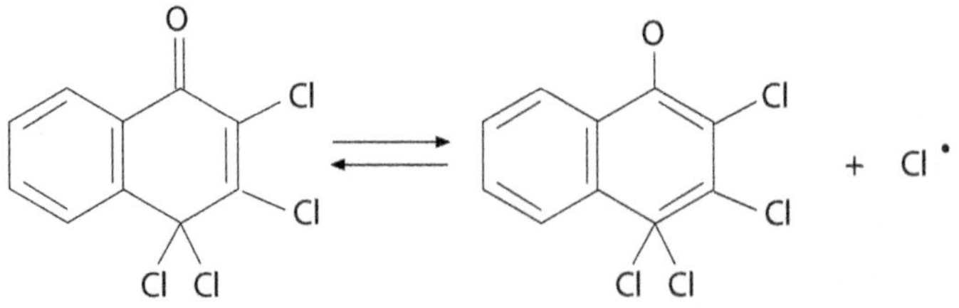

FIGURE 2.9   Photochromism of 1–oxo–2,3,4,4–tetrachloro–1, 4–dihydronaphthalene.

the molecule rotates around the 11-12 bonds and the trans-isomer is formed (pre–lumirhodopsin as shown in Figure 2.10).

Other examples are the photochromism of stilbene (Figure 2.11) and the photochromism of polymeric poly (4-methacryl aminoazobenzene) (Figure 2.12) [39–41].

2.2.2.5 Photochromism Based on Tautomerism

Tautomerism refers in general to the reversible interconversion of isomers, which differ in the position of a hydrogen atom. In the case of photochromic

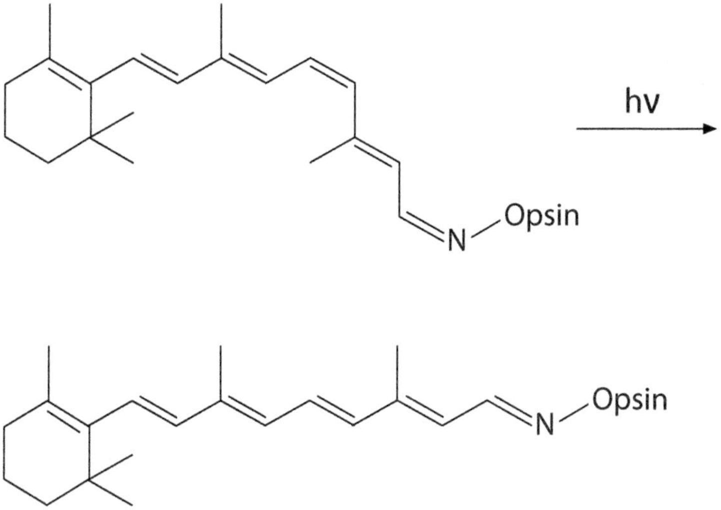

FIGURE 2.10 Photochromic reaction of rhodopsin.

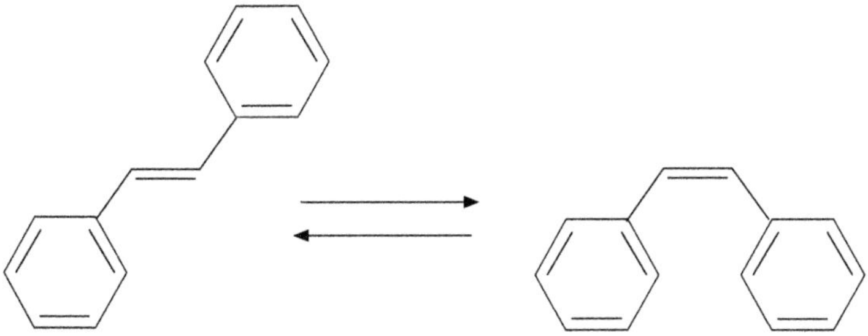

FIGURE 2.11 Photochromic reaction of Stilbene.

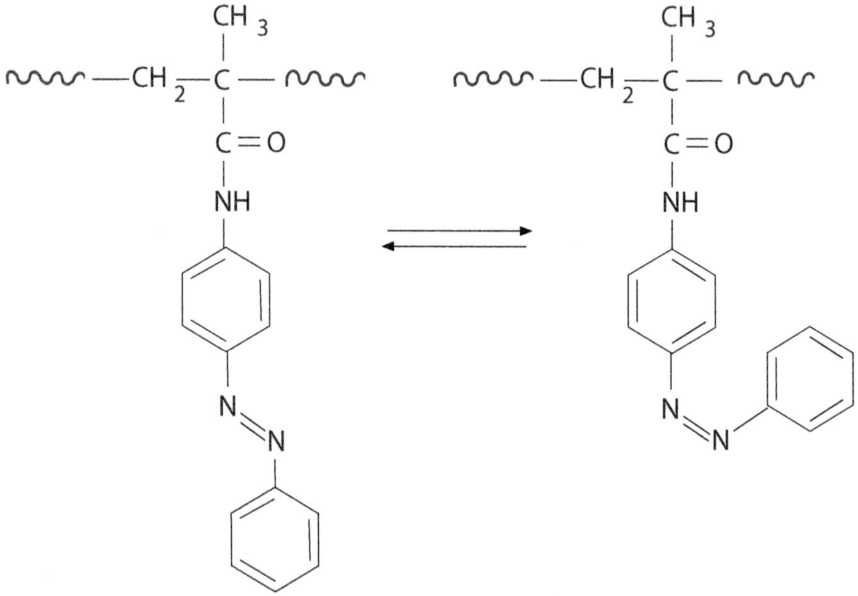

FIGURE 2.12   Photochromic reaction of poly(4–methacryl aminoazobenzene).

tautomerism, this interconversion occurs after irradiation with light. A major type of photochromic tautomerism is hydrogen transfer. A typical example of photo-tautomerism involves metal-free phthalocyanine. In the excited form at low temperature, two hydrogen atoms are shifted between four central carbon atoms [42] (see Figure 2.13).

Another example is hydrogen tautomerism observable in (E)-2-((phenylimino) methyl) phenol (Figure 2.14). The enol-form is pale yellow and after irradiation, the keto-form appears, which is reddish or brown [1].

2.2.2.6   Photodimerization

There are many organic compounds that during excitation by UV irradiation create excimers, in which one molecule is in the ground state and the second is in the excited state. This specific arrangement was measured by spectroscopic methods [43], and a maximum conversion rate of 75% was determined. The creation of a complex of one photo-excited molecule and

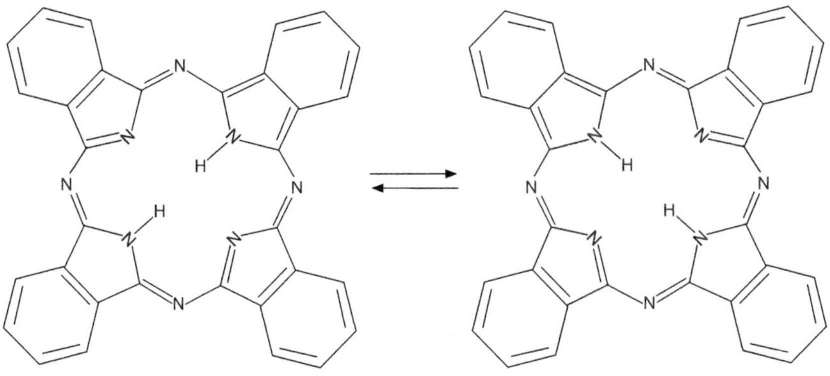

FIGURE 2.13 The tautomerism of phthalocyanine.

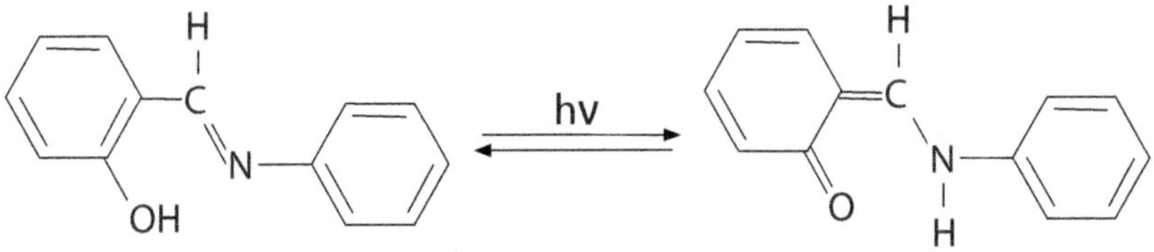

FIGURE 2.14 Photochromic reaction of (E)-2-((phenylimino) methyl) phenol.

one ground state molecule occurs during photoreaction. Different tautomeric sites lead to an additional disorder parameter adding further complexity in the photodimerization reaction. Bond energies are so high as to create a stable dimmer in the solid-state. An example is the pair of σ covalent bonds between two neighboring parallel molecules in anthracene [43, 44] (see Figure 2.15). Anthracene absorbs light in the 240–290 nm and 310–390 nm regions, and a fluorescence emission can be observed between 390 nm and 450 nm. Upon irradiation in the near ultraviolet (around 360 nm), a dimerization reaction can also occur. Note that rather intense illumination conditions and high anthracene concentrations are required for this second-order reaction to proceed at a measurable rate. Upon irradiation, side photoproducts such as photo-oxides can eventually be formed and interfere with the dimerization reaction.

Photochromic materials can be classified based on the back reaction. If back reaction was caused through irradiation with light, it is called P-type

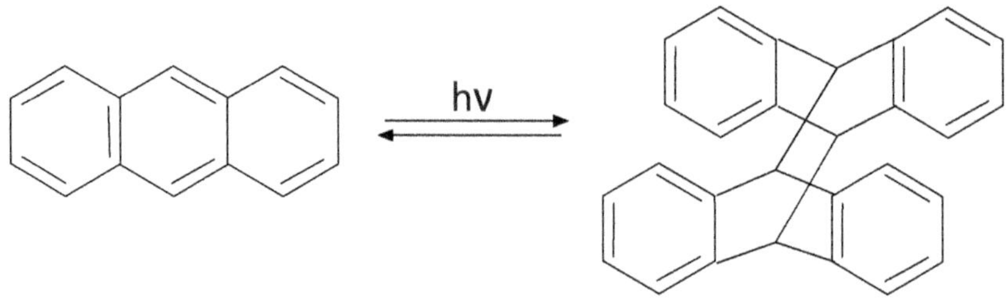

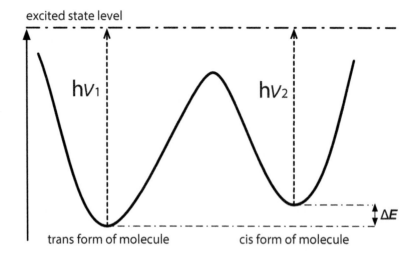

FIGURE 2.15   Photodimerization of anthracene.

photochromic materials (Figure 2.16). Similarly, if the back reaction was caused thermally, it is called T-type (Figure 2.17), which is used widely and commercially. The P-type photochromic materials have two thermally stable isomers and the transformation between the two isomers is reversible upon irradiation. Hence, this system shows good thermal stability, resistance to fatigue and are important as photo-switches [45].

The ring-closing reaction from the colorless open-ring isomer to the colored closed-ring isomer is induced by irradiation with UV light, whereas the reverse ring-opening reaction is prompted by excitation with visible light

FIGURE 2.16   Photochromic reactions of P-type compounds (both forms of the molecule are stable).

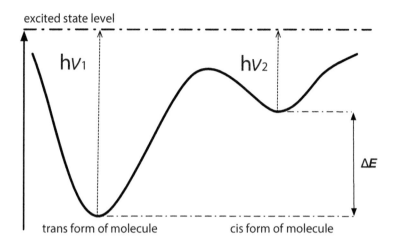

FIGURE 2.17 Photochromic reactions of T-type compounds (cis form of molecule is unstable).

[46–49]. Due to the above reaction, huge spectral shifts take place between the shorter wavelengths absorbing the open structure and the longer wavelength for the closed form. The structural changes on ring closure can affect the properties such as fluorescence, refractive index, polarizability, and electrical conductivity [50]. T-type photochromic material having only one unstable isomer, which can able to isomerizes into a stable isomer, isn't necessary additional irradiation causing reversion of system as in case of P-type photochromic materials. This reaction was induced either by thermal or photoirradiation with visible light spectrum [51, 52].

2.2.2.7 T-Type Photochromism

In the last five decades, many T-type photochromic compounds have been investigated (e.g., anils, perimidine spirocyclohexadienones, spirodihydro-indolizines, etc.). Among these, few have been successfully commercialized [53–55]. However, a commercially interesting photochromic colorant must satisfy the following requirements [1, 3]:

- the residual color is low during the weak visible absorption;
- quick response to an increase in illumination;

- have a strongly absorbing activated state;
- has good photostability (minimum of 2 years);
- resist the tendency to "fatigue," during activation;
- not affected by the temperature of its environment.

2.2.2.7.1 Spiropyrans

Spiropyrans are the most extensively studied photochromic compounds because of their potential application in many fields, namely memory disk, optical switches, sensors and photochromic inks, and dyes for plastic and textile applications [56]. Spiropyrans is the first commercialized T-type photochromic compounds in the modern world. The general structure of photochromic spiropyrans containing a second ring system attached to pyran in a spiro manner at the second position [1]. Photochromic properties of spiropyrans can occur through the ring opening of pyran compounds, which is nothing but a molecular rearrangement (Figure 2.18). The ring opening of pyran is caused by the irradiation with UV light sources; similarly, the removal of light source can become an opposite effect (colored to colorless). During the ring opening of pyrans (under UV irradiation) shifting the balance of the equilibrium, which means the open merocyanine form exists as cis-cis/trans-trans mixture that is at equilibrium. It causes the increase of the concentration of photo merocyanine (colorless) form which has been observed as an intensification of color [1]. The equilibrium moves toward to the colorless pyran form of the colorant, which is seen as fading. Generally, the thermal bleaching rates for spiropyrans low because it is very sensitive to temperature so the photochromic effect will be weaker at higher temperature. The molecular structure and photochromic mechanism for spiroxazines and spiropyrans are similar. During 1970s, spiroindolinonaphthoxazine ring

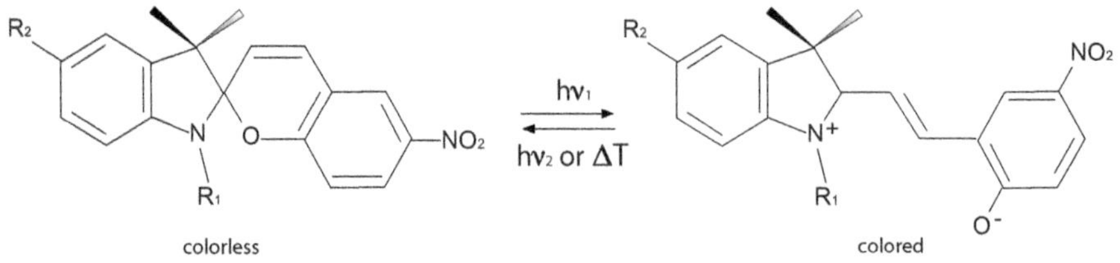

FIGURE   2.18 Photochromic   (heterolytic   cleavage/photocyclization)   reaction   of spiropyrans.

system has been developed after spiropyrans [23, 57–65]. In the last four decades, several studies have focused on indolinospirodipyrans, which are readily available as well as have diverse potential applications. In the beginning era, the practical application of spiropyrans was concentrated on photochromic dyes and inks for the plastic ophthalmic sunglass; however, the photo degradation of indolinospirodipyrans is not up to the level (rapid photodegradation), and thus, it is not suitable for this application. Recently, this issue can be solved by incorporating the spiropyran with polymers, which have become most promising photochromic materials in the field of photochromic dyes, optical storage, and sunglass [56].

2.2.2.7.2 Spiroxazines

The chemical structures of spiroxazines are similar to spiropyrans, except that the pyran is replaced by the oxazine group. The spiroxazines molecules contain a condensed ring substituted 2H-(1,4)oxazine in which the number 2 carbon of the oxazine ring is involved in a spiro linkage. In 1970s, the first spiroxazine photochromic compound such as ring system was studied and reported by Ono and Osada [66] and Arnold and Volmer [67]. The spiroxazines compounds remained unnoticed before 1982 due to the growth of spiropyran [1]. Generally, the spiroxazine gives relatively fast-fading blue photo coloration, but it can be enhanced by modifying the structure (Figure 2.19). The parent structure of indolinonaphthoxazine will turn to blue during the UV irradiation and rapidly fade back to colorless (after removal of the activation radiation, i.e., light source). This is caused due to ring opening of merocyanine structure, which absorbs dominant wavelength of 600 nm [53]. In 1990, the commercial plastic photochromic ophthalmic lenses were successfully

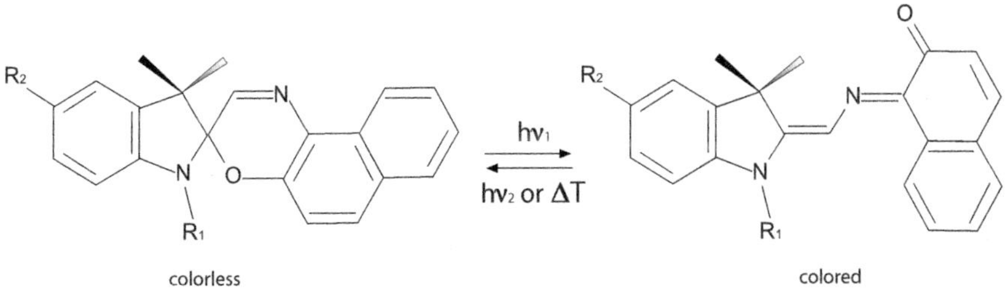

colorless   colored

FIGURE 2.19 Photochromic (heterolytic cleavage/photocyclization) reaction of spiroxazines.

produced using spiroxazines. Later, other applications were developed such as photochromic inks and dyes and cosmetic products. The synthesis of large number of modified spiroxazines has been developed, which create the possibility to utilizing the spiroxazines in many other fields.

The application of spiroindolinonaphthoxazines as disperse dyes to polyester, nylon, and acrylic fabric was studied by Billah et al. [68], Spiroindolinonaphthoxazines have a similar structure with traditional disperse dyes, in that they are small-to-medium-size neutral molecules with a balance of hydrophilic and hydrophobic characteristics. The dyeing procedure for this compound is explained in Chapter 3. Based on the previous report [69–71], the color build up for nylon was significantly higher than that for polyester and acrylic, and the dyeing result for acrylic and polyester are the same. Apart from that, the dye uptake appears to be relatively low and is probably concentrated near to the fabric surface. Also, they found that the photochromic intensity can be increased after a mild washing treatment, which may be because of enhanced photochromic dye migration due to aqueous alkaline surfactant treatment, colorant fiber interactions, and disaggregation of the colorant within the fabric that provides more favorable conditions for conversion to the merocyanine form [61].

2.2.2.7.3 Naphthopyrans

Naphthopyrans are also known as benzochromenes or simply as chromene. Typical structures are related to 2-H-1-benzopyran. In 1966, Becker and Michl published the first compounds of photochromic 2H-1-benzopyrans [72]. However, Becker investigated more than 25 compounds and found that some compounds change their color at room temperature. In 1971, Becker [86] published in his patent that the material changes to orange with ultraviolet irradiation below about 40°C and then bleaches to colorless at room temperature. Around 1990, the photochromism of benzo and naphthopyrans was sparkled by commercialization in the field of plastic ophthalmic industries [1]. Similarly, the photochromism of spiropyrans and benzo and naphthopyrans is the same and involves the breaking of the oxygen-carbon bond of pyran. The photochromic effects for naphthopyrans is light induced ring opening and the structure as it is shown in Figure 2.20 [1, 24]. During irradiation, the colorless or faintly colored molecules undergo an electrocyclic pyran ring opening with the cleavage of the

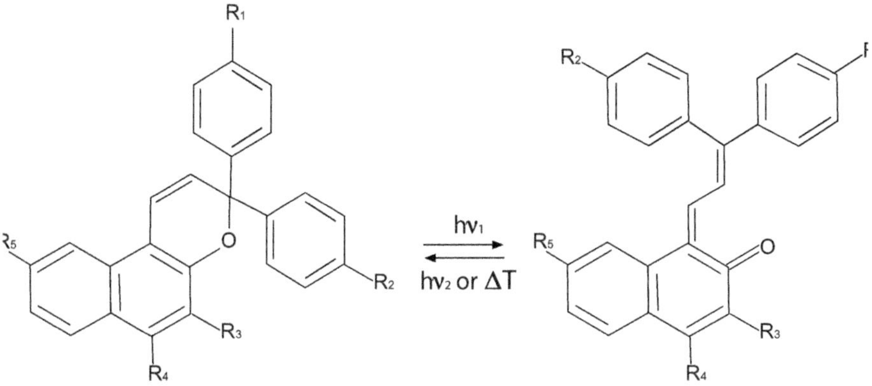

FIGURE 2.20 Photochromic reaction of naphthopyrans.

carbon to oxygen bond. The structural reorganization allows the photogenerated species to adopt more planar structures. The greater conjugation is responsible for absorption in the visible region of the spectrum. The colored geometric isomers gradually electrocyclize to regenerate the colorless form on the cessation of the irradiation [73]. Initially naphthopyrans have limited of orange/yellow colors, but recently it is extended to other colors. These families offer considerable scope for the structural modification on this photochromic property because many convenient modifications of the structure are possible. However, careful design is needed because substituent choice usually affects both kinetics and color. It can enable to a more extensive gamut of colors that spans across the visible spectrum from yellows to oranges, reds, purples and blues, but also features more neutral colors such as olive, brown, and grey. The stability of such dyes is generally as good as any other class, while their photochromism tends to be more independent of temperature than that of spiroxazines [1,74,75]. Naphthopyrans are widely used to make ophthalmic plastic lenses and solar protection glasses, fuel markers, and security markers [72].

2.2.2.8   P-Type Photochromism

In the last three decades, significant efforts were made to develop P-type photochromic materials due to their potential applications in textiles and other fields [76, 77]. The two most important groups of P-type photochromic dyes are fulgides and diarylethenes, as shown in Figures 2.21 and 2.22, respectively [78]. The photochromic mechanism for both fulgides and dia- rylethenes is a cyclization (an electrocyclic reaction) in which ring-opened species are a colorless or weakly colored species, whereas the ring-closed species are colored species. For T-type photochromic materials, the photo- chromic mechanism was due to ring opening of compounds, and for P-type, it is ring closing of compounds [1, 3, 74].

2.2.2.8.1 Fulgides

The fulgide and fulgimide family constitutes an important class of P-type photochromic compounds; they belong to organic type of photochromic compounds. In 1905, Stobbe synthesized the photochromic fulgides, namely phenyl-substituted bismethylene succinic anhydrides [1, 79, 80]. In 1968,

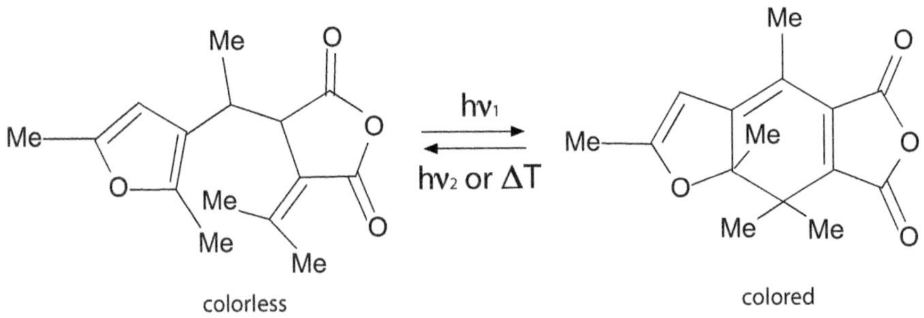

FIGURE 2.21 Photochromic (photocyclization) reaction of fulgides.

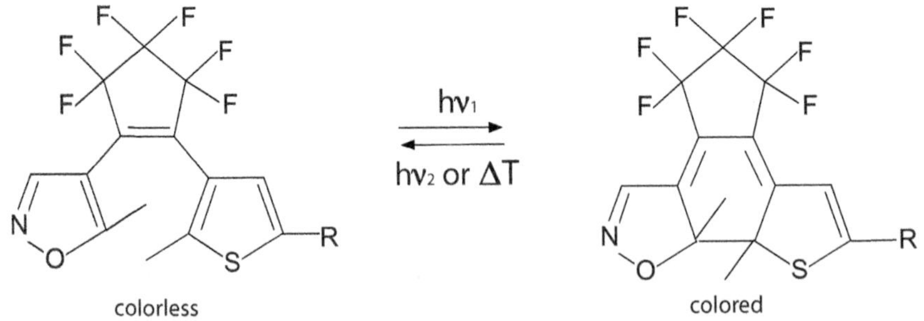

FIGURE 2.22 Photochromic (photocyclization) reaction of diarylethenes.

various fulgides and their derivatives were synthesized, and the first compound of the fulgide family was developed by Hart et al. [81] as succinimide, which was called a fulgimide; it is one of the most important and practical derivatives of fulgides [1]. Fulgides and fulgimides exhibit great photochromic properties such as readily distinguishable absorption spectrum for each form, efficient photoreactions, and thermal and photochemical stabilities; so, these materials can be used in the field of memory disk, optical switches, sensors, and dyes/inks [82].

2.2.2.8.2   *Diarylethenes*

Diarylethenes is the one of the P-type photochromic compound and has heterocyclic five-membered rings as the aryl groups, such as thiophene or benzothiophene rings, which undergoes thermally irreversible and fatigue-resistant photochromic reactions. The coloration and the discoloration processes are driven photochemically [1]. Theoretical consideration based on the molecular orbital theory revealed that the thermal stability of both isomers of diarylethenes is attained by introducing aryl groups and that have low aromatic stabilization energies. Hence, most studies have focused on increasing the thermal stability of the closed form of diarylethenes. When the aryl group was furan or thiophene, the closed forms were thermally stable, and the discoloration reaction was thermally prohibited. On the other hand, diarylethenes exhibited thermally reversible reactions when the aryl group was phenyl or indole [1]. Generally, thermal stability depends mostly on the nature of the aryl groups. Usually, the open-ring isomers are colorless compounds and the closed-ring compounds are colored form (see Figure 2.22). However, the color is depending on their chemical structure which means that the extended conjugation along with the molecular backbone. Therefore, many of the diarylethenes have the photochromic behavior in solution and solid state [83, 84].

2.3   MATERIALS WITH THERMOCHROMIC COLORANTS

Thermochromism is a phenomenon in which certain dyes made from organic or inorganic compounds and from liquid crystals change color reversibly when their temperature is changed. Day, in his work, defined thermochromism as noticeable reversible color change brought about by the boiling point of each liquid, boiling point of solvent in the case of solution, or the melting point for solids [85]. The change may be reversible, i.e., the original color may return upon cooling, or irreversible. Similar changes in color, reversible or irreversible, are often observed when a thermochromic substance or system is cooled down [86].

Some researchers are of opinion that only the reversible changes should be included in term of thermochromism. This point of view is logical because the irreversible changes in color caused by heating are, for the most part, merely the results of true chemical reactions, and thus do not require special terminology. Closer studies, however, reveal that the situations are in many cases not so simple and clear cut [87]. The mechanism responsible for thermochromism of organic compounds varies with molecular structure of the compound. The color change is due to equilibrium changes, either between two molecular species (e.g., acid-base, keto-enol or lactim-lactam tautomerism); or between different crystal structures; or between stereoisomers.

Thermochromic phenomena can be observed in solids, liquids, solutions, dispersions, and gases. The change in the color occurs at a specific temperature, which is called "*the thermochromic transition temperature.*" According to the investigations, thermochromism occurs based on different mechanisms in which some of the following materials are used:

- inorganic compounds;
- organic compounds;
- polymers; and
- sol-gels.

Inorganic thermochromic compounds consist of many metals and inorganic compounds with thermochromic behavior in solid or solution phase; these compounds do not fulfill the demands of textile industry. These materials obviously exhibit their thermochromic effect in solution and in high temperatures; therefore, they are not suitable for textile applications. Temperature-indicating paints, crayons, etc. are widely used in industrial applications for observing heat patterns and for detecting high and low temperature points on surfaces of heat engines, pipelines, and refrigeration fins. With an irregular and nonuniform temperature rise, a number of colored bands separated by isothermal lines will appear on the workpiece surface, allowing the thermal record to be made of the temperature gradient across the surface [87].

Thermochromism in organic compounds occurs due to different mechanisms. Three main mechanisms of thermochromism in organic compounds have been proven as below:

- variation in crystal structure (equilibrium between crystal structures).
- stereoisomers.
- molecular rearrangement (equilibrium between two different species of the material; acid-base, keto-enol, etc.).

Thermochromic pigments have reached widespread use through different industries, including textile industry, military applications, and plastic industry. Types of pigments are known that change color either reversibly or irreversibly. In order to be functional, thermochromic pigments are produced in a microencapsulated form. However, they are still problematic in some ways. In some cases, low thermal stability and easy extraction from the products, which results in toxicity have been reported.

Thermochromic dyes are typically produced in two types:

- liquid crystal type; and
- leuco type

The leuco dye-based thermochromic pigments generally change from colored to colorless or to another color with an increase in temperature. The cholesteric liquid crystals exhibit "color play" by passing through the whole spectrum with an increase in temperature. They can be painted on a surface or suspended in a fluid and used to make the distribution of temperature visible. Normally clear, or slightly milky in appearance, liquid crystals change in appearance over a narrow range of temperatures called the color-play bandwidth (the temperature interval between first red and last blue), centered around the nominal event temperature (mid-green temperature).

2.3.1 THERMOCHROMIC LIQUID CRYSTALS

A liquid crystal (LC) is a state that exists between the solid and isotropic liquid phase (mesophase). While exhibiting the fluid characteristic of an isotropic liquid, liquid crystal phases possess one- or two-dimensional ordering that imparts crystalline properties such as birefringence. Compounds forming liquid crystal phases are described as mesomorphic or said to possess mesogenic properties. Liquid crystal phases may be divided into two broad categories:

- a lyotropic liquid crystal phase is formed by a mixture of amphiphilic molecules and a solvent (result from the action of solvents); and
- a thermochromic liquid crystal (TLC) phase (results from the action of temperature) is intrinsic to a given compound and appears as a function of temperature.

Low-molecular-weight TLCs (as opposed to liquid crystalline polymers) are classified according to their structure and shape: calamatic, discotic, polycatenar, and amphiphilic (Figure 2.23). Calamatic mesogens, which have an overall rod-like shape, are the most commonly known; discotic mesogens have a disc-like shape; polycatenar mesogens, which have more than two alkyl side-chains, are neither rod- nor disc-shaped but may form liquid crystal phases typical of both categories.

In general, calamatic mesogens may form a nematic (N) phase and one or more smectic phases. The two most commonly observed smectic phases are the smectic A (SmA) and smectic C (SmC) phases. A typical thermochromic phase sequence is shown in Figure 2.24. Hence, cooling from the isotropic liquid state can result in the formation of a N phase which, upon further cooling, can form more highly ordered SmA and SmC phases, and finally a crystalline phase.

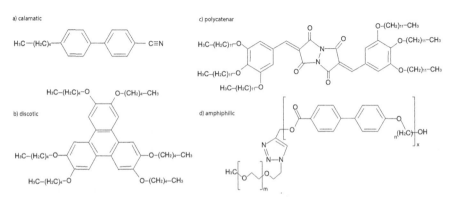

FIGURE 2.23 Examples of (a) calamatic, (b) discotic, (c) polycatenary, and (d) amphiphilic mesogens.

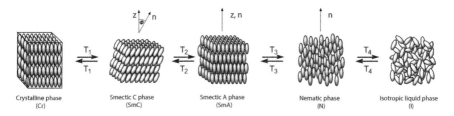

FIGURE 2.24 Schematic representation of crystalline (Cr), smectic C (SmC), smectic A (SmA), nematic (N), and isotropic (I) phases of calamatic liquid crystals.

Liquid crystals with chiral center, known as cholesteric liquid crystals, change color due to influence of temperature. Initially, the products were derivatives of cholesterol as given in their name. However, synthetic chiral molecules known as chiral nematic liquid crystals are being used presently. The cholesteric liquid crystals against a black background manipulate the incident light and reflect selected wavelengths of light that vary with change in temperature. The particular reflected wavelengths depend on the pitch length of the helix formed by the liquid crystals, a parameter that changes with temperature. For integration and application purposes, the leuco dye-based and liquid crystal thermochromic pigments are normally microencapsulated. TLCs show colors by selectively reflecting incident white light. Conventional temperature-sensitive mixtures, in thin films, reflect bright, almost pure colors, turning from colorless (black, on a black background) to red at a given temperature and, as the temperature is increased, pass through the other colors of the visible spectrum in sequence (orange, yellow, green, blue, violet) before turning colorless (black) again at a higher temperature. The color changes are reversible, and on cooling, the color change sequence is reversed. A typical wavelength (color)/temperature response is shown in Figure 2.25.

Thermochromic pigments have been commercialized since the late 1960s and used, for example, in thermographic recording materials. The thermochromic pigments are also used as temperature indicators such as measuring the body temperature, in food containers to determine the temperature or history of the food storage, in medical thermography for diagnosis purposes, in thermal mapping of engineering materials to diagnose faults in product design and in mechanical performance, in the cosmetic industry for moisturizing and as a carrier for vitamins, etc. These materials also have been used in memory devices, in batteries for life indication, and in the architecture field for decoration or for its functionality. These materials have been reported as

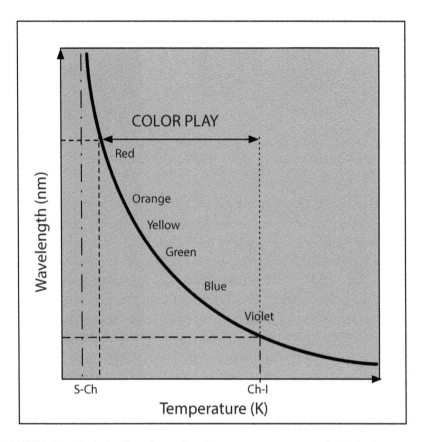

FIGURE 2.25   Typical reflected wavelength-temperature response of a TLC mixture.

novelty materials in toys, ornaments, kettles, umbrellas, toilet seats, etc. In textiles, these pigments have been applied by printing, coating, by a method involving cationization of cotton fabric, and extrusion. They have been used in T-shirts, children clothing, jeans, electronic heat profiling circuitry, incorporated in man-made cellulose fibers, and acrylic fibers. Thermochromic materials have been used in textiles but not to a vast extent [85].

Due to their inherently oily form, pure TLC are very difficult to work, and their thermal performance degrades rapidly due to chemical contamination and UV radiation exposure. This problem is obviously solved by two manufacturing processes, namely microencapsulation and polymer dispersion, which improve protection of the raw TLC. The optimum optical effect is obtained over a black background. Other background colors cause shift in a*b* values, as shown in Figure 2.26.

2.3.1.1 Microencapsulation (ME)

Microencapsulation is a chemical process that takes raw TLC material and encases it in protective 1- to 10-micron-diameter capsules, as shown in Figure 2.27. Custom-made formulations consisting of TLC-filled microcapsules suspended in a water-based binder material are commercially available for widespread application. The microencapsulation process offers very good chemical contaminant resistance and good radiation protection, and it can also make the TLC surface less sensitive to the lighting-viewing geometry. However, special care is necessary to properly prepare and apply ME-TLCs in order to avoid problems associated with over-attenuation of the reflected light coming from the TLCs and segregation of the binder and the microcapsules.

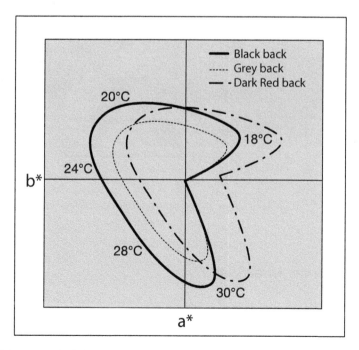

FIGURE 2.26 Typical shape of TLC color change on CIELAB chromaticity plane, and the effect of different backgrounds.

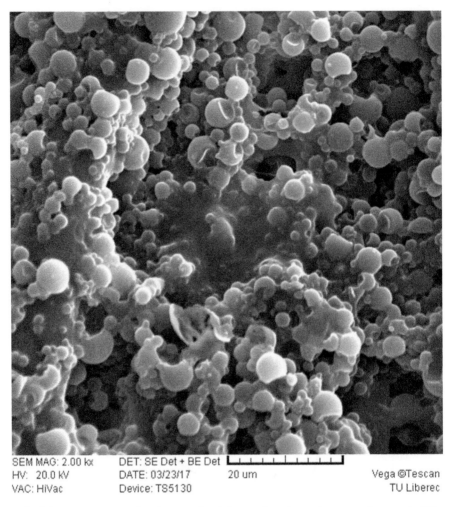

SEM MAG: 2.00 kx   DET: SE Det + BE Det
HV: 20.0 kV   DATE: 03/23/17   20 um   Vega ©Tescan
VAC: HiVac   Device: TS5130   TU Liberec

FIGURE 2.27   An example of encapsulated thermochromic coating on a microscopic slide.

2.3.1.2   Polymer Dispersion (PD)

The process chemically disperses pure monomer-based TLC material into a solid polymer-based matrix. A key benefit of this process is that the virtually transparent polymer material only causes mild attenuation of the reflected light from the TLC (see Figure 2.28). This feature preserves more of the brilliant color response characteristics of the raw TLC material.

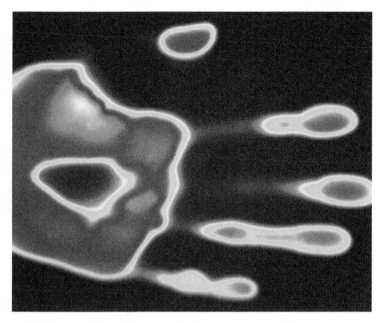

FIGURE 2.28 Example of hand mark on the PD-TLC material.

On the other hand, PD-TLCs are fairly limited in their usability for two reasons:

1. they are not typically available in a sprayable medium, limiting their use to relatively flat surfaces;
2. edge effects due to chemical contamination can destroy a PD-TLC surface when a cutout portion of a manufactured sheet is used.

TLC mixtures have a characteristic red start or mid-green temperature and color-play bandwidth. The bandwidth is defined as the blue start temperature minus the red start temperature. The color play is defined by specifying either the red start or mid green temperature and the bandwidth. For example, R30C6W describes a liquid crystal with a red start at 30°C and a bandwidth of 6°C, i.e., a blue start 6°C higher, at 36°C. Other possibility is the description range of temperature alone, such as liquid crystal ink at 24°C to 29°C [88].

Both the color-play bandwidth and the event temperature of a liquid crystal can be selected by its proper chemical composition. The event temperatures of liquid crystals range from –30°C to 115°C with color-play bands from 0.5°C to 20°C, although not all combinations of event temperature and color-play bandwidth are available. Liquid crystals with color-play bandwidths of 1°C or less are called narrow-band materials, while those whose bandwidth exceeds 5°C are referred to as wide-band. The type of material to be specified for temperature indicating should depend very much on the type of available image interpretation technique: human observers, image processing, or spectrophotometers. The uncertainty associated with direct visual inspection is about one-third the color-play bandwidth, given an observer with normal color vision; this means about ±0.2°C to 0.5°C. The uncertainty of image processing interpreters using wide-band liquid crystals is of the same order as the uncertainty assigned to human observers using narrow-band materials and depends on the pixel-to-pixel uniformity of the applied paint and the size of the area averaged by the interpreter. It is frequently used for transformation into HSI color space. Local calibration can be achieved for high resolution systems with uncertainty pf ±0.05°C, 0.3% to 7.9% of the calibrated temperature range. Using a spectrophotometric system, the resolution is better than ±0.1°C.

2.3.2 THERMOCHROMISM OF LEUCO DYES

The leuco dye (LD)-based thermochromic pigments consist of a color former, a developer, and a solvent in a specific combination [89]. The color former is a pH-sensitive dye, mostly belonging to the spirolactone or fluoran class, as shown in Figure 2.29.

The color developer is a proton donor and normally is a weak acid such as bisphenol A; octyl p-hydroxybenzoate; methyl p-hydroxybenzoate; 1,2,3-triazoles; 4-hydroxycoumarin derivatives; laurylgallate; ethylgallate; p-hydroxybenzoic acid methyl ester; and various other phenols, aromatic amines, carboxylic acids, and Lewis acids (see Figure 2.30).

The solvent is a low-melting hydrophobic, long aliphatic chain fatty acid, amide, or alcohol, and its melting point is used to associate the system with a specific temperature at which the color former and developer interact [5, 90].

Leuco dyes color change is provided therefore via two competing reactions, namely one between dye and developer and the other between solvent and developer. The first reaction prevails at low temperatures where the solvent is in solid form; therefore, dye-developer reaction prevails which

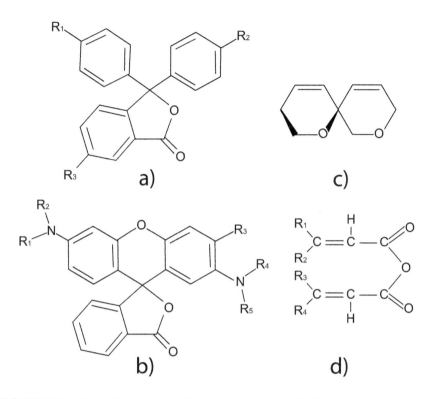

FIGURE 2.29 Typical leuco dyes include (a) spirolactones, (b) fluorans, (c) spiropyrans, and (d) fulgides.

gives rise to the formation of dye-developer complexes. In most cases, these complexes are colored. At a higher temperature, the solvent melts and the solvent-developer interaction becomes dominating; thus, dye-developer complexes are destroyed, and the system converts into its colorless state (see Figure 2.31).

Provided that the mixture is formulated correctly, a striking color change from colored to colorless occurs upon heating the composition above its melting point, and the original color returns when the material solidifies through cooling. Additional components can be added to the thermochromic composite. Stabilizers can be added to enhance the light fastness of the color formers. Regular textile dyes can be added to create a base color and thereby a thermochromic transition, between this base color and the combined color, can be attained.

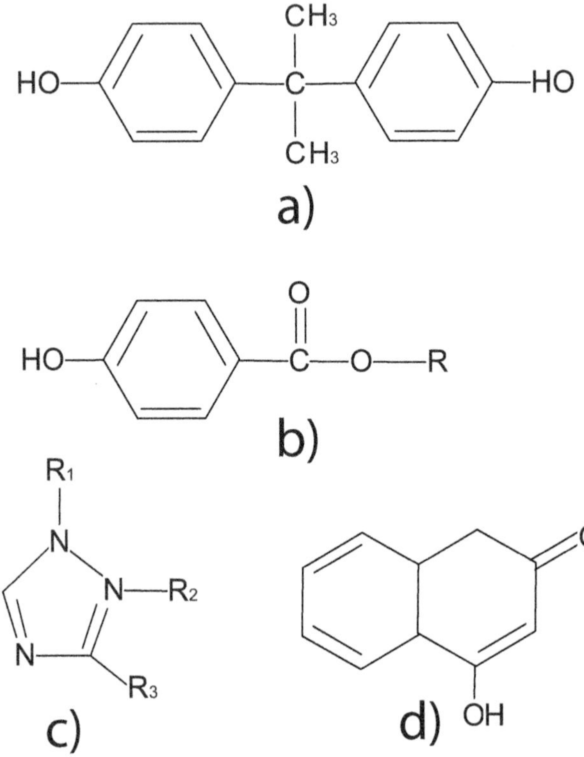

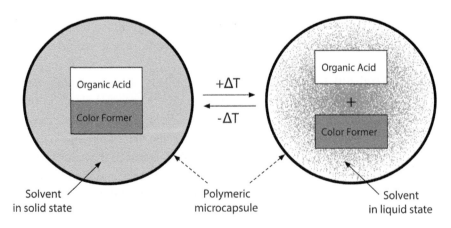

FIGURE 2.30 Typical color developers include weak acids such as (a) bisphenols, (b) alkoxy-p-hydroxybenzoates, (c) 1,2,3-triazoles, and (d) 4-hydroxycoumarin.

FIGURE 2.31 Scheme of the LD thermochromic change.

The temperature-driven phase change mechanism implies that it is important to have the appropriate balance between the solubility and solvating power of the components in the composite as well as the correct balance between the acidity (tendency to donate a proton) of the developer and basicity (tendency to accept a proton) of the dye. It is important that the developer possesses just enough solubility in the molten solvent to cause complete dissolution (full developer–solvent interaction) and yet have a poor enough solubility in the cooled composition to maximize phase separation of the developer and solvent so that developer–dye interactions prevail. The dye should have sufficient solubility in the solvent as well as the correct basicity to give color formation with high contrast. Furthermore, the dye must be compatible (give good association) with the developer. Otherwise, phase separation of these two components, in the solid, color developed state, could occur and result in the loss of color over time [91].

Because these systems involve changes in phase between colored solid and colorless liquid states, applications must generally employ microencapsulation or lamination to protect the composites and safeguard their thermochromic properties. This restricts the techniques available for applying the thermochromic material to textiles to that of pigment printing of fabrics or incorporation into synthetic fibers during their manufacture.

Unlike the TLCs, the LDs have very robust color possibilities. They are relatively inexpensive and substantially easier to process and apply [92]. However, LDs are not nearly as accurate at changing color at a precise temperature when compared to LCs. Color will begin to change over a 2–5°C range although certain LDs have been processed to have more or less of a range. Further LDs exhibit a unique property called "hysteresis" [93]. This property, which is defined as a lag or delay in response to the change of stimuli (in this case, it is the returning temperature decrease), causes the return color change (now nearly colorless) to maintain its colorless state until it further cools 2–5°C below than when it originally cleared during the warming process. The desired range of hysteresis is another variable characteristic that can be customized by the encapsulation processor (Figure 2.32).

Leuco dye-based thermochromic inks of various transition temperature are commercially available, from −15°C to 65°C, but most current applications are limited to three standard temperature ranges [91]:

- body-heat activated (~ 31°C); and
- warm (~ 43°C).

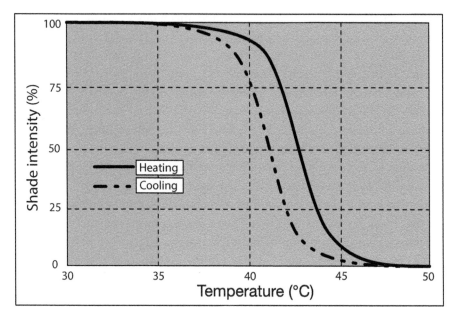

FIGURE 2.32 Hysteresis of leuco dyes.

Cold region is applied for checking the freshness of drinks, i.e., whether it is appropriately cooled. Body-activated region can serve to cover some information by a layer that could become transparent by the heat of the body (see Figures 2.33 and 2.34).

Warm region is applied to indicate the customers if the food is too hot to be eaten. Thermochromic printing inks of all basic types are available; water-based and photocuring inks for printing on paper, plastics, and textile.

2.3.2.1 Thermochromic Polymers

Thermochromism can appear in all different classes of polymers like thermoplastics, duroplastics, coatings, or gels [94]. The polymer is an embedded additive or a supramolecular system built by the interaction of the polymer with an incorporated additive, which can cause the thermochromic effect. Polymers that show conjugation such as poly(3-alkylthiophenes) and poly(3-alkoxythiophenes) may show thermochromism along with other chromism in the solid and solution form. In macromolecular systems, a hypsochromic

FIGURE 2.33   Baby T-shirt before and after thermal exposure.

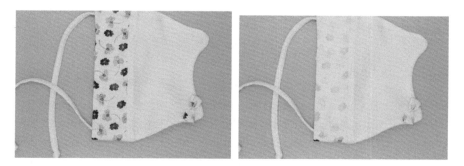

FIGURE 2.34   Baby cap before and after thermal exposure.

reversible color change occurs, which is also known as negative thermochromism. For example, poly[3-oligo(oxyethylene)-4-methylthiophene] is violet at room temperature and yellow at 80°C [95]. The effect occurs due to a planar molecular structure at room temperature, which is effectively conjugated. On increasing temperature, the structure becomes disordered, twisted, and non-planar and becomes less effectively conjugated as shown in Figure 2.35.

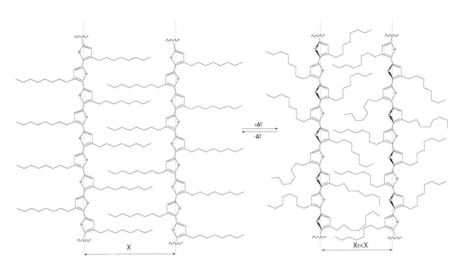

FIGURE 2.35 Polythiophene side chain melting.

If a solubilizing group (such as the ethanesulphonate group in the case of water) is present or induced in the polymer chain, then the color change can occur in different solvents. By controlling side chain length and flexibility, the temperature range in which color change may take place can be varied.

In the context of sun protection, thermochromic polymer systems obtained by doping the polymer matrix with thermochromic additives are of special interest.

2.3.2.2 Sol-gel

Sol–gel glasses have attracted much attention in recent times as encapsulation matrices for molecular entities [96]. Organic reversible thermochromic complexes have become increasingly important in recent years in the study and are used for textile and smart coating. Organic complexes with a leuco dye, a weak acid (electron acceptor), and a solvent are commonly used in consumer applications [89, 97, 98].

An example of the other type of temperature-sensing system comprises a porous organosilica sol–gel with encapsulated pyridinium N-phenoxide betaine dye [99]. While typical thermochromic response of materials constitutes a temperature-modulated change in equilibrium between two colored forms (or between colored and colorless forms), the thermochromic responses of pyridinium N-phenoxide betaine dyes constitute a temperature-dependent change in absorption maxima [99]. As such, these dyes provide a means for fabrication of optical color sensors for temperature.

Generally, numerous applications for the materials have been claimed in the patent and scientific literature over the years, although not all have been successfully commercialized. A representative selection of applications is summarized below:

a.   General temperature indication/digital thermometers;
b.   Battery testers and other voltage measuring devices;
c.   Temperature indicators for medical applications;
d.   Medical thermography;
e.   Radiation detection;
f.   Esthetic.

2.4   MATERIALS WITH CHEMOCHROMIC COLORANTS

Chemochromic materials are defined as materials that change color in the presence and by the effect of chemical agents. The chemical agents that these show color effect under their presence (for example, acids, alkalis) can be classified into these categories of chromogenic materials:

- Gasochromic materials;
- Halochromic materials;
- Solvatochromic materials; and
- Hygrochromic materials.

2.4.1   GASOCHROMIC MATERIALS

Gasochromic materials are commonly used in the form of optical gas sensor. Similarly, other fiber optic sensors have intrinsic advantages over the electronic counterpart due to their small footprint, low cost, rapid sensing speed, real-time monitoring capability, high sensitivity and selectivity, possibility of distributed measurement, and immunity to electromagnetic interferences. There are two groups of fiber-optic sensors according to constructions: extrinsic and intrinsic sensors. For extrinsic sensors, fibers are simply used to guide light to and from the region where the light beam is influenced by the measurand. In this case, the fiber is not used for the sensing function. On the other hand, for intrinsic sensors, the interaction with the analyte occurs within the optical fiber element; that is, the optical fiber structure is modified and the fiber itself plays an active role in the sensing function. Many parameters like phase, polarization, and intensity of the output light may be modulated. Gasochromic material-based optical gas sensors are produced commonly in the form of "optrode." The word optrode is a combination of the words "optical" and "electrode" and refers to fiber optic devices that measure the concentration of a specific chemical or a group of chemicals [100, 101]. The basic structure of an optrode is composed of a source fiber and a receiver fiber that are connected to a sensing layer as illustrated in Figure 2.36.

An alternative gas sensor was designed by Courbat and others [102]. The authors presented the development of colorimetric films for the detection of ammonia in a waveguide configuration. These films were based on a pH indicator embedded in a polymeric matrix and subsequently spin-coated onto a glass slide and coupled to external LED and photo-detector components. Later, the authors [103] proposed a modified low-cost and low-power colorimetric gas sensor based on plastic technologies and additive processes targeting applications in autonomous and wireless systems, which is based on plastic foil – PEN or PET – as the substrate and directly used as a planar optical waveguide to simplify its fabrication. Until today, many materials have demonstrated gasochromic behavior. Among the notable ones, tungsten trioxide (WO_3) is a well-known metal oxide for its applications in gas sensing industry. WO_3 has been treated as a promising candidate for gasochromic platform-based sensing device in many studies [104]. It has been widely reported that when WO_3 is coated with a catalytic thin coating of noble metals (e.g., Au, Pt and Pd), and exposed to hydrogen gas, it will undergo a

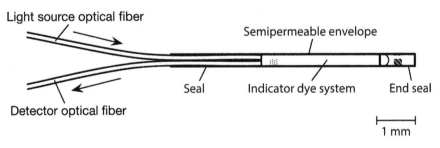

FIGURE 2.36 Typical design of an optrode – optical sensor with a color changeable dye system.

reversible coloration process [105–107]. Although the origin of this coloration mechanism is still controversial, recently, the double injection model is widely accepted [108]. Generally speaking, gasochromic smart windows, which fall into the subclass of chromic devices, have very similar operation principles to electrochromic smart windows, except that their external stimulus is reducing gases instead of applied voltages.

Hsu and others [109] reported decreasing response of WO_3 films with H_2 based on growing temperature as shown in Figure 2.37.

Another application of gasochromic sensors is reported by Viková [110, 111]. One of the possibility of evaluation of barrier properties of protective clothes is its testing during acting model gas. Special protective clothes for dangerous environment expects the determination of limits of their safety and reliability. The commonly used method is based on the system of local passive dosimeters, which are placed on specific positions on the human body. This method is complicated, costly, and time-consuming. Moreover, this method gives only approximate information of place of gas penetration through protective clothes. Chemochromic indicator in the form of textile underwear was developed as an alternative at the National Authority for Nuclear, Biological and Chemical Protection, Czech Republic. The intensity of color change allows quantitative evaluation of amount of penetrated gas as well as the area of place where color change happened. This will provide information about the weakness of protective clothes construction, as shown in Figure 2.38 or insufficient management of wearing of protective clothes (see Figure 2.39).

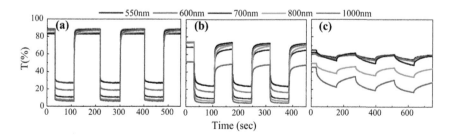

FIGURE 2.37 The dynamic responses of the Pt/WO_3 thin films toward 100% of H_2 gas measured include the wavelength of 550, 600, 700, 800, and 1000 nm, (a) room temperature, (b) 250°C, and (c) 500°C grown samples (reprinted from Chia-Hao Hsu et al. [109] with permission © 2013 Elsevier.)

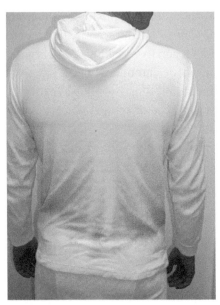

FIGURE 2.38 Example of color change after model gas penetration through protective clothes.

FIGURE 2.39 Example of color change after model gas penetration through the zipper of protective clothes.

2.4.2 HALOCHROMIC MATERIALS

Halochromic materials change color according to the changes in pH of the surrounding medium and are frequently called as pH indicators. They are usually weak acids or bases, which when dissolved in water dissociate slightly and form ions. A specific class herein is halochromic textiles, i.e., fibrous materials that change color with pH. Such halochromic textiles play an important role in the continuous monitoring and visual reporting of the pH with applications in various fields, such as wound treatment and protective clothing. Several studies demonstrated that halochromic dyes can be applied on conventional fabric with normal techniques and that the response of these sensors depends on the density of the fabric. As is well known, due to a large sorption area, pH-sensitive nanofibrous nonwovens have high sensitivity and a fast response time, and they are mostly fabricated by introducing a pH-responsive dye via dye-doping of the feed mixture before fabrication. Dye leaching, however, is a major problem [112]. The coloration of halochromic nonwoven fabric made of nanofibers can instead be obtained directly during the formation of the fibers and covalent linking of the dye prevents dye leaching. Through blend electrospinning of polyamide-6 (PA6) with a dye-functionalized copolymer, large sheets of uniform, halochromic nanofibrous material can be fabricated showing a fast pH-sensitive color change [113]. Polymeric entanglements within the nanofibers are proposed to immobilize the dye-functionalized copolymer in the PA6 matrix, resulting in drastically reduced dye leaching. Nevertheless, above neutral pH, such halochromic system is sensitive on hydrolysis as shown in Figure 2.40.

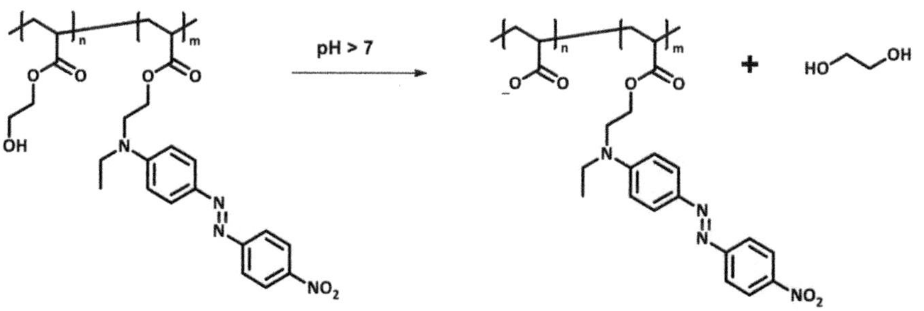

FIGURE 2.40 Proposed hydrolysis mechanism of poly(HEA-co-DR1-A) under basic conditions (reprinted from Steyaert et al. [113] with permission © 2015, Royal Society of Chemistry.); poly(HEA-co-DR1-A) means copolymer of HEA that is 2-hydroxyethylacrylate, DR1 that is Disperse Red 1, and A means acrylate.

On the other side, as an advantage, it is possible to point out that significant differences in electrospinning behavior or fiber morphology were found for the blank, dye-doped or blend nanofibers, as illustrated by Figure 2.41.

Such stable nanofibrous, PA6-based, halochromic materials are particularly interesting in the design of new colorimetric sensors applicable in several sectors, including the biomedical field, agriculture, safety, and technical textiles. Figure 2.42 shows the basic idea of previously

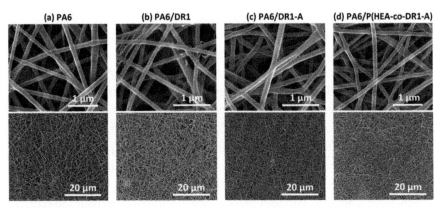

FIGURE 2.41　SEM images of the electrospun samples: (a) PA6 – 136 ± 19, (b) PA6/DR1 – 123 ± 18, (c) PA6/DR1-A – 120 ± 14, and (d) PA6/P(HEA-co-DR1-A) – 138 ± 25 (reprinted from Steyaert et al. [113] with permission © 2015, Royal Society of Chemistry.)

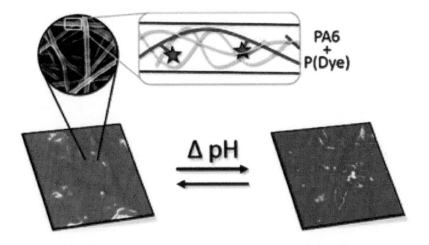

FIGURE 2.42　Scheme of dye immobilization in halochromic nanofibers through blend electro spinning of a dye-containing copolymer and polyamide-6 (reprinted from Steyaert et al. [113] with permission © 2015, Royal Society of Chemistry.)

mentioned fiber sensor, which is sensitive to change in pH in strongly acidic environment.

2.4.3 SOLVATOCHROMIC MATERIALS

The term solvatochromism is used to describe the change in position, intensity, and shape of the UV–VIS absorption spectrum of the chromophore in the solvent of different polarity [114]. Positive solvatochromism refers to the absorption of the chromophore, which has a neutral ground state show a bathochromic

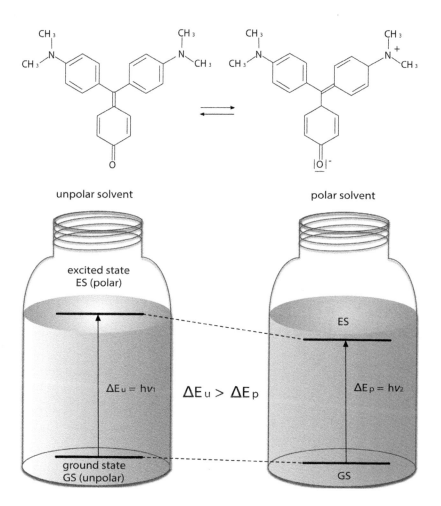

FIGURE 2.43   Scheme of positive solvatochromism of 4,4'-bis-dimethylamino-fuchson.

(red direction) shift with increasing solvent polarity. On the other hand, when the chromophore has a charge-separated ground state, it leads to negative solvatochromism that refers to a hypsochromic (blue direction) shift of their absorption with increasing solvent polarity. According to the zwitterionic nature of the chromophore, a strong negative solvatochromism effect can be observed as a function of the medium polarity [115]. Moreover, in the case of negative solvatochromism, increasing the solvent polarity would stabilize the ground state of the zwitterionic chromophores. Now, let us compare the solvent effects on absorption and fluorescence. The energy levels are shown for the unsolvated ground and excited states of this dye molecule. It clearly indicate how these levels (and the gap) are affected by polar solvation in the case of absorption. In case of positive solvatochromism, the difference between ground and excited state is higher than that in polar solvent as shown in Figure 2.43.

The resulting absorption curves of 4,4'-bis-dimethylamino-fuchson in toluene and methanol, with visible bathochromic shift of absorption maxima, are presented in Figure 2.44.

Similarly, we can explain negative solvatochromism as was mentioned before, contrary to positive version (Figure 2.45).

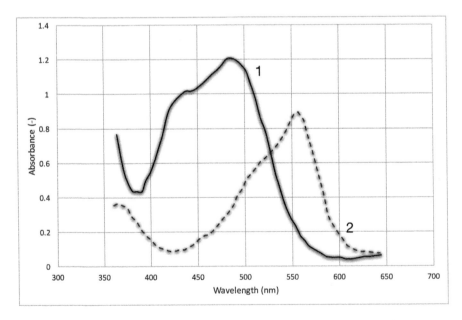

FIGURE 2.44 Absorption curves of 4,4'-bis-dimethylamino-fuchson in toluene (1) – polarity index 2.4 and methanol (2) – polarity index 5.1.

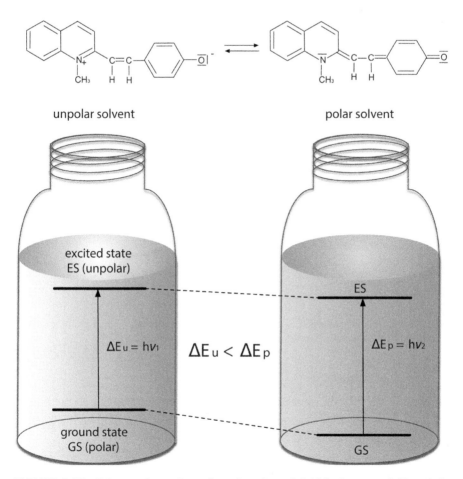

FIGURE 2.45 Scheme of negative solvatochromism of 2-(4'-hydroxystyryl)-N-methyl-chinolinium-betains.

The resulting absorption curves of 2-(4'-hydroxystyryl)-N-methyl-chinolinium-betains in chloroform and water, where visible hypsochromic shift of absorption maxima is presented in Figure 2.46.

In solvatochromic materials, we can also add different detectors of chemical agents, which contain pigments selectively soluble in these agents. Paper chemical agent detectors for certain chemical warfare purposes were developed during the 1960s and have been in widespread use in military applications. In their most sophisticated form, these detectors consist of a paper base in which three dyes have been incorporated, each dye being sensitive to a particular family of liquid chemical warfare agent droplets. These

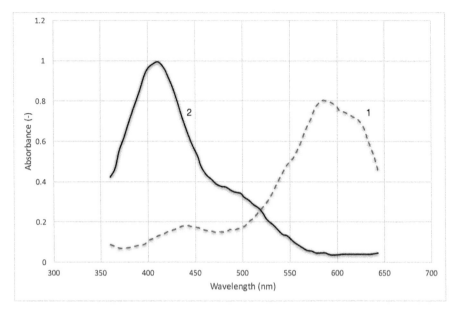

FIGURE 2.46 Absorption curves of 2-(4'-hydroxystyryl)-N-methyl-chinolinium-betains in chloroform (1) – polarity index 4.1 and water (2) – polarity index 10.2.

droplets react with the specific dyes to produce color changes on the paper. According to the color change produced, the type of chemical warfare agent can be identified.

Other forms of paper chemical agent detectors contain only one dye. In these cases, not as much information on the type of chemical warfare agent can be obtained from the reaction with the paper detector. A paper chemical agent detector strip is attached to the clothing or equipment of service personnel so that these personnel can immediately determine whether they or their equipment have been exposed to liquid agent contamination. Also, the paper can be used to determine whether an unknown liquid on clothing, equipment, or terrain is a chemical warfare agent. The preferred method of manufacturing these detectors is to include the dyes in the papermaking slurry so that they are integral with the paper as shown in Figure 2.47.

The result is qualitative, but the detector paper has a sensitivity of about 20 µL of liquid. Some substances can act as interferences and produce false positives, such as insecticides, antifreeze, and petroleum products. The dyes must be solids with a minimum melting point that will

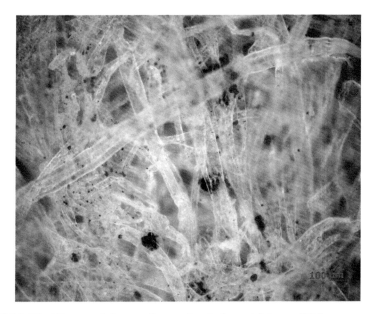

FIGURE 2.47   Microscopic image of paper chemical agent detectors PP-3.

keep them solid during a paper drying process and must give the correct strong color on reaction with chemical warfare agent droplets. There
are various other requirements, including compatibility in terms of color
produced with other dyes in the detector. As currently utilized, the detectors incorporate three dyes, one each for the detection of H-type agents
(mustard), G-type nerve agents, and V-type nerve agents. Such three dyes
are   frequently   2,5,2',5'-Tetramethyltriphenylmethane-4,4'-diazo-bis-β-
hydroxynaphthoic anilide commonly called E, 4-[(4-(phenylazo)phenyl)
azo]-phenol is sold under the common name C.I. Disperse Yellow 23, and
ethyl-bis-(2,4-dinitrophenyl)acetate commonly called as EDA (Figures
2.48–2.50).

Unfortunately, EDA was found to be strongly mutagenic [116]. Despite
the fact that soldiers attach these papers routinely to their uniforms and
equipment have a very low exposure to the impregnated dyes, the problem
is crucial for people who are involved in production of these paper chemical agent detectors. Based on this issue, a suitable replacement of EDA
was proposed as 3',3",5',5"-tetrabromophenolphtalein ethyl ester, which is
referred as TBPE [117]. Other possibility is production of textile chemical

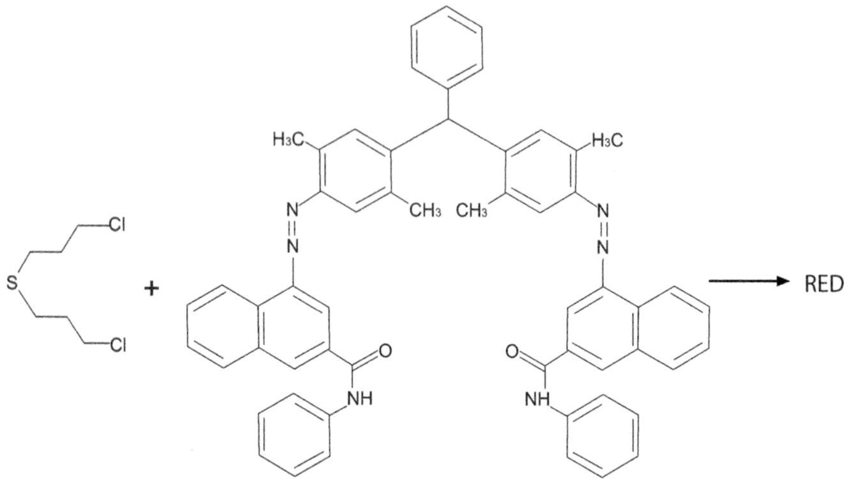

FIGURE 2.48 Reaction of 2,5,2',5'-tetramethyltriphenylmethane-4,4'-diazo-bis-β-hydroxynaphthoic anilide with H-type agents creates red color.

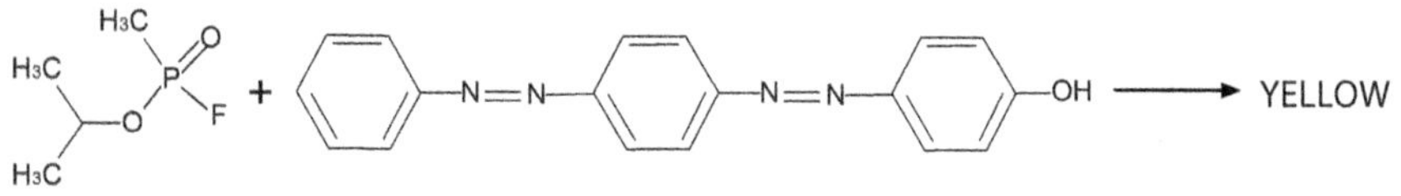

FIGURE 2.49 Reaction of 4-[(4-(phenylazo)phenyl)azo]-phenol with G-type nerve agents creates yellow color.

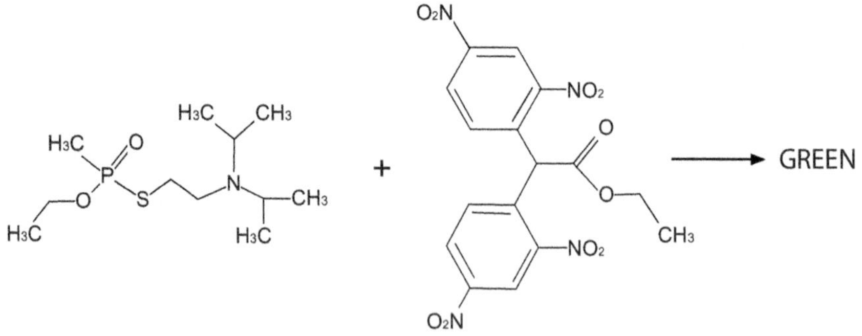

FIGURE 2.50 Reaction of ethyl-bis-(2,4-dinitrophenyl)acetate with V-type nerve agents creates green color.

agent detector by the method of layer coating, which gives a short reaction time of detector [118]. An example of textile alternative of three-way paper chemical agent detector is shown in Figure 2.51.

2.4.4 HYGROCHROMIC MATERIALS

Hygrochromic materials change color in response to the presence of moisture or to contact with water. Humidity sensors are used extensively in industry as well as for environmental monitoring. Their widespread applications cover a broad range of domestic, medical, and industrial applications. For example, in food packaging, excess moisture in meat packaging can accelerate food spoilage, and as a consequence, desiccants are often included in packaging to extend the shelf-life [119].

It is possible to classify the humidity-sensitive materials into four groups, namely electrolytes, organic polymers, porous ceramics [120], and zeolites [121]. Among these materials, functional ceramics have played a major role because of their intrinsic characteristics. They are superior in mechanical strength, exhibit high chemical resistance in most environments, and

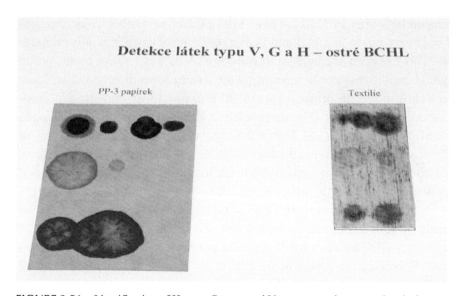

Detekce látek typu V, G a H – ostré BCHL

PP-3 papírek

Textilie

FIGURE 2.51 Identification of H-type, G-type, and V-type agents by paper chemical agent detectors PP-3 and its textile alternative.

show good reproducibility of the electrical properties. However, the sensor devices are expensive and are sometimes affected by exposure to liquid water or organic solvent vapors. As an alternative to the measurement of electrical properties, optical sensors have become increasingly important as they can be miniaturized, are relatively cheap, and simple to set up. Among these are optical sensors for humidity monitoring that rely on the reaction of metal ion complexes with water, resulting in color change, such as cobalt (II) chloride, copper chloride, and zinc chloride, or on luminescence changes in calcein or metallo-porphyrines [122]. Another approach is based on the response of polarity-sensitive dyes to water vapor [123]. European Union classified cobalt chloride as a category 2 carcinogen, while repeated exposure to copper chloride can lead to copper poisoning. Zinc chloride is also known to be very harmful to the environment. These materials are very toxic to aquatic organisms and may cause long-term adverse effects in the aquatic environment. While this does not require elimination of use, some customers have decided to eliminate their use based on the chemicals listing as a "Substance of Very High Concern" (SVHC). More environment friendly alternatives with equivalent performance have been identified [124–126].

Matsushima et al. have proposed using thiazine and flavylium salts (parent of the anthocyanidines) in gels as simple colorimetric humidity and temperature sensors [127]. These salts exhibit reversible color changes from blue (dry) to purple (humid), as a result of a change in relative humidity. This has been attributed to the absorption of water vapor by the gel in humid conditions which encourages the dye to form dimers and so leads to a shift in λ_{max} absorbance (ca. 10–20 nm) to a lower wavelength. Spin-coating alternative of this novel relative humidity sensor was developed by Mills and others [128]. Figure 2.52 shows the response and recovery of the absorbance for a typical indicator produced from methylene blue/urea/hydroxyethylcellulose (MB/urea/HEC) film upon exposure to repeated cycles of humid and dry air.

As demonstrated in Figure 2.53, a typical relative-humidity sensor was made by spin coating an ink comprising 5 mg MB, 100 mg urea in 2 g of a 5% w/v HEC aqueous solution for 30 s at 3500 rpm on to a 25-mm glass disc. Following the drying of the film at 70°C for a few minutes, the final product is an opaque pink film (ca. 1.7 mm thick) under ambient conditions (40–60% RH) with a λ_{max} at 570 nm, which upon exposure to high

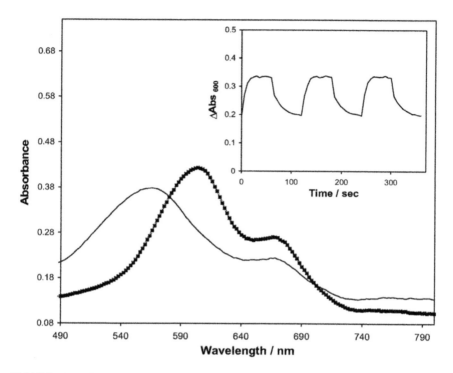

FIGURE 2.52 Spectral changes of a typical MB/urea/HEC relative-humidity indicator film before (—) and after (■) exposure to 100% RH air. Inset diagram is a plot of the change in absorbance MB $\lambda_{maxhumid}$ (600 nm) for the indicator film on exposure to 3 cycles of 1 min 100% RH air and 1 min dry air. Abs_{600} recorded every 6 s. MB is methylene blue, and HEC is hydroxyethylcellulose (reprinted from Mills et al. [128] with permission of Royal Society of Chemistry).

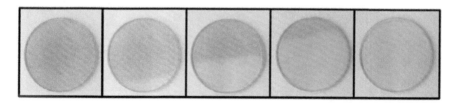

FIGURE 2.53 Photographs of a typical MB/urea/HEC relative-humidity indicator changing color from pink (left) to blue (right) upon exposure to 100% RH air, from the bottom right (reprinted from Mills et al. [128] with permission of Royal Society of Chemistry).

relative-humidity conditions (>85% RH) turns rapidly blue ($\lambda_{max} \approx 600$ nm) and clear. This color change process occurs rapidly, and the response and recovery time is 10 s and 60 s, respectively.

Optical chemical sensors must perform two functions: they must interact with analyzed chemicals and subsequently report such interactions. Due to some interactions between chemicals (analyte) and the chemochromic dye, it is necessary to use set of dyes in the form of sensor array for improvement differentiation. The other problem is resolution of compounds that are not spectroscopically active. As solution of this problem is a chemochromic dye with complexing agents allowing creation of suitable chromophore. For gas phase sensing, a colorimetric sensor array is simply digitally imaged before and during exposure to any volatile analyte. The imaging is mostly commonly achieved with an ordinary flatbed scanner, but one may also use digital cameras, portable handheld readers, and even cell phones; constancy of illumination is, of course, important. An example of a 36-spot sensor array for use with gas phase analytes is shown in Figure 2.54.

From the digital images, a difference map (Figure 2.54) is easily generated by digital subtraction, pixel by pixel, of the image of the array before and after exposure: red value after exposure minus red value before, green minus green, and blue minus blue. Averaging of the centers of the spots avoids artifacts from nonuniformity of the dye spots, especially at their edges. The other advantage of using the differences in RGB colors is that it tends to cancel out discrepancies in printing because the color differences

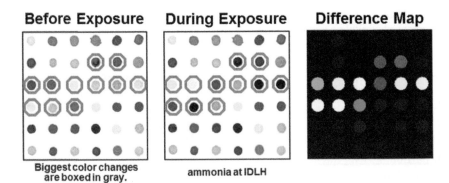

FIGURE 2.54 Image of the 36-dye colorimetric sensor array (left) before exposure and (middle) during exposure to ammonia at its IDLH (immediately dangerous to life or health concentration), (right) subtraction of the two images yields a difference vector in 108 dimensions (i.e., 36 changes in red, green, and blue color values); this vector is usefully visualized using a difference map, which shows the absolute values of the color changes. For purposes of display to increase the color palate, the color range of difference maps are usually expanded (reprinted from Askim et al. [129] with permission of Royal Society of Chemistry).

are only a weak function of variation of the dye concentration or spot intensity from array to array [129].

An efficient chemochromic system is found in olfactory systems. The olfactory system permits differentiation among a huge number of chemical compounds and complex mixtures over an enormous range of concentrations. This kind of molecular recognition could not utilize the usual model of bio specificity, i.e., the lock-and-key mechanism of enzyme–substrate interaction. The olfactory receptors represent the exact opposite of that kind of specificity and show highly cross-reactive, nonspecific interactions with odorants. Molecular recognition instead occurs through the pattern of response from hundreds of different types of olfactory receptor epithelia cells (each of which expresses only a single one of the hundreds of olfactory receptors found in our genome), as analyzed by the olfactory bulb and the brain [129]. Based on their recognition element properties, the sensors used in an array will span a range of molecular specificity. Importantly, by proper choice of dyes and substrate, the array is essentially nonresponsive to changes in humidity. A selection of the difference maps of a representative subset of 24 volatile organic compounds is presented in Figure 2.55.

Since each volatile organic compound produces a different mixture of oxidized derivatives, the array response to these more reactive volatile by-products provides a unique, but much more sensitive, signature for the initial volatile compound. Twenty commonly found volatile organic compound pollutants in indoor air were examined as representative analytes, and all were discriminable by hierarchical cluster analysis, as shown in Figure 2.56, both at their immediately dangerous to life or health and at their permissible exposure limit.

2.5 MATERIALS WITH ELECTROCHROMIC COLORANTS

Electrochromism is defined as an electrochemically induced color change. In particular, it is the change in optical absorption bands brought on by an electrochemical redox reaction in a material. A discussion of electrochromism first appeared in the literature in the early 1960s [130]. Since then, electrochromism has been demonstrated in a variety of materials, both inorganic and organic, and has been utilized to develop contrast-based display and various other technologies. The three fundamental types of electrochromic materials were first defined in Ref. [131] and subsequently adopted in Refs. [132, 133]:

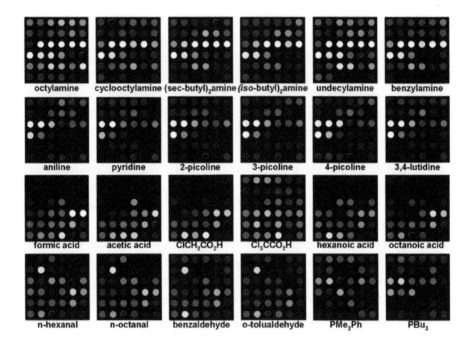

FIGURE 2.55 Colorimetric array response to volatile organic compounds visualized as color difference maps. The figure shows 24 representative volatile organic compounds after equilibration at their vapor pressure at 295 K (reprinted from Askim et al. [129] with permission of Royal Society of Chemistry).

Type I – Materials soluble in both their oxidized and reduced forms. Methyl viologen (1,1'-dimethyl-4,4'-bipyridinium di-cation, MV) is an example of a type I material. Its structure is shown in Figure 2.57.

Type II – Materials that are soluble in one redox state, but become insoluble in another, existing as a solid deposit on the electrode surface. Heptyl viologen (1,1'-diheptyl-4,4'-bipyridinium di-cation, HV) in an aqueous solution behaves in this manner. HV is soluble as a di-cation salt, $HV^{2+} 2X^-$, but forms a deposit when reduced to the radical-cation salt, $HV\bullet^+ X^-$. The structure of HV is shown in Figure 2.58.

The abovementioned viologens can undergo two successive electron-transfer reactions from the di-cation (Figure 2.59A) to produce a radical cation (Figure 2.59B) and then a neutral species (Figure 2.59C), yielding differently colored species at each step. The colors formed depend upon on the substituents.

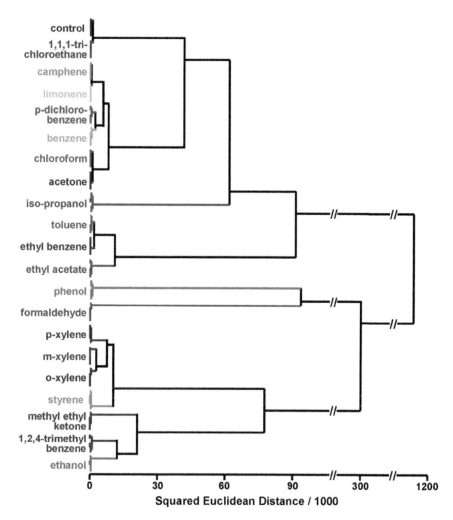

FIGURE 2.56 Hierarchical cluster analysis dendrogram for 20 commonly found indoor pollutant VOCs at their IDLH concentrations and a control. All experiments were run in quintuplicate with 30 mg chromic acid on silica as the pre-oxidation reagent; no confusions or errors in classification were observed in 105 trials (reprinted from Askim et al. [129] with permission of Royal Society of Chemistry).

Type III – Solid films or surface bound materials that remain bound whilst their redox state is changed. Examples of such materials include inorganic species (e.g., Prussian Blue, PB; tungsten trioxide, WO_3), where coloration change is due to the new electronic transitions raised along the intercalation process:

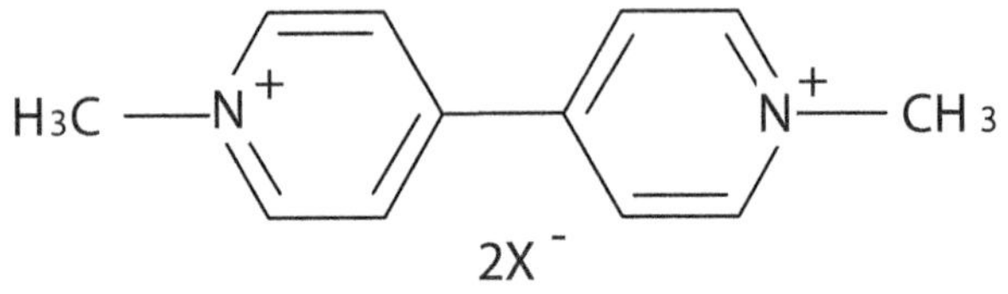

H3C
N+
N+
CH3
2X-

FIGURE 2.57 Methyl viologen, type I electrochromic material, where X– is an appropriate counter ion.

H3C
CH2
6
N+
N+
CH2
6
CH3
2X-

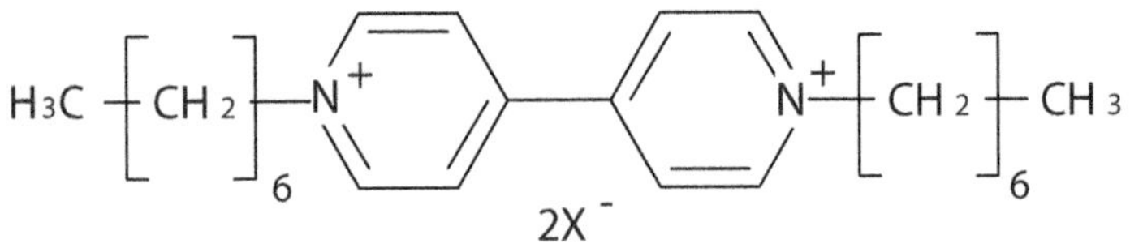

FIGURE 2.58 n-Heptyl viologen, type II electrochromic material, where X– is an appropriate counter ion.

R'
N+
N+
R
+ e-
- e-
R'
N
N+
R
+ e-
- e-
R'
N
N
R
A)
B)
C)

FIGURE 2.59 The three common redox states of viologens, showing the two successive electron transfer reactions. A, di-cation; B, radical cation; C, neutral species.

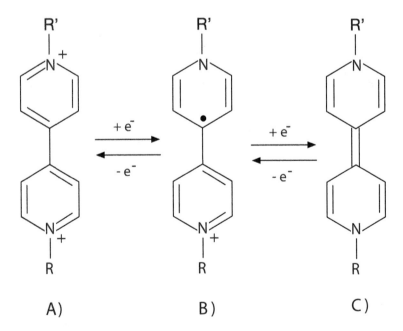

The electrochromic effect is strongly dependent on crystal stoichiometry with significantly different properties being reported for amorphous and crystalline films as visible on reactions of Prussian blue. Figure 2.60 is a visible gasochromism as mentioned in the previous section, which is similar to electrochromism that the optical properties of a number of MO_n can be switched reversibly and persistently in the presence of catalysts when exposed to the reducing or oxidizing gases.

Type III materials are also conducting polymers. Such systems are commonly investigated as thin films or deposits on the electrode's surface. Upon doping, a new range of states appear in the gap region (polarons and bipolarons), leading to new electronic transitions that produce the optical change (see Figure 2.61).

$$MO_n + xI^+ + xe^- \rightleftharpoons I_xMO_n$$

transparent   colored

$$[Fe^{III}Fe^{III}(CN)_6] + e^- \rightleftharpoons [Fe^{III}Fe^{II}(CN)_6]^-$$

Prussian brown   Prussian blue

$$3[Fe^{III}Fe^{II}(CN)_6]^- \longrightarrow [Fe^{III}_3\{Fe^{III}(CN)_6\}_2\{Fe^{II}(CN)_6\}]^- + 2e^-$$

Prussian blue   Prussian green

$$[Fe^{III}Fe^{II}(CN)_6]^- + e^- \longrightarrow [Fe^{II}Fe^{II}(CN)_6]^{2-}$$

Prussian blue   Prussian white (transparent)

$$H_2 \rightleftharpoons H^+ + e^-$$

$$WO_3 + xH^+ + xe^- \rightleftharpoons H_xWO_3$$

transparent   colored

FIGURE 2.60 Electrochemical reactions of transitional metal oxides. M is a transitional metal ion; I+ is a positive ion; the quantity x becomes the stoichiometric parameter of the product and can vary between 0 and 1.

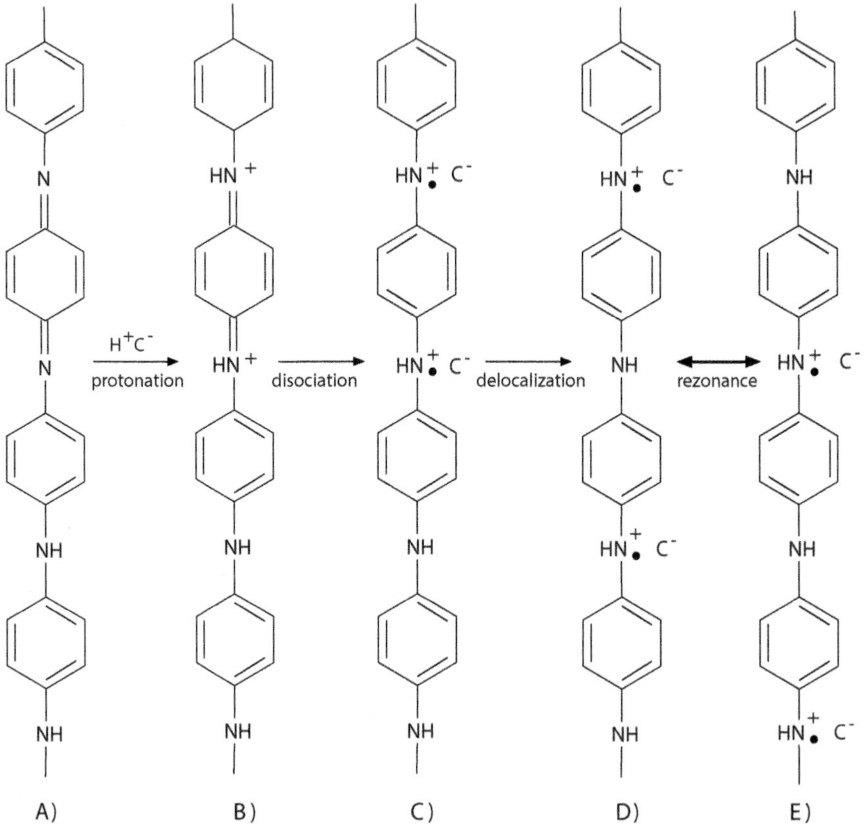

FIGURE 2.61 The doping of polyemeraldine with protons to form the conducting state of polyaniline. (A) polyemeraldine, (B) bi-polaron form, (C) polaron form, (D) delocalized polaron form, and (E) resonance forms of delocalized polaron lattice.

An important property of conducting polymers is that they could be chemically modified easily to provide practically all the colors that we desire (see Figure 2.62).

Polyaniline, polypyrrole, and poly(3,4-ethylenedioxythiophene) are among the most widely used conducting polymers. Compared to metal oxides, they are inexpensive, easy to process, have high coloration efficiency, great amount of availability of different colors, and low fabrication cost. However, they exhibit poor stability. Apart from their use in electrochromism, there are other applications such as organic light emitting diodes, solar cells, batteries, sensors, etc. The electronic properties of any specific polymer are

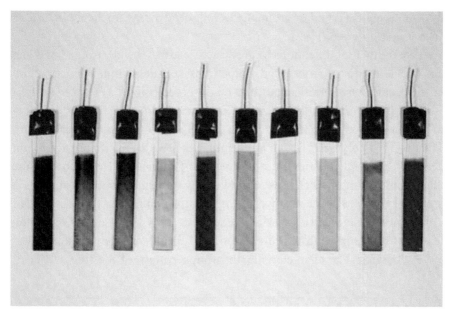

FIGURE 2.62 The photograph of a series of neutral EDOT and BEDOT-arylene variable color electrochromic polymer films on ITO/glass illustrating range of colors available (reprinted from Mortimer et al. [133] with permission © 2006 Elsevier.)

unique. This means that tailoring of the polymers functionality — usually on the monomer — can allow for property manipulation, e.g., band gap control and thus color control. Figure 2.63 shows some examples of how different monomers can be synthesized based on the same core structure by changing

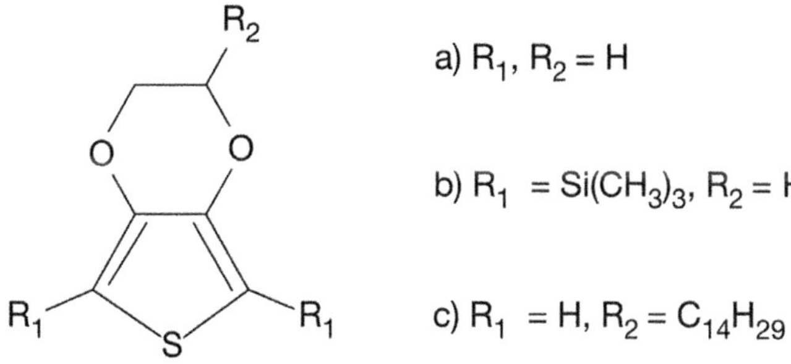

a) $R_1, R_2 = H$

b) $R_1 = Si(CH_3)_3, R_2 = H$

c) $R_1 = H, R_2 = C_{14}H_{29}$

FIGURE 2.63 Some examples of a monomer's core structure, e.g. EDOT, (a) EDOT, (b) EDOT-TMS, and (c) EDOTOC$_{14}$H$_{29}$.

the functionality (R and R' groups), (3,4-ethylene dioxythiophene, EDOT, as shown in Figure 2.63) [92].

These slight modifications to the structure affect the electronic band gap and thus alter the color observed. Sometimes, seemingly minor functionality changes can cause significant effects on observed color.

Figure 2.64 shows the spectroelectrochemical series for an alkylene-dioxy-substituted thiophene polymer, poly(3,4-(ethylene-dioxy)thiophene). The undoped polymer's strong absorption band, with a maximum at 621 nm (2.0 eV), is characteristic of a π-π^* inter-band transition. Upon doping, the inter-band transition decreases, and two new optical transitions (at ca 1.25 and ca 0.80 eV) appear at lower energy, corresponding to the presence of a polaronic charge carrier (a single charge of spin 1/2). Further oxidation leads to the formation of a bi-polaron, and the absorption is enhanced at lower energies. The characteristic absorption pattern of the free carrier of

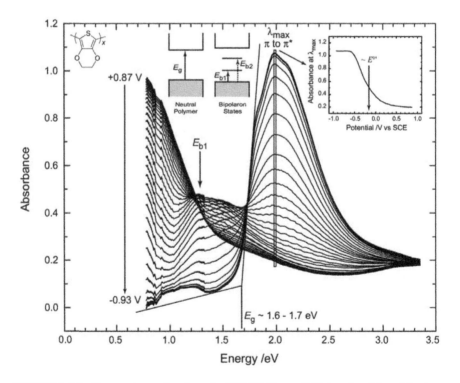

FIGURE 2.64 Spectroelectrochemistry for PEDOT film on ITO (reprinted from Mortimer et al. [133] with permission © 2006 Elsevier.)

the metallic-like state then appears when the bi-polaron bands finally merge with the valence and conduction bands [133].

The ability to tune the color of the neutral polymer by control over the co-monomer concentration during electrochemical copolymerization has also been shown by utilizing the monomers EDOT and BEDOT-NMeCz. As shown in Figure 2.65, by varying the ratios of co-monomer concentrations, colors ranging from yellow to red to blue can be reached in the neutral polymer film. In all copolymer compositions, the films pass from a green intermediate state to a blue fully oxidized state [133].

The discovery of new electrochromic materials and creating new combinations of electrochromic materials for use in novel, operational devices is fundamental to research in this field. Current research focuses on this with less attention spent on the theory of materials and device operation. Commercially, electrochromic materials can be used to make a variety of different devices such as optical information displays/storage, mirrors, electrochromic shutters, windows for energy and light control in buildings and cars, military protective eyewear, camouflage materials, controllable canopies for aircraft, spacecraft thermal control, secondary battery materials, the optical iris of a camera, and even items such as electrochromic sunglasses [134, 135].

2.6   MATERIALS WITH LUMINESCENT COLORANTS

Luminescent materials (or phosphors) are generally characterized by the emission of light with energy beyond thermal equilibrium. Therefore, the nature of luminescence is different from blackbody radiation. As a consequence, external energy has to be applied to luminescent materials to enable them to generate light. Luminescence can occur as a result of very different kinds of excitation, like photo- or electroluminescence. In practice, the luminescence generally is due to excitation with X-rays, cathode rays, UV, or even visible light [136].

Persistent phosphor materials having a long afterglow after the optical excitation has ceased to be used since a long time. Recently, they have gained renewed interest as several newly developed materials show visible luminescence many hours after excitation, thus opening up many novel applications. One of the most important of these is powerless safety illumination in case

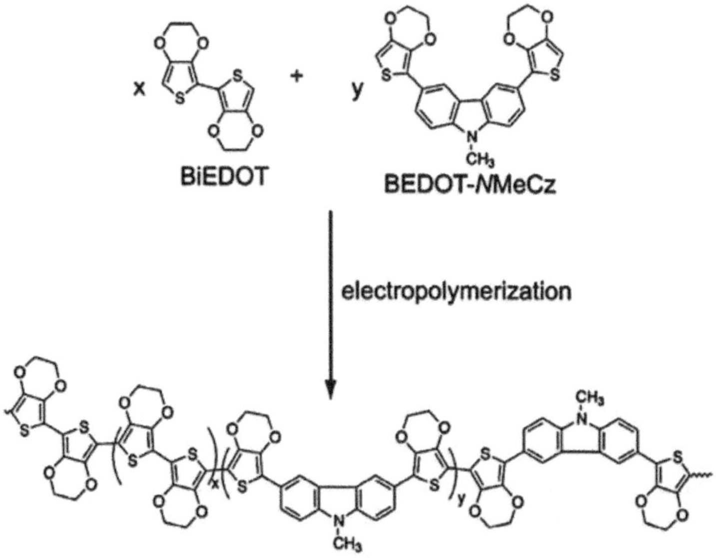

Comonomer Solution Composition	Neutral Polymer λ$_{max}$ (nm)	Neutral Electrochromic Response (Photograph)
100% BiEDOT	577	
90:10	559	
80:20	530	
70:30	464	
50:50	434	
30:70	431	
20:80	429	
10:90	420	
100% BEDOT-NMcCz	420	

FIGURE 2.65 Representative copolymer structure and electrochromic properties of electrochemically prepared copolymers of varied compositions (reprinted from Mortimer et al. [133] with permission © 2006 Elsevier.)

of electric power failures [137, 138]. The usual way in which these phosphors are evaluated, is by measuring their photometric output as a function of time. The afterglow intensity after, for example, 10 min is then quoted as a figure of merit. Alternatively, an extinction time is defined as the time until the luminance has decayed to 0.32 mcd.m-2, which is roughly 100 times the eye sensitivity in dark-adapted condition [139]. Both methods use the photometric quantity candela, which is defined in the context of high-intensity color vision, where only the cones in the human retina are assumed to contribute to vision. However, below about 1 cd.m-2, rods in the retina start to contribute to the perception of luminance, and below 1 mcd.m-2, rods take over completely. Rods have a much higher sensitivity than cones, and their sensitivity curve is shifted toward the blue compared to cones. This results in the Purkinje effect, named after the 19th century Czech physiologist Jan Evangelista Purkyně (Purkinje), who observed that red objects that appeared brighter than blue or green objects in daylight, appeared darker than the blue and green objects at dawn [140]. Therefore, a new, unified luminance definition, as developed by Rea et al. [141], is used to more accurately describe the low-level performance of light sources such as persistent phosphors. We will see that this sheds an entirely new light on the relative performance of phosphors with different emission spectra. The effect is illustrated using a number of selected benchmark commercial persistent phosphor powders. Several relaxation phenomena in complex condensed-matter systems have been found to follow the stretched-exponential decay law [142]:

$$\phi(t) = \exp\left[-(t/\tau)^{\beta}\right] \quad 0 < \beta < 1 \tag{2.1}$$

The parameters β and t depend on the material and the specific phenomenon under consideration, and they can be a function of external variables such as temperature [143]. The book "An Overview: Physics through the 1990s" published in 1986 stated that there seems to be a universal function that slow relaxations obey. If the system is driven (or normally fluctuates) out of equilibrium, it returns according to the function $\exp[-(t/\tau)\beta]$. Unfortunately, this is not a mathematical expression that is frequently encountered in physics, and hence, very little idea exists of what the underlying mechanisms are [144]. Scher et al. [145] pointed out that there are several derivations of the stretched exponential for systems in three dimensions, involving diverse concepts such as percolation, hierarchical relaxation of constraints, and multi-polar interaction transitions. Some authors chose to describe the stretched exponential decay with $\beta<1$ as the result of superposition of many exponential decays [146].

One of the first natural phosphorescence was observed in 1568 by Cellini [147]. Copper-doped zinc sulfide (ZnS:Cu) has been known since the beginning of the 20th century as a long-term green light-emitting material; however, the glow it provides in applications and the continuity of such glow is limited. The visible phosphorus effect cannot be maintained more than a few hours and the phosphorescent glow is lost easily. Therefore, to maintain glow, sometimes radioactive elements are added to ZnS:Cu phosphor-based pigments allowing them to emit the energy by the radioactive substance. However, the procedure of processing and disposal of radioactive elements are very difficult, leading to many problems. Consequently, the use of such some pigments is also limited [148].

Earth alkali sulfides such as CaS and SrS are the next generation long-lasting phosphors. They are among those sulfide phosphors that were so-called Lenard's phosphors studied since 1930s [149]. They were re-discovered by Lehmann et al. in the early 1970s. The phosphors of the type were alkali earth sulfides such as CaS:Bi3+, CaS:Eu2+, CaS:Ce3+, etc. During World War II, long persistence of these phosphors was considered for military application due to their photostimulation properties. In term of long persistence, their major advantages were not only the long persistence time but also that they could be excited under natural solar light [150].

The most efficient persistent luminescent materials include the alkaline earth aluminates and magnesium disilicates doped with Eu2+ and co-doped with other rare earth ions, $CaAl_2O_4$:Eu2+,Nd3+, $SrAl_2O_4$:Eu2+,Dy3+, $Sr_4Al_{14}O_{25}$:Eu2+,Dy3+, and $Sr_2MgSi_2O_7$:Eu2+,Dy3+ [151]. The best materials emit in visible even in excess of 24 h in the dark. Although many new phosphors were developed within the current 10 years, only some of them can be excited under visible natural light. The others required UV or deep UV excitation to trap electrons. The issue whether the long persistence can be obtained by natural light excitation is very important, and actually, it sometimes limits the application of long-lasting phosphors [150].

Technologically important forms of luminescence may be split up into several categories such as photoluminescence, cathodoluminescence, and electroluminescence. In textile applications, an important aspect is fashion and design where different colors of textile substrate are used. Retro-reflex-ive/luminescent strips are placed on fluorescent textiles such as warning vest or protective clothes (first aid, firemen's...) that are relatively rigid and less permeable. The wear comfort of sportswear is an important qual-ity criterion that affects performance, efficiency, and well-being. Hence, fabric breathability (moisture and air permeability) and thermal properties should be tailored in order to meet the requirements of sportswear. Protec-tive design brings new point of view in the world of personal protective clothes, where high visible strips are obvious. An important question is how long the phosphorescent design need to be addressed satisfactorily. Results of work by Vik et al. [146] shows that based on different colors, the predicted difference in decay time could be more than 100% and phos-phorescence spectral effectiveness function allows computing of relative response of individual materials used as background of phosphorescent pigment and relating decay time of phosphorescent pattern (see Figure 2.66). Based on the simple calibration set, it is now possible to obtain similar phosphorescence characteristics on different colored materials.

In this case, it is necessary to remember that the photometric quanti-ties, as defined for photopic vision, are highly inaccurate at intermediate light levels. More specifically, low-wavelength emission is underestimated

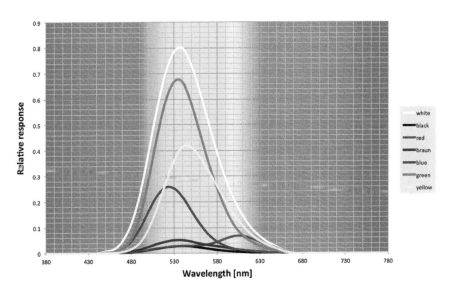

FIGURE 2.66   Relative phosphorescence response curves of colored textile substrates.

in brightness, while red emission is seriously overestimated. Indeed, due to the nature of human vision – shifting from cone to rod vision, with a corresponding change in absolute sensitivity and wavelength-dependent sensitivity curve – low intensity red light is perceived much weaker than green or blue light with equal photometric luminance.

KEYWORDS

- **chemochromism**
- **chromic materials**
- **electrochromism**
- **luminescence**
- **photochromism**
- **thermochromism**

REFERENCES

1. Crano, J. C., & Gugliemetti, R. J., (1999). *Organic Photochromic and Thermochromic Compounds*, Kluwer Academic Publishers, New York, vols. 1 and 2, pp. 2.
2. Eggins, B. R., (2002). *Chemical Sensors and Biosensors*, John Wiley & Sons: Chichester. pp. 5–9.
3. Dürr, H., & Bouas-Laurent, H., (2003). *Photochromism, Molecules and Systems*, Elsevier, Amsterdam. pp. 5–13.
4. Kim, S. H., (2006). *Functional Dyes*, Elsevier, Amsterdam, pp. 85–88.
5. Bamfield, P., & Hutchings, M. G., (2010). *Chromic Phenomena, Technological Applications of Color*, 2nd edn., RSC Publishing, Cambridge.
6. Ferrara, M., & Bengisu, M., (2014). *Materials that Change Color, Smart Materials, Intelligent Design*, Springer Cham. pp. 3–8.
7. Monk, P. M. S., Mortimer, R. J., & Rosseinsky, D. R., (2007). *Electrochromism and Electrochromic Devices*, Cambridge University Press Cambridge.
8. Abe, J., Yamashita, H., & Katsuya, M., (2015). Photochromism, in: *Encyclopedia of Polymeric Nanomaterials*, Kobayashi, S., & Müllen, K., ed., Springer-Verlag Berlin, pp. 1572–1583.
9. Börjesson, K., Herder, M., Grubert, L., Duong, D. T., Salleo, A., Hecht, S., Orgiu, E., & Samori, P., (2015). Optically switchable transistors comprising a hybrid photochromic molecule/n-type organic active layer, *J. Mater. Chem. C.*, 4156–4161.
10. Christie, R. M., (2015). *Color Chemistry*, 2nd edn., The RSC Cambridge. pp. 90–91.
11. Feczko, T., & Voncina, B., (2013). Organic nanoparticulate photochromes, *Curr. Org. Chem.*, *17*(16), 1771–1789.

12. Pardo, R., Zayat, M., & Levy, D., (2011). Photochromic organic–inorganic hybrid materials, *Chem. Soc. Rev.*, *40*, 672–687.

13. Galston, A. W., (1950). Riboflavin, light, and the growth of plants, *Science, 111*, 619–624.

14. Bouas-Laurent, H., & Dürr, H., (2001). Organic Photochromism, *Pure Appl. Chem.*, *73*, *4*, 639–665.

15. Smets, G., Braeken, J., & Irie, M., (1978). Photochemical effects in photochromic systems, *Pure Appl. Chem.*, *50*, 845–856.

16. Alfimov, M. V., Fedorova, O. A., & Gromov, S. P., (2003). Photoswitchable molecular receptors, *J. Photochem. Photobiol. A: Chemistry*, *158*(2–3), 183–198.

17. Kelly, T. R., De Silva, H., & Silva, R. A., (1999). Unidirectional rotary motion in a molecular system, *Nature*, *401*, 150–152.

18. Ahmed, S. A., (2006). Photochromism of dihydroindolizines Part VI: synthesis and photochromic behavior of a novel type of IR-absorbing photochromic compounds based on highly conjugated dihydroindolizines, *J. Phys. Org. Chem.*, *19*, 402–414.

19. Li, J., Speyer, G., & Sankey, O. F., (2004). Conduction switching of photochromic molecules, *Phys. Rev. Lett.*, *93*(24), 248302.

20. Janus, K., Koshets, I. A., Sworakowski, J., & Nespurek, S., (2002). An approximate non-isothermal method to study kinetic processes controlled by a distribution of rate constants: the case of a photochromic azobenzene derivative dissolved in a polymer matrix, *J. Mater. Chem.*, *12*, 1657–1663.

21. Nespurek, S., Toman, P., & Sworakowski, J., (2003). Charge carrier transport on molecular wire controlled by dipolar species: Towards light-driven molecular switch, *Thin Solid Films*, *438–439*, 268–278.

22. Nassau, K., (2001). *The Physics and Chemistry of Color,* 2nd edn., John Wiley and Sons, Inc., New York, pp. 133.

23. Minkin, V. I., Bren, V. A., & Lyubarskaya, A. E., (1990). In: *Organic Photochromes*, Eltsov, A. V., ed., Plenum Publ. Corp., New York, pp. 218–244.

24. Frick, M., (2008). Photochromic textiles – Production and property (in German), PhD Dissertation, Stuttgart University, Germany. pp. 4.

25. Brown, G. H. ed., (1971) *Techniques of Chemistry*, vol. III, Photochromism, John Wiley & Sons, New York. pp. 55–288

26. Gerhardt, G. E., & Township, W., (1973). Single and double energy transfer in triplet-triplet photochromic composition, *US. Patent 3, 725*, 292.

27. Mancheva, I., Zhivkov, I., & Nespurek, S., (2005). Kinetics of the photochromic reaction in a polymer containing azobenzene groups, *J. Opt. Adv. Materials*, *7*, 253–256.

28. Kühn, D., Balli, H., & Steiner, U. E., (1991). Kinetic study of the photodecoloration mechanism of an inversely photochromic class of compounds forming spiropyran analogues, *J. Photochem. Photobiol. A: Chemistry*, *61*(1), 99–112.

29. Liu, R. S. H., & Hammond, G. S., (2003). Photochemical reactivity of polyenes. from dienes to rhodopsin, from microseconds to femtoseconds, *Photochem. Photobiol. Sci.*, *2*, 835–844.

30. Ostrovskii, M. A., & Fedorovich, I. B., (1978). Photochemical transformations of the visual pigment rhodopsin, *J. Quantum Electron*, *8,* 1274–1278.

31. Suzuki, T., & Callender, R. H., (1981). Primary photochemistry and photoisomerization of retinal at 770K in cattle and squid rhodopsins, *Biophys. J.,* *34*(5), pp. 261–270.

32. Nishioku, Y., Hirota, N., Nakagawa, M., Tsuda, M., & Erazima, M., (2001). The energy and dynamics of photoreaction intermediates of' Octopus rhodopsin studied by the transient grating method, *Anal. Sci.*, *17*, 323–325.

33. Kandori, H., Shichida, Y., & Yoshizawa, T., (2001). Photoisomerization in rhodopsin, *Biochemistry*, *66*(11), 1197–1209.

34. Yan, M., Manor, D., Weng, G., Chao, H., Rothberg, L., Jedju, T. M., Alfano, R. R., & Callender, R. H., (1991). Ultrafast spectroscopy of the visual pigment rhodopsin, *Proc. Natl. Acad. Sci. USA, 88*, 9809–9812.

35. Liu, R. S. H., & Colmenares, L. U., (2003). The molecular basis for the high photosensitivity of rhodopsin, *Proc. Natl. Acad. Sci. USA*, *100*(25), 14639–14644.

36. Han, M., Groesbeek, M., Sakmar, T. P., & Smith, S. O., (1997). The C9 methyl group of retinal interacts with glycine-121 in rhodopsin, *Proc. Natl. Acad. Sci. USA, 94*, 13442–13447.

37. Nakanishi, K., (1977). Photochemical studies of visual pigments, *Pure & Appl. Chem.*, *49*, 333–339.

38. Birge, R. R., Gillespie, N. B., Izaguirre, E. W., Kusnetzow, A., Lawrence, A. F., Singh, D., Song, Q. W., Schmidt, E., Stuart, J. A., Seetharaman, S., & Wise, K. J., (1999). Biomolecular Electronics: Protein-based associative processors and volumetric memories, *J. Phys. Chem. B.*, *103*, pp. 10746–10766.

39. Lovrien, R., & Waddington, J. C. B., (1964). Photoresponsive systems I. Photochromic macromolecules, *J. Amer. Chem. Soc.*, *86*, pp. 2315–2322.

40. Cojocariu, C., & Rochon, P., (2004). Light-induced motions in azobenzene containing polymers, *Pure Appl. Chem., 76*(7–8), pp. 1479–1497.

41. Tatewaki, H., Baden, N., Momotake, A., Arai, T., & Erazima, M., (2004). Dynamics of water-soluble stilbene dendrimers upon photoisomerization, *J. Phys. Chem. B., 108*, 12784–12789.

42. Wehrle, B., & Limbach, H. H., (1989). NMR study of environment modulated proton tautomerism in crystalline and amourphous phatlocyanine, *Chem. Physics, 136*, pp. 223–247.

43. Dürr, H., (1990). A new photochromic system – potential limitatir and perspectives, *Pure & Appl. Chem.*, *62*(8), pp. 1477–1482.

44. Cicogna, F., Ingrosso, G., Lodato, F., Marchetti, F., & Zandomeneghi, M., (2004). 9-anthroylacetone and its photodimer, *Tetrahedron, 60*, pp. 11959–11968.

45. Zhang, J. Z., Schwartz, B. J., King, J. C., & Harris, C. B., (1992). Ultrafast studies of photochromic spiropyrans in solution, *J. Am. Chem. Soc., 114*, pp. 10921–10927.

46. Sun, Z., Li, H., Liu, G., Fan, C., & Pu, S., (2014). Photochromism of new unsymmetrical diarylethenes based on the hybrid of azaindole and thiophene moieties, *Dyes & Pigments, 106*, pp. 94–104.

47. Viková, M., & Vik, M., (2011). Alternative UV sensors based on color-changeable pigments, *Adv. Chem. Eng. Sci., 1*, pp. 224–230.

48. Tamai, N., & Miyasaka, H., (2000). Ultrafast dynamics of photochromic systems., *Chem. Rev., 100*, pp. 1875–1890.

49. Preigh, M. J., Stauffer, M. T., Lin, F. T., & Weber, S. G., (1996). Anodic oxidation mechanism of a spiropyran, *J. Chem. Soc., Faraday Trans., 92*, pp. 3991–3996.

50. Patel, P. D., Mikhailov, I. A., & Belfield, K. D., (2009). Theoretical study of photochromic compounds, Part 2 : Thermal mechanism for byproduct formation and fatigue resistance of diarylethenes used as data storage materials, *Int. J. Quantum Chem., 109*, pp. 3711–3722.

51. Montalti, M., Credi, A., Prodi, L., & Gandolfi, M. T., (2006). *Handbook of Photochemisty*, Third ed., CRC Taylor & Francis, London, UK.

52. Adamo, C., & Jacquemin, D., (2013). The calculations of excited-state properties with time-dependent density functional theory, *Chem. Soc. Rev., 42*, pp. 845–856.

53. Cheng, T., Lin, T., Brady, R., & Wang, X., (2008). Fast response photochromic textiles from hybrid silica surface coating, *Fibers Polym., 9*, pp. 301–306.

54. Albini, A., & Germani, L., (2010). Photochemical methods, in: *Handbook of Synthetic Photochemistry*, Albini, A., & Fagnoni, M., ed., WILEY-VCH Verlag GmbH & Co. KGaA, Weinheim, pp. 1–24.

55. García-Amorós, J., & Velasco, D., (2012). Recent advances towards azobenzene-based light-driven real-time information-transmitting materials, *Beilstein J. Org. Chem., 8*, pp. 1003–1017.

56. Zhang, C., Zhang, Z., Fan, M., & Yan, W., (2008). Positive and negative photochromism of novel spiro [indoline-phenanthrolinoxazines], *Dyes & Pigments, 76*, pp. 832–835.

57. Christie, R. M., (2013). Advances in dyes and colorants, in *Advances in the Dyeing and Finishing of Technical Textiles*, Gulrajani, M. L., ed., pp. 1–37.

58. Dawson, T. L., (2010). Changing colors: Now you see them, now you don't, *Color. Technol., 126*, pp. 177–188.

59. Lisyutenko, V. N., Barachevskii, V. A., Pankratov, A. A., & Konoplev, G. G., (1989). Excitation-energy relaxation in photochromic Indoline spiropyrans, *Teori Eksp. Khimiya*, pp. 420–427.

60. Jiang, G., Song, Y., Guo, X., Zhang, D., & Zhu, D., (2008). Organic functional molecules towards information processing and high-density information storage, *Adv. Mater., 20*, pp. 2888–2898.

61. Oda, H., (1998). A novel approach for improving the light fatigue resistance of spiropyrans, *J. Soc. Dye. Color. 114*, pp. 363–365.

62. Semyakina, G. M., Samedova, T. G., Popova, N. I., Zakhs, E. R., Malkin, Y. N., Martynova, V. P., et al., (1983). Regulation of photochromic properties of fast-relaxing spiropyrans, *Izv. Akad. Nauk SSSR*, pp. 1277–1282.

63. Ohnishi, Y., Yoshimoto, S., & Kimura, K., (2001). Novel compounds producing a photochromic spiropyran on heating, *J. Photochem. Photobiol. A. Chem., 141*, pp. 57–62.

64. Li, X., Li, J., Wang, Y., Matsuura, T., & Meng, J., (2004). Synthesis of functionalized spiropyran and spirooxazine derivatives and their photochromic properties, *J. Photochem. Photobiol. A Chem., 161*, pp. 201–213.

65. Lukyanov, B. S., & Lukyanova, M. B., (2005). Spiropyrans: Synthesis, properties, and application, *Chem. Heterocycl. Compd., 41*, pp. 281–311.

66. Ono, H., & Osada, T., (1971). Photochromic compound and composition containing the same, *US. Patent 3562172*.

67. Arnold, G., Vollmer, H. P., Wilhelm, A., & Paal, G., (1971). Spiro [benzothiazole(or indoline)-2,3'-[3H]naphth[2,1-b][1,4]oxazines], Ger. Patent 1,947,714, *Chem. Abstr., 75*, 28320b.

68. Billah, S. M. R., Christie, R. M., & Shamey, R., (2008). Direct coloration of textiles with photochromic dyes. Part 1: Application of spiroindolinonaphthoxazines as disperse dyes to polyester, nylon and acrylic fabrics, *Color. Technol., 124*, pp. 223–228.

69. Lee, S. J., Son, Y. A., Suh, H. J., Lee, D. N., & Kim, S. H., (2006). Preliminary exhaustion studies of spiroxazine dyes on polyamide fibers and their photochromic properties, *Dyes & Pigments, 69*, pp. 18–21.

70. Son, Y. A., Park, Y. M., Park, S. Y., Shinand, C. J., & Kim, S. H., (2007). Exhaustion studies of spirooxazine dye having reactive anchor on polyamide fibers and its photochromic properties. *Dyes & Pigments, 73,* pp. 76–80.

71. Suh, H. J., Keum, S. R., Koh, K., & Kim, S. H., (2007). Anchoring of photochromic spirooxazine into silica xerogels, *Dyes & Pigments, 72,* pp. 363–366.

72. Becker, R. S., & Michl, J., (1966). Photochromism of synthetic and naturally occurring 2H-chromenes and 2H-pyrans, *J. Am. Chem. Soc., 88,* pp. 5931–5933.

73. Jacobson, R. E., (1989). Photochromic imaging, in: *Photopolymerisation and Photoimaging Science and Technology,* Allen, N. S., ed., Springer Netherlands, Dordrecht, pp. 149–186.

74. Cheng, T., Lin, T., Brady, R., & Wang, X., (2008). Fast response photochromic textiles from hybrid silica surface coating, *Fibers Polym., 9,* pp. 301–306.

75. Billah, S. M. R., Christie, R. M., & Morgan, K. M., (2008). Direct coloration of textiles with photochromic dyes. Part 2: The effect of solvents on the color change of photochromic textiles, *Color Technol., 124,* pp. 229–233.

76. Ortica, F., (2012). The role of temperature in the photochromic behavior, *Dyes & Pigments, 92,* pp. 807–816.

77. Little, A. F., & Christie, R. M., (2010). Textile applications of photochromic dyes. Part 2: Factors affecting the technical performance of textiles screen-printed with commercial photochromic dyes, *Color. Technol., 126,* pp. 164–170.

78. Liu, G., Liu, M., Pu, S., Fan, C., & Cui, S., (2012). Photochromism of new unsymmetrical isomeric diarylethenes bearing a pyridine group, *Tetrahedron, 68,* pp. 2267–2275.

79. Stobbe, H., (1905). The color of the Fulgenic acid and Fulgides, *Berichte der Dtsch. Chem. Gesellschaft, 38,* pp. 3673.

80. Stobbe, H., (1907). A product of the action of light on diphenyl sulfide and the polymerization of phenylpropiolic acid, *Berichte der Dtsch. Chem. Gesellschaft, 40,* pp. 3372–3382.

81. Hart, R. J., Heller, H. G., & Salisbury, K., (1968). The photochemical rearrangements of some photochromic fulgides, *Chem. Commun.* (London), pp. 1627–1628.

82. Yao, B., Wang, Y., Menke, N., Lei, M., Zheng, Y., Ren, L., Chen, G., Chen, Y., & Fan, M., (2005). Optical properties and applications of photochromic fulgides, *Mol. Cryst. Liq. Cryst., 430,* pp. 211–219.

83. Nakamura, S., Yokojima, S., Uchida, K., Tsujioka, T., Goldberg, A., Murakami, A., Shinoda, K., Mikami, M., Kobayashi, T., & Kobatake, S., (2008). Theoretical investigation on photochromic diarylethene: A short review, *J. Photochem. Photobiol. A. Chem., 200,* pp. 10–18.

84. Van der Molen, S. J., Van der Vegte, H., Kudernac, T., Amin, I., Feringa, B. L., & Van Wees, B. J., (2006). Stochastic and photochromic switching of diarylethenes studied by scanning tunnelling microscopy, *Nanotechnology, 17,* pp. 310–314.

85. Day, J. H., (1963). Thermochromism, *Chem. Rev., 63*(1), pp. 65–80.

86. Antonov, L., (2014). *Tautomerism, Methods and Theories,* Wiley-VCH Weinheim. pp. 203–263.

87. Sone, K., & Fukuda, Y., (1987). *Inorganic Thermochromism,* Springer-Verlag Berlin. pp. 2–128.

88. Christie, R. M., & Bryant, I. D., (2005). An evaluation of thermochromic prints based on microencapsulated liquid crystals using variable temperature color measurement, *Color. Technol., 121,* pp. 187–192.

89. Burkinshaw, S. M., Griffiths, J., & Towns, A. D., (1998). Reversibly thermochromic systems based on pH-sensitive functional dyes, *J. Mat. Chem.*, *8*, pp. 2677–2683.

90. Chowdhury, M. A., Joshi, M., & Butola, B. S., (2014). Photochromic and thermochromic colorants in textile applications, *J. of Eng. Fibers and Fabrics*, *9*(1), pp. 107–123.

91. Kulcar, R., Friskovec, M., Hauptman, N., Vesel, A., & Klanjsek G. M., (2010). Colorimetric properties of reversible thermochromic printing inks, *Dyes and Pigments*, *86*(3), pp. 271–277.

92. Towns, A., (1999). The heat is on for new colors, *J. Soc. D. Col.*, *115*(7/8), pp. 196–199.

93. Ogrodnik, W., (2008). Use of color-changing pigment to detect wire and cable hazards, *Wire Journal International*, *41*(4), pp. 150–155.

94. Seeboth, A., & Lotzsch, D., (2008). *Thermochromic Phenomena in Polymers*, Smithers Rapra Technology Ltd., Shropshire, pp. 17–71

95. Lucht, B. L., Euler, W. B., & Gregory, O. J., (2002). Investigation of the thermochromic properties of polythiophenes dispersed in host polymers, *Polymer Preprints*, *43*(1), pp. 59–60.

96. Avnir, D., (1995). Organic chemistry within ceramic matrices: doped sol–gel materials, *Acc. Chem. Res.*, *28*, pp. 328–334.

97. Masashi, T., Masashi, S., Katsuji, M., & Tadataka, Y., (1998). Thermochromism of dyes on silica gel, *Dyes and Pigments*, *39*, pp. 97–109.

98. Sergio, H. A., (1996). Solid-state electronic absorption, fluorescence and 13C CPMAS NMR spectroscopic study of thermo- and photo-chromic aromatic Schiff bases, *J. Chem. Soc., Perkin Transactions*, *2*, pp. 2293–2296.

99. Reichardt, C., (1992). Solvatochromism, thermochromism, piezochromism, halochromism, and chiro-solvatochromism of pyridinium N-phenoxide betaine dyes, *Chem. Soc. Rev.*, *21*, pp. 147–153.

100. Israel, B., & Walt, R., (2002). Optrode-based fiber optic biosensors. In: *Optical Biosensors: Present and Future,* Second Edition, Ligler, F. S., & Tait, C. R.; Elsevier Science, Amsterdam, pp. 5–55.

101. Yin, S., Ruffin, P. B., & Yu, F. T. S., (2008). *Fiber Optic Sensors*, Sec. ed., CRC Press, Boca Raton.

102. Courbat, J., Briand, D., Damon-Lacoste, J., Wöllenstein, J., & De Rooij, N. F., (2009). Evaluation of pH indicator-based colorimetric films for ammonia detection using optical waveguides, *Sensors and Actuators B: Chemical*, *143*(1), pp. 62–70.

103. Courbat, J., Linder, M., Dottori, M., Briand, D., Wöllenstein, J., & De Rooij, N. F., (2010). *Inkjet Printed Colorimetric Ammonia Sensor on Plastic Foil for Low-Cost and Low-Power Devices*, IEEE 23rd International Conference on Micro Electro Mechanical Systems (MEMS), Wanchai, Hong Kong, pp. 883–886.

104. Korotcenkov, G., (2007). Metal oxides for solid-state gas sensors: What determines our choice? *Materials Science and Engineering B.*, *139*(1), pp. 1–23.

105. Granqvist, C. G., (2000). Electrochromic tungsten oxide films: review of progress 1993–1998, *Solar Energy Materials and Solar Cells,* *60*(3), pp. 201–262.

106. Chan, C. C., Hsu, W. C., Chang, C. C., & Hsu, C. S., (2011). Hydrogen incorporation in gasochromic coloration of sol–gel WO_3 thin films, *Sensors and Actuators: B Chemical,* *157*(2), pp. 504–509.

107. Chan, C. C., Hsu, W. C., Chang, C. C., & Hsu, C. S., (2010). Preparation and characterization of gasochromic Pt/WO3 hydrogen sensor by using the Taguchi design method, *Sensors and Actuators B Chemical,* *145*(2), pp. 691–697.

108. Faughnan, B. W., Crandall, R. S., & Heyman, P. M., (1975). Electrochromism in WO$_3$ Amorphous Films, *RCA. Review, 36*(1), pp. 177–197.

109. Hsu, C. H., Chang, C. C., Tseng, C. M., Chan, C. C., Chao, W. H., Wu, Y. R., Wen, M. H., Hsieh, Y. T., Wang, Y. C., Chen, C. L., Wang, M. J., & Wu, M. K., (2013). An ultrafast response gasochromic device for hydrogen gas detection, *Sensors and Actuators B., 186*(9), pp. 193–198.

110. Viková, M., Slabotinsky, J., Vik, M., & Musilova, M., (2003). Possibility of colorimetric identification of dangerous gas penetration through protective clothes, International conference TEXSCI 03, Liberec, Czech Republic, pp. 203–207.

111. Viková, M., (2004). Colorimetric identification of dangerous gas penetration through protective clothes via intelligent underwear, *Quality Textiles for Quality Life*, 1–4.

112. Van der Schueren, L., De Meyer, T., Steyaert, I., Ceylan, O., Hemelsoet, K., Van Speybroeck, V., & De Clerck, K., (2013). Polycaprolactone and polycaprolactone/chitosan nanofibres functionalised with the pH-sensitive dye Nitrazine Yellow, *Carbohydr Polym., 91*(1), 284–293.

113. Steyaert, I., Vancoillie, G., Hoogenboom, R., & De Clerck, K., (2015). Dye immobilization in halochromic nanofibers through blend electrospinning of a dye-containing copolymer and polyamide-6, *Polymer Chemistry, 6*(14), 6, 2685–2694.

114. Reichardt, Ch., & Welton, T., (2011). *Solvents and Solvent Effects in Organic Chemistry*, 4th edn., Wiley-VCH Verlag, Weinheim. pp. 4–140

115. Wetzler, D. E., Chesta, C., Fernandez-Prini, R., & Aramendia, P. F., (2001). Dynamic solvatochromism in solvent mixtures, *Pure Appl. Chem., 73*(3), 405–409.

116. Thoraval, D., Bets, R. W., Bovenkamp, J. W., & Dix, J. K., (1988). Development of paper, chemical agent detector, 3-way liquid containing non-mutagenic dyes. *I-Replacement of the Yellow Dye Thiodiphenyl-4,4'-diazo-bis-salicylic Acid (A2)*, Defense Research Establishment Ottawa, Report No. 962.

117. Thoraval, D., & Bovenkamp, J. W., (1989). Paper chemical agent detectors, *EP 0334668 A1, 27*.

118. Vik, M., & Viková, M., (2011). *Identification Methods for Evaluation of Amount of Dangerous Substances in Air and on the Surface* (in Czech), Report for National Authority for Nuclear, Biological and Chemical Protection Czech Republic.

119. Esse, R., & Saari, A., (2008). In: *Smart Packaging Technologies*, Kerry, J., & Butler, P., (ed.), John Wiley & Sons, West-Sussex, pp. 130–149.

120. Yamazoe, N., & Shimizu, Y., (1986). Humidity sensors: *Principles and Applications, Sensors and Actuators, 10*, 379–398.

121. Sohrabnezhad, S., Pourahmad, A., & Sadjadi, M. A., (2007). New methylene blue incorporated in mordenite zeolite as humidity sensor material, *Materials Letters, 61*, 2311–2314.

122. Mohr, G. J., & Spichiger-Keller, U., (1998). Development of an optical membrane for Humidity, *Mikrochim. Acta, 130*, 29–34.

123. Sata, T., (2004). *Ion Exchange Membranes, Preparation, Characterization, Modification and Application*, The Royal Society of Chemistry, Cambridge. pp. 276–280.

124. Song, S., Cho, I. H., Kim, J., & Park, D. A., (2005). Composition for non-toxic, non-hazardous, and environmentally friendly humidity-indicating agent and its application, *US Patent 20050106735*.

125. Knyrim, J., & Dick, S., (2011). Halogen and heavy metal-free humidity indicating composition and humidity indicator card containing the same, *US Patent 20110171745*.

126. Benas, M., Fujisawa, K., Madsen, R. R., Mathur, R., & Yeakley, T., (2010). *Cobalt Dichloride Free Humidity Indicator Cards*, Texas Instruments Application Report SLVA410.

127. Matsushima, R., Nishimura, N., Goto, K., & Kohno, Y., (2003). Vapochromism of ionic dyes in thin films of sugar gels, *Bulletin of the Chemical Society of Japan, 76*(6), 1279–1283.

128. Mills, A., Grosshans, P., & Hazafy, D., (2010). A novel reversible relative-humidity indicator ink based on methylene blue and urea, *Analyst, 135*, 33–35.

129. Askim, J. R., Mahmoudi, M., & Suslick, K. S., (2013). Optical sensor arrays for chemical sensing: the optoelectronic nose, *Chem. Soc. Rev., 42*, 8649–8682.

130. Platt, J. R., (1961). Electrochromism, a possible change of color producible in dyes by an electric field, *The Journal of Chemical Physics, 34*(3), 862–863.

131. Chang, I. F., Gilbert, B. L., & Sun, T. I., (1975). Electrochemichromic systems for display applications, *J. Electrochem. Soc., 122*(7), 955–962.

132. Monk, P. M. S., Mortimer, R. J., & Rosseinsky, D. R., (2007). *Electrochromism and Electrochromic Devices*, Cambridge University Press, Cambridge.

133. Mortimer, R. J., Dyer, A. L., & Reynolds, J. R., (2006). Electrochromic organic and polymeric materials for display applications, *Displays, 27*, 2–18.

134. Sapp, S. A., Sotzing, G. A., & Reynolds, J. R., (1998). High contrast ratio and fast-switching dual polymer electrochromic devices, *Chem. Mater., 10*, 2101–2108.

135. Kang, J., Paek, S., Hwang, S., & Choy, J., (2008). Optical iris application of electrochromic. thin films, *Electrochem. Commun., 10*, 1785–1787.

136. Buxbaum, G., & Pfaff, G., (2005). *Industrial Inorganic Pigments*, Wiley-VCH Verlag, Weinheim.

137. Doi, T., (2004). *Study on the Outdoor visual Information Indicating System Raised by Wide Power Failure Due to the Next Nankai Earthquake*, 13th World Conference on Earthquake Engineering Vancouver, BC, Canada.

138. Saito, M., Adachi, N., & Kondo, H., (2007). Full-color illumination that needs no electric power, *Opt. Express, 15*, 1621–1626.

139. Wördenweber, B., Wallaschek, J., Boyce, P., & Hoffman, D., (2007). *Automotive Lighting and Human Vision*, Springer-Verlag Berlin. pp. 60–93.

140. Vik, M., Shamey, R., & Purkyne Jan, E., (2016). In: *Encyclopedia of Color Science and Technology*, Luo, R., ed., Springer Reference: New York, pp. 1090–1093.

141. Rea, M. S., Bullough, J. D., Freyssinier-Nova, J. P., & Bierman, A., (2004). A proposed unified system of photometry, *Light Res. Technol., 36*, 85–111.

142. Chen, R., (2003). Apparent stretched-exponential luminescence decay in crystalline solids, *J. Lumin., 102–103*, 510–518.

143. Wang, B., Zhuo, Z., & Lu, Z., (2006). The influence of temperature on the afterglow feature of $SrAl_2O_4$Eu, Dy Phosphors, *Journal of Wuhan University of Technology – Mater. Sci.*, ed., *21*(3), 120–122.

144. Physics Survey Committee, Board on Physics and Astronomy, Commission on Physical Sciences, Mathematics, and Resources, National Research Council: *Physics through the 1990s An Overview*, National Academy of Sciences, Washington, 1986.

145. Scher, H., Shlesinger, M. F., & Bendler, J. T., (1991). Time-scale invariance in transport and relaxation, *Phys. Today, 44*, 26–34.

146. Vik, M., Viková, M., & Kasparova, M., (2014). Decay of Phosphorescent warning design on textile substrates, *Applied Mechanics and Materials, 440*, 112–117.

147. Pátek, K., (1962). *Luminescence* (in Czech), SNTL Prague, pp. 29–78.
148. Blasse, G., & Grabmaier, B. C., (1994). *Luminescent Materials*, Springer-Verlag Berlin. pp. 10–70.
149. Kitai, A., (2008). *Luminescent Materials and Applications*, John Wiley & Sons: Chichester. pp. 94–100.
150. Ronda, C., (2008). *Luminescence*, Wiley-VCH Verlag, Weinheim, pp. 4–247.
151. Chen, Y., Liu, B., Kirm, M., Qi, Z., Shi, C., True, M., Vielhauer, S., & Zimmerer, G., (2006). Luminescent properties of blue-emitting long afterglow phosphors $Sr_{2-x}Ca_x MgSi_2O_7:Eu^{2+}, Dy^{3+}$ ($x = 0, 1$) *J. Lumin.*, *118*, 70–78.

CHAPTER 3

PRODUCTION OF CHROMIC MATERIALS

ARAVIN PRINCE PERIYASAMY and MARTINA VIKOVÁ

CONTENTS

Abstract 109
3.1   Introduction 110
3.2   Production of Chromic Materials 111
3.3   Interesting Videos 147
Keywords 147
References 148

ABSTRACT

This chapter deals with the application of color changing materials on textiles. Color changeable materials are defined as a reversible transformation in a chemical species between two forms having different absorption spectra by photoirradiation, which is caused by external stimuli. Since 1960, some color changeable materials have been commonly used, such as colorless spectacles that turn to grey in sunshine to become sunglasses. However, the development of color changeable textiles has not led to significant industrial production mainly due to technical difficulties in the application of the dyes and their performance. The growing interest in color changing textiles is due to the high commercial potential applications, especially in the clothing sector, nanofibers, and smart textiles. This chapter describes the applications of color changing materials to textile materials. There are many color changing materials. Photochromic and thermochromic are the types of functional colorants that are predominantly used in the clothing, textiles, and smart materials made from textile materials to determine the environmental variation by changing the color (acting as sensory materials). Later, we discuss different techniques used to prepare the color changeable textiles, namely screen printing, mass coloration, exhaust dyeing, melt-blown technology, digital printing, and electrospinning. Amongst these, the sol-gel and electrospinning methods are very useful to prepare color changeable textiles with respect to lower temperature. The main advantage of these methods is that the photochromic compounds can be embedded in the pores of textile materials.

3.1   INTRODUCTION

Textiles are one of the basic needs of human life after food and shelter. It can be applied in various forms from clothing to high-tech applications such as protective textiles, medical textiles, geo textiles, and sport textiles. Imparting colors in textiles have opened the door for inventive applications on textile materials. The pleasure delivered by colors may have been behind the thinking of ancient Egyptians who applied color to their clothing as long ago as 2500 BC, which led to the consequent development of colorants from natural dyes derived mainly from plants through to the modern synthetic dyes that are used for introduction of different colors on various textile materials. The introduction of colors to textiles presently is a well-established technology.

Some materials can change their optical properties when exposed to a specific stimulus such as temperature, electrical field, light, solvents, mechanical pressure of friction, etc. The concept to produce color changeable textile materials is based on applying the color changeable materials to them. However, these colorants possess the ability to undergo reversible color change under the influence of various external stimuli. Photochromic and thermochromic pigments are commercialized since the late 1960s and used, for example, in photochromic sunglass and thermochromic recording materials.

Thermochromic materials are a class of color changing materials that change the color with respect to the reaction of heat, especially through the application of thermochromic dyes whose colors change at particular temperatures. In textiles, two types of thermochromic systems have been applied successfully. They are liquid crystal type and molecular rearrangement type. In both cases, the dyes are entrapped in microcapsules and applied to the garment fabric like a pigment in a resin binder. Thermochromic transition temperature is a specific temperature at which the color change occurs depending upon the thermochromic materials. At this transition temperature, the color changes occur very rapidly. Thermochromic dyes are often used for rigid applications such as thermometers, security printings, cosmetics (nail polish), medical thermography, and food packaging. In textiles, thermochromic materials are used for home textiles and fashion, for example, table cloth, which when kept in contact with a hot dish, changes the color. Similarly, curtains change color during sunny days or depending on the outdoor temperature. Thermochromism through molecular rearrangement in dyes has aroused a degree of commercial interest. However, the overall mechanism underlying the changes in color is far from being clear and is still very much open to speculation. These mechanisms are described in Chapter 2.

3.2 PRODUCTION OF CHROMIC MATERIALS

3.2.1 CHROMIC POLYMERS

Photochromic polymer is nothing but a chromophore incorporated in the backbone or side groups of the polymer structure, which changes its various properties during the external stimuli. These chromophores are based on photochromic compounds that undergo reversible transformation of two isomers in one or both directions by photoirradiation. During the photoirradiation, these polymers can change their physical, chemical, and optical properties in a reversible manner. These changes may be like polymer chain conformation, shape of polymer gels, surface wettability, membrane permeability, pH, solubility, sol-gel transition temperature, and phase separation temperature. In 1967, Lovrrien carried out the first attempt of polymer chain with photoirradation properties by incorporating the azochrysophenine (photochromic colorant) into poly(methacrylic acid). From this result, he reported that the viscosity strongly influenced the change in the intermolecular interaction between the photochromic dye and the polymer. So, it is necessary to control the viscosity of the solution to obtain desirable photochromic properties [1].

Later, azobenzene was incorporated in the backbone of various polyamides as shown in Figure 3.1.

In 1970, Agolini and Gay [3] studied the photochromic polyamide with azobenzene. They reported that at higher temperature (200°C) under UV irradiation, the cis isomer content in the photostationary state is low. In 2001, Finkelmann et al. [4] studied the first photochromic elastomer poly[oxy(methylsilylene)] as shown in Figure 3.2. After incorporating the azobenzene into the polymeric backbone, the phase transition temperature shifts depend on the ratio of the trans and cis forms. So, photoirradiation can induce a change in the phase transition temperature. This mechanism tends to change the shape of the elastomer. Based on this mechanism, many photoresponsive polymers have been developed [5].

These photochromic polymers are used in photochromic glasses, UV sensors, optical wave guides, holographic recording media, nonlinear optics, memory devices, etc. [2]. Seboth et al. [6] developed nontoxic thermochromic polymer by incorporating anthocyanidins on poly (lactic acid) (PLA). They

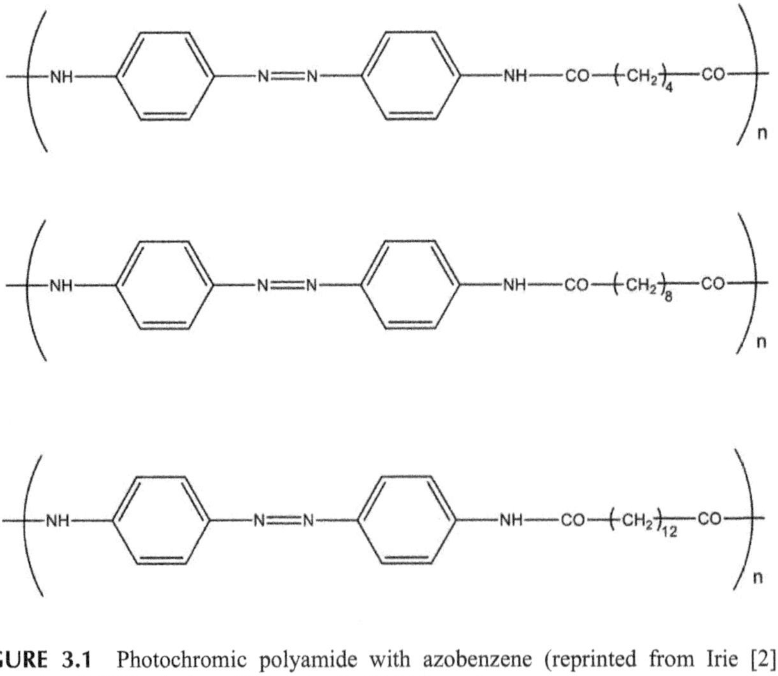

FIGURE 3.1 Photochromic polyamide with azobenzene (reprinted from Irie [2] with permission © 2013, SpringerVerlag Berlin Heidelberg.)

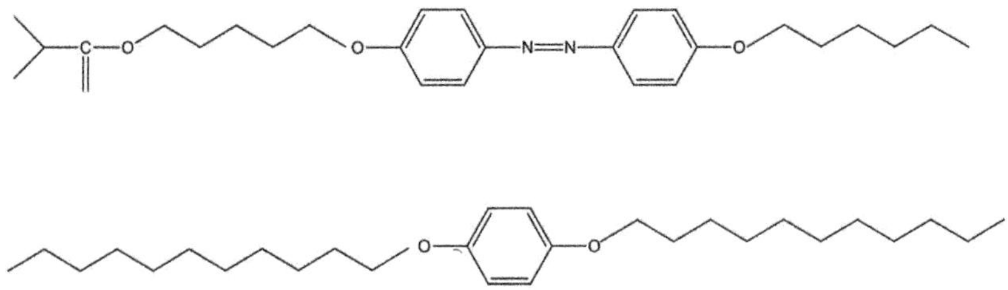

FIGURE 3.2 Photochromic elastomer of poly[oxy(methylsilylene)] (reprinted from Irie [2] with permission © 2013, SpringerVerlag Berlin Heidelberg.)

found thermochromic effects at 45°C, where the color turned red to violet. Further heating up to 60°C resulted in cloudy to clear transformation (thermotropic effect). They designed thermochromic polymers for sensor application.

Douglas A. Davis et al. [7] developed mechanoresponsive synthetic polymeric materials by directly linking force-activated mechanophores into the polymerchains of bulk polymers or by using the mechanophores as crosslinks (Figure 3.3). Experimentally, these mechanophores polymers were produced by crosslinking the poly (methyl acrylate) (PMA) with a glassy mechanophore of poly (methyl meth-acrylate) (PMMA). A commonly used monomer for the preparation of photoresponsive polymer hydrogel networks is N-isopropylacrylamide (abbreviated as NIPAAM), of which the corresponding polymer possesses well-studied thermoresponsive changes in hydrophilicity [8].

Ivanov et al. [9] described that the lower critical solution temperature (LCST) can be increased or decreased by the incorporation of hydrophilic and hydrophobic groups, respectively, in the polymer backbone. Therefore, it is possible to alternate between hydrophobic and hydrophilic isomer of the spiropyran which can be controlled not only by temperature but also by light. Therefore, it gives dual effects. Figure 3.4 shows the poly(NIPAM–X) copolymer under acidic condition (pH 4). The polymers are water soluble with hydrophilic MC, but readily precipitate with hydrophobic SP.

Wismontski-Knittel et al. [10] concluded that the spiropyran incorporated methacrylate can induce crystallization with stepwise isomerization. In other words, crystallization and isomerization mutually stimulate each other. This process looks like closing a zipper at the molecular scale; therefore,

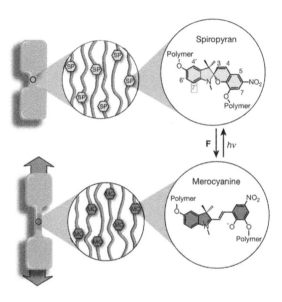

FIGURE 3.3 Schematic diagram of "dog bone" mechanophore polymers prepared from linear 80-kDa PMA with spiropyran (reprinted from Davis et al. [7] with permission © 2009, Rights Managed by Nature Publishing Group.)

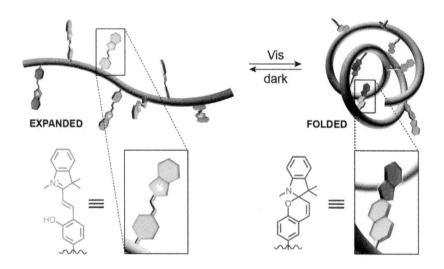

FIGURE 3.4 Light-controlled thermoresponsive polymers [14].

it aptly called "zipper crystallization." The schematic diagram of this process is shown in Figure 3.5. PEDOT is a widely used conducting polymers with a variety of applications [11]. This polymer and its alkyl derivatives showed typically cathodical electrochromic properties. Therefore, it can be used with the combination of anodical coloring materials in order to develop dual polymer devices [12, 13]. The polymer bipropylenedioxythiophene (poly(spiroBiProDOT)) also show the dual cathodical and anodical coloring properties [13–15].

3.2.2 SCREEN PRINTING ON TEXTILES

Humans have been printing designs on fabrics for centuries [16]. Designs have been found on clothes in Egyptian tombs dating to about 2500 B.C [17]. Printing is one of the most complex routine among all textile operations due of a number of variables and the need for a high degree of precision, particularly since there is no way to correct a bad print [16, 18, 19]. Technically,

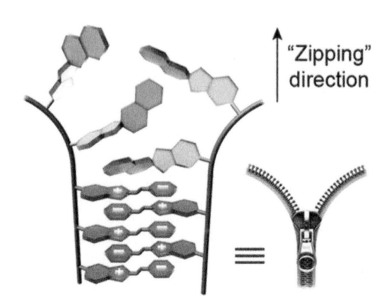

FIGURE 3.5 Schematic representation of zipper crystallization on polymer with spiropyran groups [14].

printing on textiles can be defined as the reproduction of a decoration by application of one tool loaded with coloring materials on a textile support [20, 21]. Techniques such as block printing, screen printing, or digital printing is used for printing on textiles [21]. Little et al. [22, 23] studied the use photochromic colorants as a disperse dye for textile screen printing. The printing paste was prepared by mixing a colorant dispersion and a thickening agent. The colorant dispersion was prepared by mixing a photochromic colorant in the presence of an oil-based low foam wetting agent. A dispersing agent based on disodium salt of naphthalene sulfonic acid, formaldehyde condensate, and water was used. Milling was done for 30 minutes on a roller mill by using ceramic balls in a glass jar. Sodium alginate and water were used as thickeners. The colorant dispersion was then mixed with the thickener solution to give a required colorant concentration in the printing paste. The fabric was printed, dried, and cured at 140°C for 5 min. After fixation, the surface dyes had to be removed. Hence, the reduction clearing process was carried out, and finally, the printed fabric was washed with a nonionic detergent solution (1 g/L) for 5 min, rinsed again, and dried in an oven at 60°C. Feczko et al. [24] prepared photochromic dye-based screen-printed cotton fabric by using ethyl cellulose-spirooxazine nanoparticles along with light absorbers that help to improve the photostability for photochromic behavior of the printed fabric. To improve the durability of photochromic compounds, microencapsulation was used for incorporating photochromic dyes. Due to the small size of the resultant microcapsules (average size of 3–5 μm), microencapsulated photochromic dyes can be screen printed onto fabrics. The advantage of this technique is that the encapsulating polymer layer offers a protection to the photochromic compounds [25, 26]. However, it reduced the fabric handling properties [27]. Viková [28] applied photochromic pigments on different fabrics (cotton, polyester (PET), polyamide (PA)) by using screen printing technology. Five different photochromic pigments were used, and the standard pigment printing procedure was followed in this study. After printing the fabrics, drying was carried out at 75°C for 5 min, and the curing process was done at 190°C for 20 s. The printed textile fabrics before and after UV irradiation are shown in Figure 3.6.

However, the optical yield of photochromic pigment-printed fabric was reduced with increasing concentration of the photochromic pigment; this may be due to the higher concentration of pigment on the fabric. Pure photochromic pigment powder is not sensitive to UV radiation in daylight. After

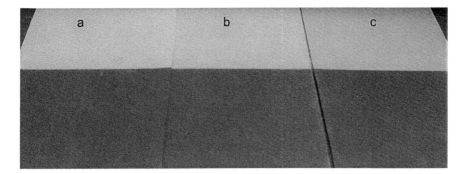

FIGURE 3.6 Printed fabrics (a=Cotton, b=Polyester and c=Polyamide) with different photochromic pigments, before (light) and after irradiation (dark side) with UV (author's own diagram).

irradiation, the photochromic pigment in the powder form does not change color. Concentrations of photochromic pigment above 50 g/kg of printing paste may well cause crystallization of the pigment into the solid-state form. Scanning electron microscopy images are shown in Figure 3.7a and 3.7b, which explain the visible crystallization of pigment at a concentration of 3 g/30 g (i.e., 3 g of the photochromic pigment in 30 g of the printing paste) [28].

In printed textiles, the scattering was increased due to the inactivated photochromic pigments. Similarly, the scattering can be decreased due to

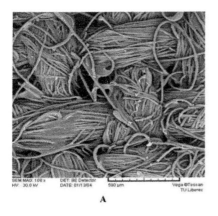

A   **B**

FIGURE 3.7 Microscopic image of a printed fabric with the photochromic pigment; A = 0.25 g/30 g; B = 3 g/30 g (i.e., 3 g of the photochromic pigment in 30 g of the acrylate printing paste) [28].

the activated pigments. This is due to the presence of pigment material in the form of capsules, as documented by microscopy images in Figure 3.8. The microscopy images confirm that the mean size of capsules around 5 μm and in all tested pigments is identical. It also seems possible to influence relatively high concentrations of pigments used for the total coverage of the textile substrate. An interesting phenomenon is reducing of the scattering coefficient relating to whole substrate that is printed by blind printing paste (a paste containing all chemicals except photochromic pigment).

Waseem [29] studied leuco dye-based thermochromic pigments applied by screen printing on textiles. Five colors, namely magenta, blue, orange,

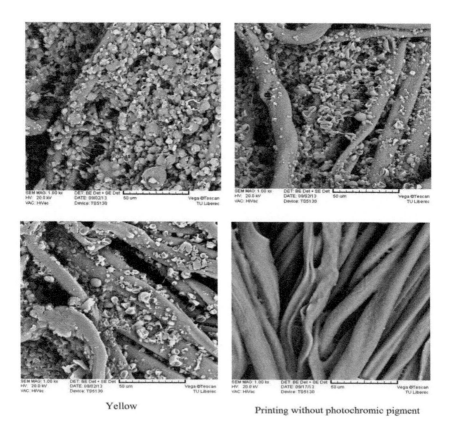

Yellow   Printing without photochromic pigment

FIGURE 3.8 Microscopy images of printing of fabrics with different photochromic pigments [28].

green, and black were used for the formulation of screen printing pastes. The color change temperature for magenta, blue, orange, and green was reported as 31°C, while for black, it was 40°C. In this study, he used a powder form of pigments that was free from water; hence, it was composed of a color former, a color developer, and a solvent contained in microcapsules. After printing, both cured and uncured samples were measured by a spectrophotometer to find the stability on the temperature. The possible color loss occurred at higher temperature. The integral value of printed sample (magenta) is shown in Figure 3.9.

Light fastness is the important property of colored textile materials. However, some dyes show poor ratings in the light fastness. It can be improved using different types of lightfast enhancers. In this study, the author used three different classes of lightfast enhancers, namely UV absorbers, hindered amine light stabilizers (HALS), and antioxidants. The concentration of lightfast enhancers was used 3% by weight (wt) of the printing paste. Generally, lightfastness enhancers provide creaminess over the colors, and after application, the printed samples were cured at 110°C for 10 min. To determine lightfastness, the author exposed the printed samples at Heraeus Xenotest 150S for different time periods such as 1, 2, 3, 4, 6, 8, 12, 18 and 24

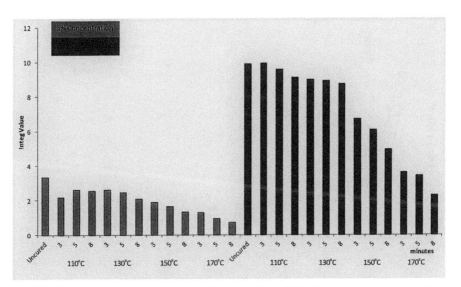

FIGURE 3.9 Integral values illustrating thermal color stability for magenta prints at various exposure temperature and time (reprinted with the permission of Waseem [29]).

hours. Normally, a blue wool sale is used to determine the light fastness (by visual comparison), However, the leuco dye-based thermochromic pigments were found to have poor lightfastness and faded very quickly when exposed to light that, instead of using the blue wool scale, the author took taken the color measurement by using the spectrophotometer; the integral values of the samples are shown in Figure 3.10 (only magenta color). The results show the antioxidant and HALS had a slight improvement in lightfastness, while the UV absorber showed a more significant improvement in lightfastness than the original sample [29].

Materials whose color reacts on contact with water are called hydrochromic or hygrochromic. Hygrochromism material can be used as a moisture indicator [30]. When applied to textile substrates, hygrochromic materials are used for hygiene, home furnishing, or clothing applications, such as umbrellas that color under raindrops or tablecloths that change their color when a drink is spilled. Several outfits created by the British designer Amy Winters use hygrochromic materials, such as the "Rainforest" dress, which is initially white and becomes colorful when wet, as shown in Figure 3.11. Hygrochromic materials can be applied to the textile substrate by coating or printing. However, they require a dry and homogeneous surface; therefore, synthetic fibers are preferred [31].

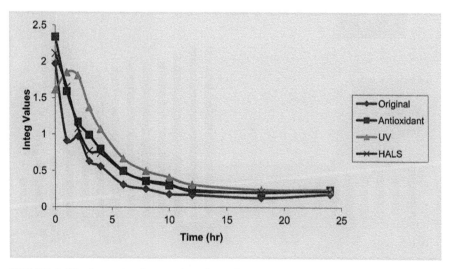

FIGURE 3.10 Integral values against time of exposure to the Xenotest of the magenta pigment mixed with lightfastness enhancers (reprinted with the permission of Waseem [29]).

FIGURE 3.11   Rainforest hygrochromic dress by Amy Winters, on the left side before wetting, on the right side after wetting (reprinted from Winters [31] with permission of Winters).

3.2.3   SOL-GEL COATING

The common name "sol-gel process" brings together a large group of methods for obtaining (synthesis) materials from solutions, in which the gel formation is present at one of the process stages [32, 33]. A sol is a dispersion of the solid particles (~0.1–1 μm) in a liquid where only the Brownian motions suspend the particles [34, 35]. However, the formulas and methods used in other industrial branches have to be adapted to the raw materials and specific textile properties. The preparatory material (or precursor) used to produce the "sol" usually consists of inorganic metal salts or metal organic components such as metal alkoxides [33, 36, 37]. These precursors are subjected to a series of hydrolysis and polymerization reactions to create a colloidal suspension (or "sol"). By further processing this suspension, this sol is transformed into a ceramic material in different forms for different applications [38, 39].

Cheng et al. [40–44] prepared silica as a matrix material for fixing the photochromic dyes (5-chloro-1,3-dihydro-1,3,3-trimethylspiro (2H-indole-2,3'-(3H)-naphth (2,1-b) (1,4) oxazine)) on to the surface of wool fabric through the sol-gel process. First, the sol was prepared by mixing the silane, aqueous nitrate-tartrate (TAS), water, and ethanol in the ratio of 1:1:10:7 (mol), respectively, with constant stirring for 24 h. The process sequence for coating is shown in Figure 3.12.

3.2.4 MASS COLORATION

Mass coloration, spun-dyeing, or dope dyeing may be defined as "a method of coloring manufactured fibers by incorporation of the colorant in the spinning composition before extrusion into filaments" [45, 46]. Viková et al. [47–49] prepared photochromic pigment-incorporated polypropylene multifilament by mixing polypropylene granulates along with photochromic pigments during the dope preparation. Several concentrations of photochromic pigments, such as 0.25, 0.5, 1.5, and 2.5 wt %, were used for this study, and

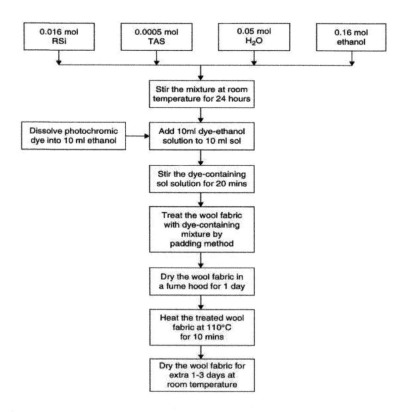

FIGURE 3.12 Typical sol-gel photochromic coating on fabric.

they were kept at 220°C as the spinning temperature; the process parameters are illustrated in Table 3.1.

Photochromic filaments with diameter ϕ=16 mm were produced using a laboratory melt spinning plant. The photochromic pigment was mechanically mixed with the polypropylene granulate, and the mixture was added in the hopper of the extruder to produce filaments as shown in Figure 3.13. Photochromic pigment-incorporated polypropylene multifilament with and without the drawing process can be carried out with specific parameters; colored polypropylene fibers were produced with different drawing ratios and different concentrations of the photochromic pigment. The filaments were drawn using a laboratory drawing machine at various drawing ratios λ with drawing temperature of 120°C [50–53]

Rubacha [54] prepared cellulose fibers with thermochromic properties by using the Lyocell process. However, this technology is based on spinning the fibers from concentrated dope solution of cellulose by using the dry-wet method in aqueous solidification bath. N-Oxide-N-methylo-morpholine (NMMO) is used as a solvent for this process; NMMO is an alternative to conventional methods of cellulose fiber production such as the viscose process or cuprammonium process, as it is free from their most important drawbacks [55]. These advantages of the NMMO process allow the process parameters to be altered within a specific range, based on the

TABLE 3.1   Various Parameters used in Melt Spinning to Produce Photochromic Filaments

Poly propylene	Metallocene Polypropylene PP HM 562 R
Temperature of melt spinning	220°C
Photochromic Pigment	5-Chloro-1,3-dihydro-1,3,3-trimethylspiro[2H-indole-2,3'-(3H)naphth[2,1-b](1,4)oxazine] (Photopia™ Blue)
Pigment concentration in fiber	0; 0.25; 0.5; 1.5; 2.5 % (on weight basis)
Drawing temperature (T_d)	120°C
Drawing ratios	1, 2, 2.5, 3, 3.5
Melt Flow Index	26.6 g/10 min
Number of holes in spinneret	13

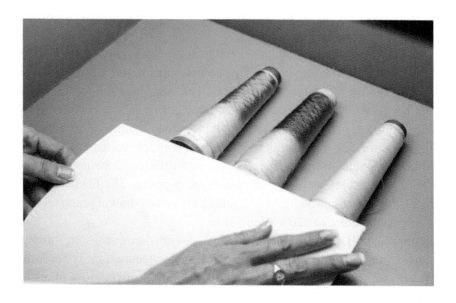

FIGURE 3.13   Photochromic polypropylene multifilaments (Author's own diagram).

fiber properties. In such cases, it is possible to obtain fibers containing considerable quantities of dyes without negative effect during the production [54].

Therefore, this study provides the information on the fibers produced from solvents with thermochromic pigmentation with specific properties and its applications. During the production of these fibers, he used Ika-Visc MKD 0.6-H60 kneader equipped with two stirrers, and a heating jacket was used for the preparation of the spinning solution. In this process, 50% aqueous solution of NMMO was used as a solvent. The fiber-forming polymers belong to cellulose, which containing 94.2% α-cellulose with moisture content of 8.3%. To prevent excessive oxidative degradation, propyl gallate (Tenox PG) was used as a stabilizer. The average of degree of polymerization (DP) was maintained at 640. Various concentrations of the thermochromic pigment were used, namely 1%, 3%, 5%, 7%, or 10% by weight. However, the commercial form of pigment contains 33% by weight of pure pigment. The dry-jet spinning process was carried out to produce the thermochromic filaments, and the production method is shown in Figure 3.14.

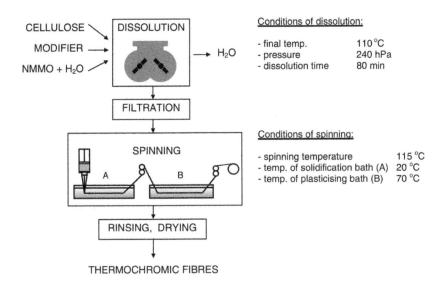

FIGURE 3.14 Scheme and process parameters for the thermochromic fibers (reprinted from Rubacha [54], with the permission of John Wiley and Sons, Copyright © 2007 John Wiley & Sons, Ltd).

The produced filaments were characterized by differential scanning calorimetry (DSC) to know their thermal properties, as mentioned in Figure 3.15. However, the color intensity of the filaments is directly proportional to the pigment concentration (Figure 3.16). The author also measured the thermochromic effects between the temperature range of 32.7°C to 32.98°C. DSC thermograms show that clear endothermic effects are linked with the color changing of decay (violet color) and exothermic effect associated with the reappearance of the color. The maxims of both effects correspond exactly to the temperatures of color changes. However, increasing the pigment concentration decreased the tenacity of the filaments, and highest tenacity loss of 50% was found at 10% of thermochromic pigment.

3.2.4.1 Thermochromic Synthetic Filaments

Waseem [29] produced different thermochromic polymers, namely polypropylene, linear low-density polyethylene, and ethylene vinyl acetate. In this work, he incorporated the thermochromic pigment at the concentration of 0.5%, 1%, and 2% by weight. He observed the color of the filaments

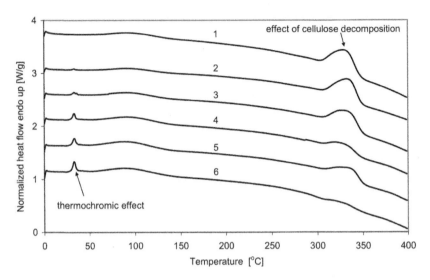

FIGURE 3.15   DSC of fibers modified with the thermochromic dye. 1: Unmodified fibers; 2–6: fibers containing 1%, 3%, 5%, 7%, and 10% by wt. of the pigment (reprinted from Rubacha [54], with the permission of John Wiley and Sons, Copyright © 2007 John Wiley & Sons, Ltd).

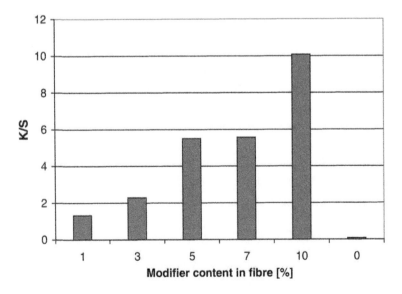

FIGURE 3.16   Dependence of color intensity on the thermochromic pigment content (reprinted from Rubacha [54] with the permission of John Wiley and Sons, Copyright © 2007, John Wiley & Sons, Ltd).

after cooling. The thermochromic pigments are sensitive to heat; so, it was assumed that the maximum color strength would be achieved at lower processing temperatures. There may be color loss at higher temperature; based on these facts, the author selected polypropylene, linear low-density polyethylene, ethylene vinyl acetate, and polycaprolactone. These polymers have different extrusion processing temperatures that cover the temperature range from 130°C to 260°C. However, in order to observe whether there was an effect of the extrusion process on the thermochromic phenomena, the extruded filaments were heated with a hair dryer and cooled a number of times. The thermochromism of pigmented filaments with 2% color concentration for all three polymers is shown in Figure 3.17.

Based on the author results [29], the UV absorber shows significant lightfastness enhancement of the thermochromic pigments, much better than using the antioxidant or HALS. The lightfastness shown by the pigments incorporated into PP and EVA are better than in LLDPE. These polymers

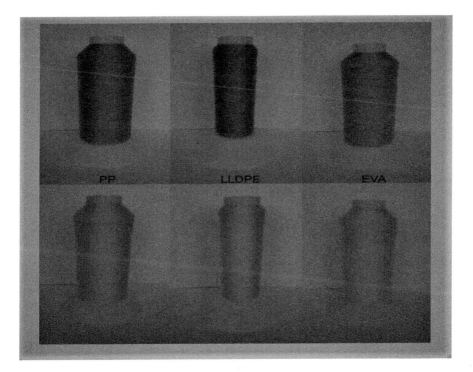

FIGURE 3.17 Thermochromic polymers containing the red pigment (reprinted with the permission of Waseem [29]).

thus provide some protection for the pigments, which may be due to UV absorption properties, as shown in Figure 3.18.

3.2.4.2 Coaxial Monofilament

In 2012, Laforgue et al. [56] produced the polypropylene coaxial multi-layer monofilament with thermochromic inks under a Canadian research lab (Industrial Materials Institute, National Research Council Canada). However, the core layer could be electrically actionable to trigger the color change of the external layer. The produced filament is shown in Figure 3.19. This filament core is covered by a polymer containing 25 wt % of a thermochromic component. The master batch of polypropylene (PP) contained 20 wt % multiwalled carbon nanotubes. The multifilaments were produced using a SandCastle co-extraction line equipped with a specially designed single-hole die, which allow to extrude the multifilaments with three coaxial layers. Three different extruding temperatures were used: 220°C for core layer (PP-MWNT), 210°C for intermediate layer (PP-TiO$_2$), and 165°C for sheath layer (PP-TCM-TiO$_2$). Finally, the monofilaments were collected on a spool with minimal stretching. Further, it can be post stretched at ambient temperature.

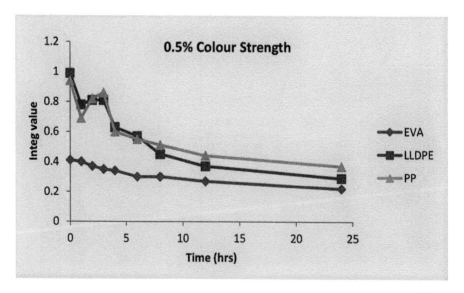

FIGURE 3.18 Lightfastness of extruded 0.5% red filaments (reprinted with the permission of Waseem [29]).

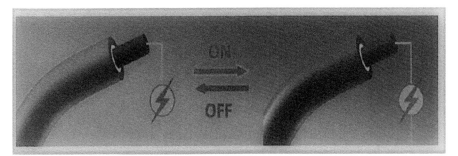

FIGURE 3.19 Electro-thermochromic filament (reprinted from Laforgue et al. [56] with the permission of ACS Publications, Copyright © 2012, American Chemical Society).

They produced the prototype woven fabric (64 cm²) by using 130 stretched monofilaments. The two upper layers of each monofilament were then selectively stripped off at the extremities, and the core layer of each monofilament was connected to an aluminum strip by using silver epoxy. All monofilaments were connected in a parallel configuration, which may decrease the ohmic resistance of the overall electric circuit. The textile was then connected to a power source to test its resistive heating properties. During the electric volt application, the fabric was heated up efficiently and evenly when the electrons were passed through the core layer of the filaments. Figure 3.20 shows the optical images of electrothermochromic fabric with different voltage applications.

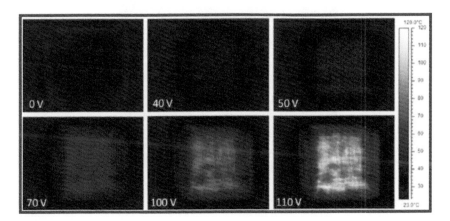

FIGURE 3.20 Infrared camera images of the 64-cm² fabric prototype at different applied voltages (reprinted from Laforgue et al. [56] with the permission of ACS Publications, Copyright © 2012, American Chemical Society).

The results show no color change up to 20 V; the temperature was increased with increasing the electric voltage and it reached 120°C at 120 V. The real color (green to beige) begins to change when the fabric reaches 40°C, as shown in Figure 3.21. Similarly, when the voltage was turned off, the fabric progressively restores its original green color. The cold-to-hot color transition time of the fabric was controlled from 2 to 30 s. The authors suggested that this fabric can be applied in various fields such as different camouflages, dynamic mural textiles, heating textiles with a visual temperature state indication, and fashionable articles [56].

3.2.4.3 Silica *Nanoparticles Functionalized* with *Thermochromic* Dye

Ribeiro et al. [57] synthesized and functionalized silica nanoparticles with a thermochromic dye by two different methods and immobilized them on cotton fabric by using a thermochromic dye-functionalized silica.

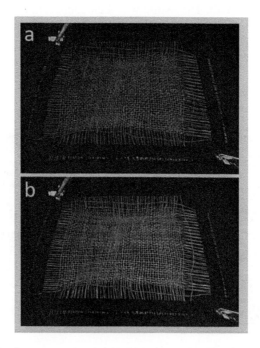

FIGURE 3.21 Textile prototype (a) in its cold color (0 V, ambient temperature) and (b) in its hot color (60 V, 50°C) (reprinted from Laforgue et al. [56] with the permission of ACS Publications, Copyright © 2012, American Chemical Society).

The silica nanoparticles were synthesized under alkaline conditions with the presence of surfactant and functionalized with the thermochromic pigment by post-synthesis (post-grafting) method; pigment-incorporated nano-silica is shown in Figure 3.22.

3.2.5 EXHAUST DYEING

Exhaust dyeing is one of the most popular dyeing methods for dyeing on textile goods [58]. The exhaust dyeing process is carried out in the various forms of textiles such as staple fiber, yarn, and fabric [59, 60]. Dye solution or dye bath is produced by dissolving the dyestuff according to the required

FIGURE 3.22 Photographs of the dye-functionalized silica (SiO$_2$) and the corresponding thermochromic textile at room temperature (left) and when heated (right) (reprinted from Ribeiro et al. [57], with the permission of Springer Publications, Copyright © 2013, Springer Science + Business Media New York).

liquor ratio. Then, the textile material is immersed into the dye solution. Initially, the surface of the fiber is dyed when dyes contact with the fiber, and then, the dyes are entered in the core of fiber [61, 62]. Proper temperature and time are maintained for diffusion and penetration of dye molecule in the fiber core. During the process, kinetic and thermodynamic reactions occur [62]. The reactive dyeing process is a chemical reaction occurring between the dye molecule and the fiber molecule [61, 64–66]. Generally, the same methods can be used to apply photochromic dyes to the cotton and polyester fibers, and the mechanisms are described in Schemes 3.1 and 3.2, respectively, as shown in Figure 3.23.

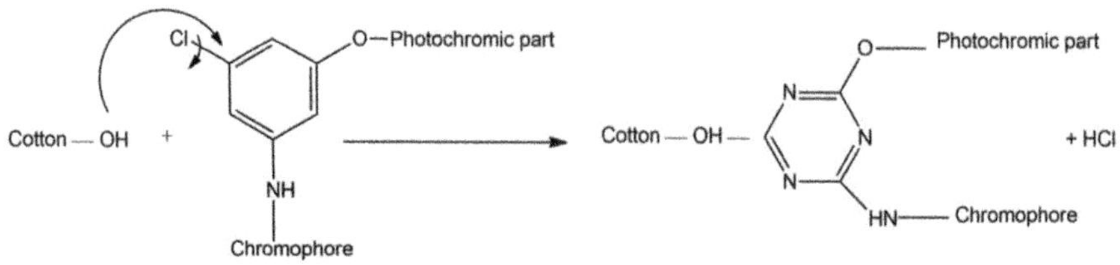

SCHEME 3.1   Mechanism of reacting cotton with the photochromic dye.

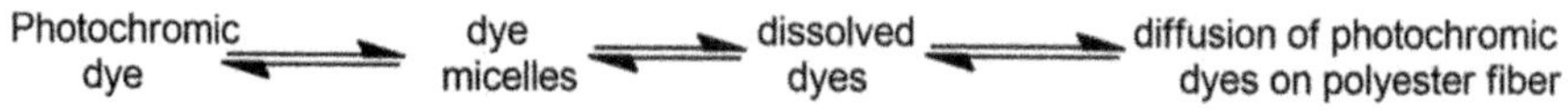

SCHEME 3.2   Process sequence for the dyeing of polyester fiber with the photochromic dye.

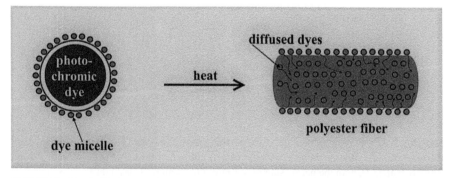

FIGURE 3.23   Diffusion mechanism of polyester fiber with the photochromic dye.

Shah et al. [67] studied the dyeing of photochromic dyes, namely mercury (II) dithizonate photochromic compound with various textile materials such as nylon, cellulose triacetate, and polyester materials. In this study, the dyeing solution was prepared by mixing of the photochromic dye with the wetting agent and the dispersing agent. Polyester and nylon dyeing was started at 60°C for 15 min with acidic pH range of 5–6. Thereafter, temperature was gradually increased to 120°C and 100°C and 1 and 1.5 h for polyester and nylon, respectively. After dyeing, the dyed fabrics were washed with soap solution (1 g/L) in alkali medium (around 2 g/L of Na_2CO_3) to remove the excess or superficial dyes that has absorbed by surface of the textile materials, and finally, the dyed fabric was dried in air [68]. Apart from these studies, Aldib and Christie was dyed on polyester fabric with the exhausting method by using of six photochromic dyes based on spironaphthooxazine and naphthopyran.

3.2.5.1 Dyeing of Synthetic Fibers (PAN, PA, and PET)

Billah et al. [68, 69] studied the direct coloration of photochromic dyes as disperse dyes. The aqueous dispersion of dyeing solution was prepared for dyeing by milling spiroindolinonaphthoxazines dyes, dispersing agent, and ceramic balls in the honey jars rotating on a set of rollers for 18 h. Subsequently, the dyeing of nylon, acrylic, and polyester was carried out in the laboratory level dyeing machine, and the liquor to material ratio was 1:50 for all samples. The initial pH was adjusted to 5.5 by using acetic acid, and the sample was loaded in the pot; the processing was continued for 10 min with increase the temperature to 60°C with gradient of 2°C min^{-1}, and the dyeing was continued for 10 min. In the later steps, the temperature was increased to 90°C with gradient of 1°C min^{-1} and the dyeing was continued for 60 min; the dyeing bath was cooled until 40°C, and the samples were removed from the pot and washed with cold water.

3.2.5.2 Dyeing of Wool

Billah et al. [70] reported that bleached wool samples were added in the stainless-steel dye pots containing the photochromic acid dye (spriooxazine-based) solution. The material to liquor ratio was 1:30 and initial pH was adjusted by diluted acetic acid. The initial dyeing process was carried out for 30 min with 40°C. Later, the temperature was increased (gradient of 1°C min^{-1}) to 60°C and the dyeing was continue for 60 min. After the dyeing process, the temperature of the dyeing machine was reduced, and the samples were removed, washed with cold water, and dried with simple air dry.

3.2.5.3 Dyeing of Nylon

Lee et al. [71] applied three spirooxazine-based photochromic dyes on polyamide (nylon) fabric by using of the exhaustion dyeing method. The dyeing mechanism of spirooxazines with nylon was described in the Figure 3.24. The dyeing process was carried out with the liquor to material ratio is 50:1. Initially, the dyeing process was started at 40°C and run for 10 min, and later increase the temperature between 80–120°C and continue the dyeing process for 60 min. At the end, the dyed samples were removed from pots, rinsed thoroughly with cold water, and dried in the open air. The results show that the exhaustion of spirooxazine (%) was increased with respect

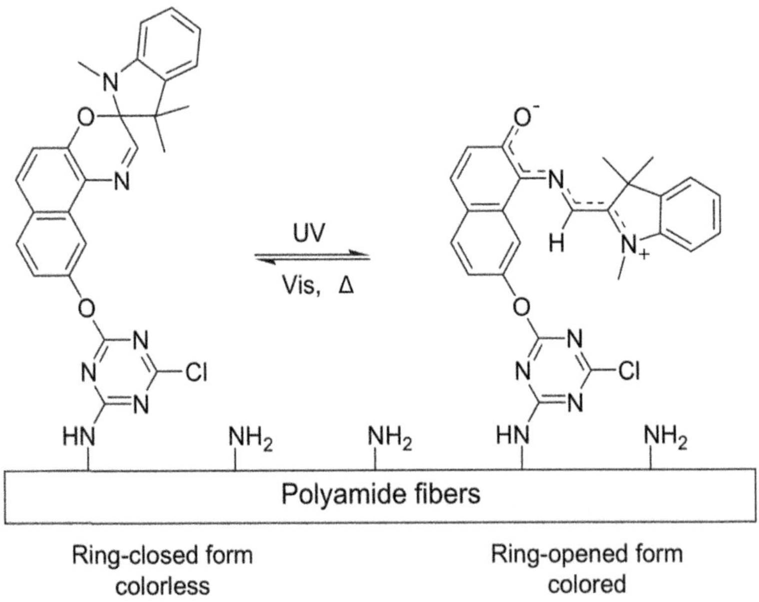

FIGURE 3.24 Bond formation of spirooxazine with polyamide fibers (reprinted from Son [72], with the permission of Elsevier Publications, Copyright © 2005 Elsevier Ltd).

to increasing the dyeing temperature, and optimum exhaustion was found between 100–110°C. This is attributed due to the higher kinetic energy of the dye molecules at elevated temperatures as well as higher fiber swelling effect. Son [72] continued their research and found that the addition of alkali during the same dyeing process increased the covalent bond fixation of dichloro-s-triazinyl group of the spirooxazine dye toward to polyamide fibers [72]. The addition of alkali (10 g/L of Na_2CO_3) enhances the initial exhaustion of spirooxazine dye and subsequent dye-fiber fixation.

Every electrochromic material exhibits at least two colors, one at the oxidized state and the other at the reduced state. When the electrochromic material has three or more stable oxidation states, each with a different color, it is called electropolychromic, or the preferred polyelectrochromic [73]. Electrochromism in commercial applications is currently limited to rigid devices, more particularly to adaptive "smart" windows and to rear-view mirrors with antiglare effect. The same principle is used for both examples: an electrochromic material is used, which is transparent at one state and opaque or darker at the other state. The color change of the device is manual or automatically controlled. When it is cloudy outside, the electrochromic window is transparent to allow the maximum light to enter, but it can turn opaque or darker when it is sunny in order to avoid glare and heating.

In 2006, Beaupre et al. [74] developed a flexible electrochromic device using textile. The flexible structure was made with a transparent electrode that was covered with a spray-coated electrochromic polymer; unfortunately, the color changing property is slow. Later, in 2011, Molina et al. [75] described the electrochromic properties of polyaniline-impregnated textile fabric for the first time. The produced fabric was immersed in the electrolytic bath and connected to a power supply makes visible color change, from light green at −1 V to dark green at +2 V. However, the author's main aim was to develop such fabrics for conductive textiles. In 2014, Yan et al. [76] developed flexible electrochromic materials by using polydimethylsiloxane (PDMS) with tungsten trioxide. These materials showed visible color change from colorless to blue, when electric volt was applied. The PDMS-based structure was combined with a cotton fabric in order to form a textile-based electrochromic device. However, the textile itself can hardly be considered as indispensable for the working of this structure. It is more a plastic-based electrochromic structure that can be fixed on textile than a textile electrochromic structure. Meunier et al. [77] developed four-layer electrochromic textile, which was composed of 100% textile spacer, impregnated with inorganic electrochromic material and sandwiched between two flexible electrodes. A Prussian Blue water solution, a type III electrochromic material, is used as both the electrochromic material and the electrolytic material. Impregnated textile spacer was placed between two electrodes, and the whole structure was sealed. Subsequently, 4.5–5.5 V of power supply was applied to the textile structure for 20 s, which initiates the visible color change from Prussian Yellow to Prussian Blue, as shown in Figure 3.25.

It was demonstrated that the electrochromic material is still fully functional when the substrate was stretched. Apart from the conductivity factor, stretching itself affects the morphology of the spray-coated electrochromic polymer films. As shown in Figure 3.26, stretching resulted in gaps in the electrochromic polymer film, thus exposing the PEDOT-PSS-soaked fibers underneath. The fibers, on the other hand, showed no color change (separation) or continuity problems upon stretching, again confirming the result of a percolated network of PEDOT-PSS particles. The intensity of the red color diminished because of the exposure of the blue underneath and returned to its original state when released [78].

The halochromic dyes can be incorporated on textile fabric by the conventional dyeing method, which is advantageous, because of simplicity and low cost as well as good penetration of dyes [68, 79]. In the ancient times,

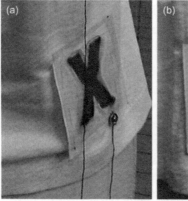

FIGURE 3.25 Electrochromic structure using Prussian Blue as the electrochromic material: (a) Prussian Yellow; (b) Prussian Blue (applied reprinting from Meunier et al. [77], with the permission of NISCAIR Publications).

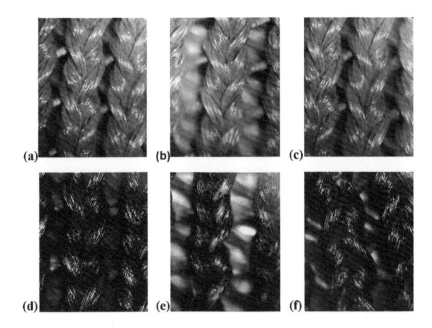

FIGURE 3.26 Optical microscope images (20×) of (a-c) conducting fabric and (d-f) electrochromic polymer-coated fabric. (a, d) Before stretching, (b, e) under ca. 20% stretching, and (c, f) after relaxation to its original state (reprinted from Ding [78], with the permission of ACS Publications, Copyright © 2010, American Chemical Society).

the pH-sensitive dye Alizarin was used, which was extracted from the roots of plants of the Rubiae family [80]. The synthetic pH indicator Congo Red was formerly applied to cellulose materials for coloring purpose [81]. However, Congo Red is not popular with respect to halochromic textiles due to carcinogenicity [82]. Another study was conducted by Hoten [83] on the application of classic cationic dyes on acrylic fibers.

The result shows specific interaction between the dye and acrylic fiber with respect to the pH, but it shows poor stability and the color change disappeared. Even though these studies do not present insights into the properties of halochromic textiles, the main aim of these work was on coloration and not on halochromism. The different types of pH indicators (azo pH indicator Brilliant Yellow and the anthraquinone pH indicator Alizarin) were applied onto the cotton fabric [84] by conventional direct dyeing method. Dyed fabric shows the visual color changes from yellow to red ((pH 4 and 7) for brilliant yellow dyes and yellow to purple for alizarin (between pH 3 and 5), thus already showing the potential of halochromic cotton fabrics. Alizarin was applied to polyamide 6 and polyamide 6.6 fabrics by the acid dyeing process. Similar to cotton, an effective pH sensor was obtained, exhibiting a reversible color change from yellow through orange to purple [85]. Staneva et al. [86] developed halochromic dye, 1-[(7-Oxo-7H- benzo[de]anthracen-3-ylcarbamoyl)-methyl]-pyridinium chloride, and they applied it to viscose rayon fabric by the conventional dyeing method. The dyed material shows good reversibility of color change.

3.2.6 PAD-DRY HEAT FIXATION DYEING OF PET

Mohanad and Christie [87, 88] optimized the solvent-based dyeing of polyester fabric carried out by the pad-dry-heat method (at 190°C for 45 s for a series of commercial photochromic dyes, with the intention of comparing the outcomes with those from a traditional aqueous disperse dyeing method. They found that the optimum solvent:fabric ratio was 2.5:1; dichloromethane was used as the solvent in their experiments, but whilst it remains a commonly used chlorinated organic laboratory solvent, it is a suspected carcinogen; thus, alternatives may need to be sought before engaging in large-scale processing. After the dyeing process, the excess dyeing solution was removed by the padding mangle followed by thermal fixation by using dry heat at 190°C for 45 s, rinsed, and subjected to a reduction clearing process at 70°C for 20 minutes. The degree of photocoloration, background color, fading characteristics, fatigue resistance, and storage stability were evaluated for the dyes. Examination of concentrated solutions of the photochromic dyestuffs in various solvents implicated the solvents in the development of background color, and it was suggested that toluene and ethyl acetate may be preferable to dichloromethane as the solvent. Results show that the degree of photocoloration of the dyed fabric was increased when the toluene or ethyl acetate was used for the application of the dye; similarly, it decreased with dichloromethane. The fatigue resistance results from aqueous exhaust dyeing were generally higher than when the dye was applied by the solvent-based method.

3.2.7 NONWOVEN FABRIC BY MELT BLOWN TECHNOLOGY

Melt blowing is a technique that directly produces the fibrous webs from polymer melt (dope) using high-velocity air or another appropriate force to attenuate the filaments where the thermoplastic fiber forming polymer is extruded through a linear die containing several hundred small orifices. This process is one of the newer techniques to produce nonwoven fabrics (Figure 3.27). This process is unique because it is used almost exclusively to produce microfibers rather than fibers the size of normal textile fibers. Convergent streams of hot air (exiting from the top and bottom sides of the die nosepiece) rapidly attenuate the extruded polymer streams to form extremely fine diameter fibers (1–5 micrometer). The attenuated fibers are subsequently blown by high-velocity air onto a collector conveyor, thus forming a fine fibered self-bonded melt blown nonwoven fabric. The melt blown process is similar to the spun bond process that converts polymer to nonwoven fabrics in a single integrated process [89–94].

Viková [28] developed photochromic melt blown nonwoven (Figures 3.28 and 3.29) fabric. In this case, the photochromic pigment was incorporated to the polypropylene fiber by mass dyeing technology. This technology was used due to the above advantages as well as its availability and relative simplicity of production in comparison to classical methods of fiber production. First, a master batch was prepared, containing a maximally technically useable concentration of pigment in the polypropylene granulate (20%). The master batch was mixed during production with clear polypropylene granulate. Two fineness of fibers (fibers with different diameter) were used to produce the nonwoven fabric with photochromic pigment concentration (1.0%, 1.8%, and 2.6%) and was found that photochromic effect wasn't sensitive on difference in fineness on measured level of color change.

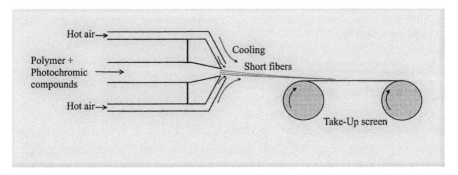

FIGURE 3.27   Production scheme for melt blown photochromic webs [95].

FIGURE 3.28   Produced photochromic nonwoven fabric under UV exposure [95].

 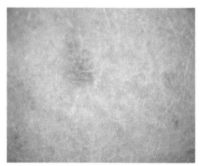

FIGURE 3.29   Nonwoven fabric exhibiting nonhomogeneity of photochromic effect under UV light exposure [95].

3.2.8   ELECTROSPINNING

Electrospinning is an electro-hydrodynamic process that uses electrical forces to produce fibers with less than 2–100 nm [96]. Predominantly, electrospinning can produce fibers from the polymer solution of natural and synthetic polymeric materials. It is a versatile and promising platform technology to produce electrospun nanofibrous materials consisting of diverse polymers and polymer composites. Electrospinning has gained much attention in the last decade, due to its wide scope to produce a variety of polymers with different diameters [97–99]. The typical electrospinning process for the production of color changeable nanofiber is shown in Figure 3.30.

Lee et al. [100, 101] incorporated SPO into PMMA through the electrospinning process, with mixing of spironaphthoxazines and D-π-A type isophorone dye; the DMF was used as solvent to make the polymer solution (dope). Polymer solution contains spironaphthoxazines, DMF, PMMA, and 2-(3-(4e((2-hydroxyethyl) (methyl) amino)styryl)-5,5-dimethylcyclohex-2-enylidene) malononitrile (vinylcyano acetate)pyran (HDMP), and it is passed toward the syringe needle. A high voltage (30 kV) is applied to the syringe needle, which helps to produce the fiber. Once the fiber is produced, it is collected on the surface of grounded aluminum foil. The diameter of the produced fiber is 400–1000 nm, and it is cylindrical in shape. Once the produced nanofiber mat was exposed to UV irradiation, it quickly changed its color from bleached position to purple. The maximum absorbance occurred

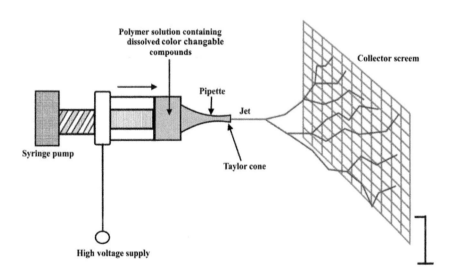

FIGURE 3.30 A typical electrospinning unit for the production of color changeable nanofibers [100].

in 500 to 550 nm, and it started to decrease above 550 nm and maximum reflectance was observed in 600 to 700 nm. A slight decoloration of SPO/ PMMA occurred at room temperature; the decoloration increased with the increase in the temperature.

Wang et al. [102] prepared photochromic nanofibers of SP-PVDF-co-HFP. Electrospinning was done with the solution containing SP-PVDF-co-HFP, N,N-DMAc, and acetone containing 1% of 1'-(3- carbomethoxypropyl) side-chain (SPEST). The polymer contains fluorine side chain that enhances the hydrophobic, thermal stability, resistance to corrosion, and UV degra-dation properties of the produced fiber. The results showed photochromic property decrease in emission at 468 nm and 482 nm and an increase the intensity in the 640 nm region. Khatri et al. [103] prepared UV-responsive photochromic nanofibers by using high voltage electrospinning machine. The polymer solution containing PVA, photochromic dyes, and ethanol was supplied through a plastic syringe attached with a capillary tip. A voltage of 12 kV was supplied to the electrospinning process to obtain nanofibers, and the produced nanofibers were deposited continuously over a rotating metal-lic drum at room temperature. Later, the electrospun nanofiber is dried to remove residual solvent, which may present in the fiber structure. The diam-eters for produced fibers were between 25 to 30 μm. Zhang et al. [104] pre-pared spiropyran-based polymeric nanowires with polystyrene (PS) as core and spiropyran-based copolymer as shell, and then the electrospinning was used to produce nanofibers. Shuipping et al. [105] prepared photochromic nanofibrous mats by electrospinning the blend solution of cellulose acetate (CA) and NO$_2$SP. First, these compounds were dissolved in acetone to make a polymer solution. During electrospinning, a positive high voltage of 10 kV was applied at the tip of a syringe needle. The resultant nanofibers were collected on a piece of an aluminum foil. The results indicate that unirradi-ated sample had only one absorption peak at 336.64 nm, while the radiated sample also showed another absorption peak at about 567 nm, which lies in the range of visible light and the increased peak due to the presence of photochromic materials.

Phase-change thermochromic materials (PCTMs) have attracted much attention for their phase-transformation reversibility and energy-storage and management properties [106, 107]. However, the fluidity characteris-tics of the PCTM after melting limits its practical applications. By using the melt coaxial electrospinning technique, Xia's group [108] developed phase-change materials encapsulated by core-shell TiO$_2$ polyvinyl-pyrrol-idone nanofibers; therefore, the combination of PCTM-encapsulated nanofibers and coaxial electrospinning fibers have further potential in the field of thermal energy management and sensors.

Li et al. [109] produced thermochromic core-shell nanofibers through melt coaxial electrospinning, and the experimental setup is shown in Figure 3.31. The heating system was provided to keep thermal atmosphere for the whole electrospinning system. The PMMA was dissolved by DMF (10–15 wt %), and then the polymeric solution (dope solution) was loaded in the outer syringe. The melting mixture (dope solution) was controlled with the flow rate of 0.2–1.0 mL/h by a syringe pump. The inner nozzle was connected to

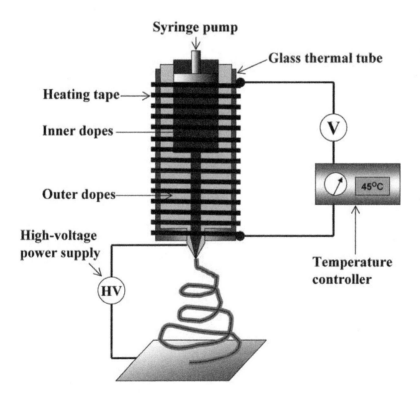

FIGURE 3.31 Melt coaxial electrospinning setup used for fabricating PMMA nanofibers loaded with PCTM CBT (reprinted from Li et al. [109], with the permission of John Wiley and Sons Publications, Copyright © 2008 Wiley Periodicals, Inc.).

the cathode of the high-voltage generator, and the metallic plate was covered with the piece of an aluminum foil and connected to the anode for collecting the substrate; electric voltage was set at 20–30 kV, and the work distance was 15–25 cm. The produced fibers were designed to be applied in thermochromic materials and sensors. These nanofibers can provide a new insight to the field of temperature sensors with good fluorescence signals.

3.2.9 DIGITAL PRINTING

Digital printing was introduced in the 1950s for paper printing. But this printing technique was extended to textiles from 1970s [110–112]. Digital printing, in simple terms, is the process of creating prints generated and designed from a computer, as opposed to analog printing, which requires printing screens [113]. From the roller printing technology of the 50s through screen printing to today's state-of-the-art inkjet printing technology by using digital CAD system, printing has evolved in stages of developments [114, 115]. In digital inkjet printing, print heads containing banks of fine nozzles fire fine droplets of individual colored inks onto pre-treated fabrics. The dyes are supplied in color cartridges by the dye maker and, once connected to the printer, are ready for instant use. Inkjet printing has a low water consumption and low energy consumption process compared with conventional printing process [116, 117].

Aldib and Christie [118, 119] prepared photochromic ink-based digital printing on polyester fabric. The photochromic dyes were dissolved in isopropanol to form a dye solution that is required to be stirred for 1 h using a magnetic stirrer to dissolve the photochromic dyes; ethylene glycol, polyethylene glycol, and polyvinylpyrrolidone were added during the stirring and continued for another 2h. After stirring, the inks were filtered through a filtered paper (pore size may be 1 µm). Later, the filtered dye solution is kept overnight and again filtered with the same quality of the paper. After second filtration, the dye solution is kept for 2 h for degassing to remove the air (air bubbles) in the solution. Otherwise, these bubbles create problems in the inkjet printing machine [120]. Then, the ready ink is applied on to the polyester fabric by using an inkjet printing machine. Once the printing is done, the printed fabric undergoes the fixation process at 190°C for 45 s. After fixation, the fabric has rinsed by cold water followed by the reduction clearing process to remove unfixed dyes present on the surface of polyester fabric. Later, the fabric is rinsed again with cold water and washed with mild soap, followed by a rinse and dry in the open air.

3.2.10 MICROENCAPSULATION OF PHOTOCHROMIC MATERIALS

Microencapsulation has been widely used in many industrial applications, including pharmaceutical and textile. The term "microencapsulation" is used to designate a category of technologies used to entrap solids, liquids, or gases inside a polymeric matrix or shell. This process involves the production of microcapsules that act as tiny containers of solids. The thickness of the shell or wall is less than 2 μm; so, it is very easy to apply in the textiles. Generally, these capsules can release the core components under controlled conditions to suit a specific purpose. The main advantage of the microencapsulation technique is that it can protect the photochromic colorants from the external environment; apart from this, it can allow us to keep several photochromic colorants in the core to develop a wide range of color effect under UV radiation. The production of microencapsulated photochromic pigments is described in Figure 3.32.

Feczko et al. [121] prepared photoresponsive polymer by microencapsulation of photochromic compounds. The nanocapsules of 5-chloro-1,3-dihydro-1,3,3-trimethylspiro[2*H*-indole-2,3'-(3*H*)naphtha[2,1-*b*](1,4)

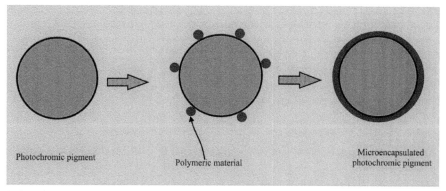

Photochromic pigment   Polymeric material   Microencapsulated photochromic pigment

FIGURE 3.32 Typical sequence for the production of microencapsulated photochromic pigments [95].

oxazine] was microencapsulated on poly(methyl methacrylate) and ethyl cellulose by emulsion solvent evaporation methods. But both the polymers were failed in practical textile application. Zhou et al. [122] prepared photochromic printed fabric by using melamine formaldehyde microcapsules. The melamine-formaldehyde microcapsules were mixed along with the print paste, and it was printed on the fabric. The results show that the improved light fastness by photochromic microcapsules as well as the lifetime of photochromic microcapsules could extend from 6–7 h to 69–75 h under continuous UV, when stirring rate, emulsification time and mass ratio of core materials/wall materials were 1000 rpm, 5 min and 1:1, respectively. Examples of invention with microencapsulated photochromic textile material are given in Table 3.2. The photochromic microcapsules changed their color very rapidly during the UV irradiation, as shown in Figure 3.33 [123].

In 1987, Toray industries developed temperature sensitive fabric by the introduction of microcapsules of diameter 3–4 mm to enclose heat-sensitive dyes that are coated with a resin homogeneously over fabric surface. The microcapsule was made of glass and contained the dyestuff, the chromophore agent (electron acceptor), and color neutralizer (alcohol, etc.) which reacted and exhibited color/decolor according to the environmental temperature. In 1996, Aitken et al. [123] studied the textile applications

TABLE 3.2   Inventions of Microencapsulated Photochromic Textiles

Invention	*Innovators*
Light- and washfast reversible photochromic fabrics coated with spironaphthoxazine derivatives, microencapsulated in hollow porous inorganic microspheres, for application in curtains.	Furuta, et al. [129, 130]
Photochromic inks for screen printed textiles, microencapsulated by a gelatin coacervation methods, formulated with an acrylic binder	Nakanishi, et al. [131]
Reversible photochromic textiles	Kamata, et al. [132]
A microencapsulated photochromic composition for textile printing paste use	Ryuichi and Yosuke [133]
Microencapsulated photochromic textiles by dyeing and printing	Yabuchi, et al. [134]

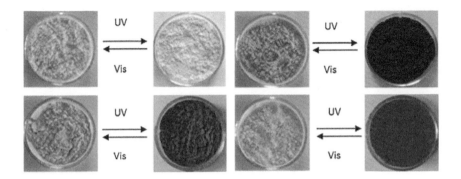

FIGURE 3.33 Color changing photochromic microencapsulation (reprinted from Zhou [122], with the permission of Elsevier Publications, Copyright © 2013 Elsevier B.V.).

of thermochromic systems. Generally, organic thermochromic systems are always dependent or related to the melting point of the incorporated solvent. The fundamental problem associated with the thermochromic microcapsules to both synthetic and natural fibers is that encapsulated microcapsules must be impermeable and insoluble in water.

3.3 INTERESTING VIDEOS

- **Photochromic pigments:** https://www.youtube.com/watch?v=C9tI_dUwmPw
- **Smart Textiles – Thermochromic paint:** https://www.youtube.com/watch?v=SYmU5i9skLg
- **Photochromic T-Shirts:** https://www.youtube.com/watch?v=nqmVqieSuCM
- **Electrochromic glasses:** https://www.youtube.com/watch?v=afLGMFUNlfY
- **Color-change electronic textile:** https://www.youtube.com/watch?v=KmdfQYD1iMc
- **Hydrochromic ink's application on umbrella:** https://www.youtube.com/watch?v=B9MjHtV4G3Y
- **Hydrochromic ink on fabrics:** https://www.youtube.com/watch?v=7wYktagIAAk

KEYWORDS

- chromic materials
- digital printing
- mass coloration
- nanofiber
- screen printing
- sol-gel coating

REFERENCES

1. Lovrien, R., (1967). The photoviscosity effect. *Proc. Natl. Acad. Sci. USA, 57*(2), 236.
2. Irie, M., (2015). Photochromic polymers, In: *Encyclopedia of Polymeric Nanomaterials*, Kobayashi, S., & Müllen, K., eds., Springer Berlin Heidelberg: Berlin, Heidelberg, pp. 1–10.
3. Agolini, F., & Gay, F. P., (1970). Synthesis and properties of azoaromatic polymers. *Macromolecules, 3*(3), 349.
4. Finkelmann, H., Nishikawa, E., Pereira, G. G., & Warner, M., (2001). A new opto-mechanical effect in solids, *Phys. Rev. Lett., 87*(1), 15501.
5. Ikeda, T., Mamiya, J., & Yu, Y., (2007). Photomechanics of liquid-crystalline elastomers and other polymers, *Angew. Chemie Int. Ed., 46*(4), 506.
6. Seeboth, A., Lotzsch, D., & Ruhmann, R., (2013). First example of a non-toxic thermochromic polymer material–based on a novel mechanism, *J. Mater. Chem. C., 1*(16), 2811.
7. Davis, D. A., Hamilton, A., Yang, J., Cremar, L. D., Van Gough, D., Potisek, S. L., Ong, M. T., Braun, P. V., Martinez, T. J., White, S. R., Moore, J. S., & Sottos, N. R., (2009). Force-induced activation of covalent bonds in mechanoresponsive polymeric materials, *Nature, 459*(7243), 68.
8. Sumaru, K., Kameda, M., Kanamori, T., & Shinbo, T., (2004). Characteristic phase transition of aqueous solution of poly (N-isopropylacrylamide) functionalized with spirobenzopyran, *Macromolecules, 37*(13), 4949.
9. Ivanov, A. E., Eremeev, N. L., Wahlund, P. O., Galaev, I. Y., & Mattiasson, B., (2002). Photosensitive copolymer of N-isopropylacrylamide and methacryloyl derivative of spyrobenzopyran, *Polymer, 43*(13), 3819.
10. Wismontski-Knittel, T., & Krongauz, V., (1985). Self-assembling of spiropyran polymers by zipper crystallization, *Macromolecules, 18*(11), 2124.
11. Groenendaal, L., Jonas, F., Freitag, D., Pielartzik, H., & Reynolds, J. R., (2000). Poly (3,4-ethylenedioxythiophene) and its derivatives: Past, present, and future, *Adv. Mater., 12*(7), 481.

12. Sapp, S. A., Sotzing, G. A., & Reynolds, J. R., (1998). High contrast ratio and fast-switching dual polymer electrochromic devices, *Chem. Mater.*, *10*(8), 2101.
13. Gazotti, W. A., Paoli, M. A. De, Casalbore-Miceli, G., Geri, A., & Zotti, G., (1999). A solid-state electrochromic device based on complementary polypyrrole/polythiophene derivatives and an elastomeric electrolyte, *J. Appl. Electrochem.*, *29*(6), 757.
14. Klajn, R., (2014). Spiropyran-based dynamic material, *Chem. Soc. Rev., 43*, 148–184.
15. Thompson, B. C., Schottland, P., Zong, K., & Reynolds, J. R., (2000). In situ colorimetric analysis of electrochromic polymers and devices, *Chem. Mater.*, *12*(6), 1563.
16. Dowds, B. F., (1970). Variables in textile screen printing, *J. Soc. Dye. Color*, *86*(12), 512.
17. Wilson, J., (2001). In: *Handbook of Textile Design*, Wilson, J., ed., Woodhead Publishing Series in Textiles, Woodhead Publishing, pp. 106–117.
18. Kozicki, M., & Sąsiadek, E., (2012). UV-assisted screen-printing of flat textiles, *Color. Technol.*, *128*(4), 251.
19. Ujiie, H., (2015). In: *Textiles and Fashion*, Sinclair, R., ed., Woodhead Publishing Series in Textiles, Woodhead Publishing, pp. 507–529.
20. Arthur, D., (2001). Broadbent. In: *Basic Principles of Textile Coloration*, Society of Dyers and Colorists, UK, pp. 493–527.
21. Miles, L. W. C., (2003). *Textile Printing: Revised Second Edition*, SDC: Bradford, UK. pp. 1–54.
22. Little, A. F., & Christie, R. M., (2010). Textile applications of photochromic dyes. Part 2: Factors affecting the technical performance of textiles screen-printed with commercial photochromic dyes, *Color. Technol.*, *126*, 164.
23. Little, A. F., & Christie, R. M., (2010). Textile applications of photochromic dyes. Part 1: Establishment of a methodology for evaluation of photochromic textiles using traditional color measurement instrumentation, *Color. Technol.*, *126*(3), 157.
24. Feczko, T., Samu, K., Wenzel, K., Neral, B., & Voncina, B., (2013). Textiles screen-printed with photochromic ethyl cellulose-spirooxazine composite nanoparticles, *Color. Technol.*, *129*(1), 18.
25. Nelson, G., (2002). Application of microencapsulation in textiles, *Int. J. Pharm.*, *242*(1–2), 55.
26. Elesini, U. S., & Urbas, R., (2016). In: *Printing on Polymers*, Izdebska, J., Thomas, S., eds., William Andrew Publishing, pp. 389–396.
27. Meirowitz, R., (2010). In: *Smart Textile Coatings and Laminates*, Smith, W. C., ed., Woodhead Publishing Series in Textiles, Woodhead Publishing, pp. 125–154.
28. Viková, M., (2011). *Photochromic Textiles*, Heriot-Watt University, Scottish Borders Campus, Edinburgh, UK. pp. 53–147.
29. Waseem, I., (2012). *An Investigation into Textile Applications of Thermochromic Pigments,* Heriot-Watt University. Edinburgh, pp. 115.
30. Lee, J., Pyo, M., Lee, S., Kim, J., Ra, M., Kim, W. Y., Park, B. J., Lee, C. W., & Kim, J. M., (2014). Hydrochromic conjugated polymers for human sweat pore mapping, *Nat. Commun.*, *5*, 3736.
31. Winters, A., (2011). *Rainbow Winters,* http://www.rainbowwinters.com/springsummer.html (accessed Dec. 2, 2016).
32. Sakka, S., (2013). In: *Handbook of Advanced Ceramics (Second Edition)*, Somiya, S., ed., Academic Press, Oxford, pp. 883–910.

33. Rondinini, S., Ardizzone, S., Cappelletti, G., Minguzzi, A., & Vertova, A., (2009). In: *Encyclopedia of Electrochemical Power Sources*, Garche, J., ed., Elsevier, Amsterdam, pp. 613–624.

34. Attia, S., Wang, J., Wu, G., Shen, J., & Ma, J., (2002). Review on sol-gel derived coatings: Process, techniques and optical applications, *J. Mater. Sci. Technol.*, *18*(3), 211.

35. Parhizkar, M., Zhao, Y., & Lin, T., (2014). In: *Handbook of Smart Textiles*, Xiaoming Tao, ed., Springer International Publishing, pp. 155–182.

36. McKinney, D., & Sigmund, W., (2013). In: *Handbook of Advanced Ceramics (Second Edition)*, Somiya, S., ed., Academic Press, Oxford, pp. 911–926.

37. Textor, T., & Mahltig, B., (2010). A sol-gel based surface treatment for preparation of water repellent antistatic textiles, *Appl. Surf. Sci.*, *256*(6), 1668.

38. Levy, D., Monte, F., Oton, J. M., Fiksman, G., Matias, I., Datta, P., & Lopez-amo, M., (1997). Photochromic doped sol-gel materials for fiber-optic devices, *J. Sol-Gel Sci. Technol.*, *8*(1–3), 931.

39. Nakazumi, H., Makita, K., & Nagashiro, R., (1997). New sol-gel photochromic thin films made by super-fine particles of organic photochromic compounds, *J. Sol-Gel Sci. Technol.*, *8*(1–3), 901.

40. Cheng, T., Lin, T., Fang, J., & Brady, R., (2007). Photochromic Wool fabrics from a hybrid silica coating, *Text. Res. J.*, *77*(12), 923.

41. Cheng, T., Lin, T., Brady, R., & Wang, X., (2008). Fast response photochromic textiles from hybrid silica surface coating, *Fibers Polym.*, *9*(3), 301.

42. Cheng, T., Lin, T., Brady, R., & Wang, X., (2008). Photochromic fabrics with improved durability and photochromic performance, *Fibers Polym.*, *9*(5), 521.

43. Cheng, T., Lin, T., Fang, J., & Brady, R., (2006). In: *Proceedings of 2006 China International Wool Textile Conference & IWTO Wool Forum, Xi'an, China, 2006*, China International Wool Textile Conference & IWTO Wool Forum: Xi/an, China, pp. 33–37.

44. Cheng, T., (2008). *Photochromic Wool Fabric by Sol-Gel Coating*, Deakin University, Australia. pp. 10–153.

45. *AATCC Technical Manual*, (2008). A glossary of AATCC standard terminology; AATCC: Research Triangle Park, vol. 83, pp. 415–423.

46. Andronova, A. P., Baranova, A. D., Timofeeva, G. I., & Aizenshtein, E. M., (1980). Mass coloration of polyethylene terephthalate produced by the continuous method, *Fibre Chem.*, *11*(5), 368.

47. Viková, M., & Vik, M., (2006). Smart textile sensors for indication of UV radiation, In: *Autex World Conference-2006*, AUTEX publications, Raleigh, North Carolina, USA, pp. 1–4.

48. Viková, M., & Vik, M., (2011). Alternative UV sensors based on color-changeable pigments, *Adv. Chem. Eng. Sci.*, *1*(4), 224.

49. Viková, M., Vik, M., & Periyasamy, A. P., (2015). In: *Workshop for PhD Students of Textile Engineering and Faculty of Mechanical Engineering*, Technical University of Liberec, Liberec, pp. 140–145.

50. Viková, M., Periyasamy, A. P., Vik, M., & Ujhelyiova, A., (2017). Effect of drawing ratio on difference in optical density and mechanical properties of mass colored photochromic polypropylene filaments, *J. Text. Inst.*, *108*(8), 1365.

51. Periyasamy, A. P., Viková, M., & Michal, V., (2016). Optical properties of photochromic pigment incorporated polypropylene filaments: Influence of pigment concentrations and drawing ratios, *Vlakna a Textil*, *23*(3), 171.

52. Periyasamy, A. P., Viková, M., & Vik, M., (2016). Problems in kinetic measurement of mass dyed photochromic polypropylene filaments with respect to different color space systems, In: *4th CIE Expert Symposium on Color and Visual Appearance*, CIE- Austria, Prague, pp. 325–333.

53. Periyasamy, A. P., Viková, M., & Vik, M., (2016). Optical properties of photochromic pigment incorporated polypropylene filaments: influence of pigment concentrations and drawing ratios, in: *24th International Federation of Associations of Textile Chemists and Colorist Congress*, University of Pardubice, Pardubice-Czech Republic, June 13–16. 2016, Book of Abstracts, pp. 152–155.

54. Rubacha, M., (2007). Thermochromic cellulose fibers, *Polym. Adv. Technol.*, *18*(4), 323.

55. Manian, A. P., Ruef, H., & Bechtold, T., (2007). Spun-dyed lyocell, *Dye. Pigment.*, *74*(3), 519.

56. Laforgue, A., Rouget, G., Dubost, S., Champagne, M. F., & Robitaille, L., (2012). Multifunctional resistive-heating and color-changing monofilaments produced by a single-step coaxial melt-spinning process, *ACS Appl. Mater. Interfaces*, *4*(6), 3163.

57. Ribeiro, L. S., Pinto, T., Monteiro, A., Soares, O. S. G. P., Pereira, C., Freire, C., & Pereira, M. F. R., (2013). Silica nanoparticles functionalized with a thermochromic dye for textile applications, *J. Mater. Sci.*, *48*(14), 5085.

58. Vigo, T. L., (1994). In: *Textile Processing and Properties Preparation, Dyeing, Finishing and Performance*, Textile Science and Technology, Elsevier, vol. *11*, pp. 112–192.

59. Vigo, T. L., (1994). In: *Textile Processing and Properties Preparation, Dyeing, Finishing and Performance*, Textile Science and Technology, Elsevier, vol. *11*, pp. 52–111.

60. Towns, A., (2014). Colorant, Photochromic, in: *Encyclopedia of Color Science and Technology*, Luo, R., ed., Springer, New York, pp. 1–9.

61. Koh, J., (2011). In: *Handbook of Textile and Industrial Dyeing*, Clark, M., ed., Woodhead Publishing Series in Textiles, Woodhead Publishing, vol. *2*, pp. 129–146.

62. Shang, S. M., (2013). In: *Process Control in Textile Manufacturing*, Majumdar, A., Das, A., Alagirusamy, R., & Kothari, V. K., eds., Woodhead Publishing Series in Textiles, Woodhead Publishing, pp. 300–338.

63. Chakraborty, J. N., (2010). In: *Fundamentals and Practices in Coloration of Textiles*, Chakraborty, J. N., ed., Woodhead Publishing India, pp. 1–10.

64. Ibrahim, N. A., (2011). In: *Handbook of Textile and Industrial Dyeing*, Clark, M., ed., Woodhead Publishing Series in Textiles, Woodhead Publishing, vol. *2*, pp. 147–172.

65. Moody, V., & Needles, H. L., (2004). In: *Tufted Carpet*, Needles, V. M. L., ed., Plastics Design Library, William Andrew Publishing: Norwich, New York, pp. 155–175.

66. Mokhtari, J., Akbarzadeh, A., Shahrestani, Z., & Ferdowsi, P., (2015). Synthesis, characterization, and evaluation of a novel spirooxazine based photochromic reactive dye on cotton, *Fibers Polym.*, *16*(11), 2299.

67. Shah, P. H., Patel, R. G., & Patel, V. S., (1985). Azodisperse dyes with photochromic mercury (ii)-dithizonate moiety for dyeing polyester, nylon and cellulose triacetate fibers, *Indian J. Fiber Text. Res.*, *10*(4), 179.

68. Billah, S. M. R., Christie, R. M., & Shamey, R., (2008). Direct coloration of textiles with photochromic dyes. Part 1: Application of spiroindolinonaphthoxazines as disperse dyes to polyester, nylon and acrylic fabrics, *Color. Technol.*, *124*(4), 223.

69. Billah, S. M. R., Christie, R. M., & Morgan, K. M., (2008). Direct coloration of textiles with photochromic dyes. Part 2: The effect of solvents on the color change of photochromic textiles, *Color. Technol.*, *124*(4), 229.

70. Billah, S. M. R., Christie, R. M., & Shamey, R., (2012). Direct coloration of textiles with photochromic dyes. Part 3: Dyeing of wool with photochromic acid dyes, *Color. Technol.*, *128*(6), 488.

71. Lee, S. J., Son, Y. A., Suh, H. J., Lee, D. N., & Kim, S. H., (2006). Preliminary exhaustion studies of spirooxazine dyes on polyamide fibers and their photochromic properties, *Dye. Pigment*, *69*(1–2), 18.

72. Son, Y. A., Park, Y. M., Park, S. Y., Shin, C. J., & Kim, S. H., (2007). Exhaustion studies of spirooxazine dye having reactive anchor on polyamide fibers and its photochromic properties, *Dye. Pigment.*, *73*(1), 76.

73. Monk, P., Mortimer, R., & Rosseinsky, D., (2007). *Electrochromism and Electrochromic Devices*, Cambridge University Press, pp. 17–292.

74. Beaupre, S., Dumas, J., & Leclerc, M., (2006). Toward the development of new textile/plastic electrochromic cells using triphenylamine-based copolymers, *Chem. Mater.*, *18*(17), 4011.

75. Molina, J., Esteves, M. F., Fernandez, J., Bonastre, J., & Cases, F., (2011). Polyaniline coated conducting fabrics. Chemical and electrochemical characterization, *Eur. Polym. J.*, *47*(10), 2003.

76. Yan, C., Kang, W., Wang, J., Cui, M., Wang, X., Foo, C. Y., Chee, K. J., & Lee, P. S., (2014). Stretchable and wearable electrochromic devices, *ACS Nano*, *8*(1), 316.

77. Meunier, L., Kelly, F. M., Cochrane, C., & Koncar, V., (2011). Flexible displays for smart clothing: part II – electrochromic displays, *Indian J. Fiber Text. Res.*, *36*(4), 429.

78. Ding, Y., Invernale, M. A., & Sotzing, G. A., (2010). Conductivity trends of PEDOT-PSS impregnated fabric and the effect of conductivity on electrochromic textile, *ACS Appl. Mater. Interfaces*, *2*(6), 1588.

79. Staneva, D., Betcheva, R., & Chovelon, J. M., (2007). Optical sensor for aliphatic amines based on the simultaneous colorimetric and fluorescence responses of smart textile, *J. Appl. Polym. Sci.*, *106*(3), 1950.

80. Canamares, M. V., Garcia-Ramos, J. V., Domingo, C., & Sanchez-Cortes, S., (2004). Surface-enhanced Raman scattering study of the adsorption of the anthraquinone pigment alizarin on Ag nanoparticles, *J. Raman Spectrosc.*, *35*(11), 921.

81. Puchtler, H., Sweat, F., & Levine, M., (1962). On the binding of Congo red by Amyloid, *J. Histochem. Cytochem.*, *10*, 355.

82. Mall, I. D., Srivastava, V. C., Agarwal, N. K., & Mishra, I. M., (2005). Removal of congo red from aqueous solution by bagasse fly ash and activated carbon: Kinetic study and equilibrium isotherm analyses, *Chemosphere*, *61*(4), 492.

83. Hoten, M., Kojima, Y., & Ito, T., (1992). Halochromism of acrylic fibers dyed with some disperse dyes, *J. Soc. Dye. Color.*, *108*(1), 21.

84. Van der Schueren, L., & De Clerck, K., (2011). Textile materials with a pH - sensitive function. *Int. J. Cloth. Sci. Technol.*, *23*(4), 269.

85. Van der Schueren, L., & Van Clerck, K. De., (2010). The use of pH-indicator dyes for pH-sensitive textile materials, *Text. Res. J.*, *80*(7), 590.

86. Staneva, D., & Betcheva, R., (2007). Synthesis and functional properties of new optical pH sensor based on benzo [de] anthracen-7-one immobilized on the viscose, *Dye. Pigment.*, *74*(1), 148.

87. Aldib, M., & Christie, R. M., (2013). Textile applications of photochromic dyes. Part 5: Application of commercial photochromic dyes to polyester fabric by a solvent-based dyeing method, *Color. Technol.*, *129*(2), 131.

88. Aldib, M., & Christie, R. M., (2013). Textile applications of photochromic dyes. Part 4: Application of commercial photochromic dyes to polyester fabric by a solvent-based dyeing method, *Color. Technol.*, *127*(2), 131.

89. Hutten, I. M., (2016). In: *Handbook of Nonwoven Filter Media (Second Edition)*, Hutten, I. M., ed., Butterworth-Heinemann, Oxford, pp. 1–52.

90. Gries, T., Veit, D., Wulfhorst, B., & Graber, A., (2015). In: *Textile Technology (Second Edition)*, Gries, T., Veit, D., & Wulfhorst, B., eds., Hanser, pp. 195–219.

91. Bhat, G. S., & Malkan, S. R., (2007). In: *Handbook of Nonwovens*, Russell, S. J., ed., Woodhead Publishing Series in Textiles, Woodhead Publishing, pp. 143–200.

92. Peacock, A. J., & Calhoun, A., (2006). In: *Polymer Science*, Peacock, A. J., Calhoun, A., eds., Hanser, pp. 285–297.

93. Mao, N., & Russell, S. J., (2015). In: *Textiles and Fashion*, Sinclair, R., ed., Woodhead Publishing Series in Textiles, Woodhead Publishing, pp. 307–335.

94. Moyo, D., Patanaik, A., & Anandjiwala, R. D., (2013). In: *Process Control in Textile Manufacturing*, Majumdar, A., Das, A., Alagirusamy, R., & Kothari, V. K., eds., Woodhead Publishing Series in Textiles, Woodhead Publishing, pp. 279–299.

95. Periyasamy, A. P., Viková, M., & Vik, M., (2017). A review of photochromism in textiles and its measurement *Text. Prog.*, *49*(2), 53.

96. Bhardwaj, N., & Kundu, S. C., (2010). A fascinating fiber fabrication technique, *Biotechnol. Adv.*, *28*(3), 325.

97. Salem, D. R., (2007). In: *Nanofibers and Nanotechnology in Textiles*, Brown, P. J., & Stevens, K., eds., Woodhead Publishing Series in Textiles, Woodhead Publishing, pp. 3–21.

98. Ko, F. K., & Gandhi, M. R., (2007). In: *Nanofibers and Nanotechnology in Textiles*, Brown, P. J., & Stevens, K., eds., Woodhead Publishing Series in Textiles, Woodhead Publishing, pp. 22–44.

99. Smit, E., Buttner, U., & Sanderson, R. D., (2007). In: *Nanofibers and Nanotechnology in Textiles*, Brown, P. J., & Stevens, K., eds., Woodhead Publishing Series in Textiles, Woodhead Publishing, pp. 45–70.

100. Lee, E. M., Gwon, S. Y., Ji, B. C., Wang, S., & Kim, S. H., (2012). Photoswitching electrospun nanofiber based on a spironaphthoxazine-isophorone-based fluorescent dye system, *Dye. Pigment.*, *92*(1), 542.

101. Lee, E. M., Gwon, S. Y., Son, Y. A., & Kim, S. H., (2012). Modulation of a fluorescence switch of nanofiber mats containing photochromic spironaphthoxazine and DA charge transfer dye, *J. Lumin.*, *132*(6), 1427.

102. Wang, M., Vail, S. A., Keirstead, A. E., Marquez, M., Gust, D., & Garcia, A. A., (2009). Preparation of photochromic poly (vinylidene fluoride-co-hexafluoropropylene) fibers by electrospinning, *Polymer*, *50*(16), 3974.

103. Khatri, Z., Ali, S., Khatri, I., Mayakrishnan, G., Kim, S. H., & Kim, I. S., (2015). UV-responsive polyvinyl alcohol nanofibers prepared by electrospinning, *Appl. Surf. Sci.*, *342*, 64.

104. Zhang, W. J., Hong, C. Y., & Pan, C. Y., (2013). Fabrication of electrospinning fibers from spiropyran-based polymeric nanowires and their photochromic properties, *Macromol. Chem. Phys.*, *214*(21), 2445.

105. Shuiping, L., Lianjiang, T., Weili, H., Xiaoqiang, L., & Yanmo, C., (2010). Cellulose acetate nanofibers with photochromic property: Fabrication and characterization, *Mater. Lett.*, *64*(22), 2427.

106. Oro, E., Gil, A., Miro, L., Peiro, G., Alvarez, S., & Cabeza, L. F., (2012). Thermal energy storage implementation using phase change materials for solar cooling and refrigeration applications, *Energy Procedia, 30*, 947.

107. Mulligan, J. C., Colvin, D. P., & Bryant, Y. G., (1996). Microencapsulated phase-change material suspensions for heat transfer in spacecraft thermal systems, *J. Spacecr. Rockets, 33*(2), 278.

108. McCann, J. T., Marquez, M., & Xia, Y., (2006). Melt coaxial electrospinning: A versatile method for the encapsulation of solid materials and fabrication of phase change nanofibers, *Nano Lett., 6*(12), 2868.

109. Li, F., Zhao, Y., Wang, S., Han, D., Jiang, L., & Song, Y., (2009). Thermochromic core–shell nanofibers fabricated by melt coaxial electrospinning, *J. Appl. Poly. Sci., 112*(1), 269.

110. Cahill, V., (2006). In: *Digital Printing of Textiles*, Ujiie, H., ed., Woodhead Publishing Series in Textiles, Woodhead Publishing, pp. 1–15.

111. Periyasamy, A. P., & Paramasivam, R., (2007). *Indian Text. J., 117*(7), 67.

112. Tyler, D. J., (2011). In: *Computer Technology for Textiles and Apparel*, Hu, J., ed., Woodhead Publishing Series in Textiles, Woodhead Publishing, pp. 259–282.

113. Nicoll, L., (2006). In: *Digital Printing of Textiles*, Ujiie, H., ed., Woodhead Publishing Series in Textiles, Woodhead Publishing, pp. 16–26.

114. Freire, E. M., (2006). In: *Digital Printing of Textiles*, Ujiie, H., ed., Woodhead Publishing Series in Textiles, Woodhead Publishing, pp. 29–52.

115. King, K. M., (2013). In: *Advances in the Dyeing and Finishing of Technical Textiles*, Gulrajani, M. L., ed., Woodhead Publishing Series in Textiles, Woodhead Publishing, pp. 236–257.

116. Cie, C., (2015). In: *Ink Jet Textile Printing*, Cie, C., ed., Woodhead Publishing Series in Textiles, Woodhead Publishing, pp. 165–178.

117. Cie, C., (2015). In: *Ink Jet Textile Printing*, Cie, C., ed., Woodhead Publishing Series in Textiles, Woodhead Publishing, pp. 15–27.

118. Aldib, M., (2013). *An Investigation of the Performance of Photochromic Dyes and their Application to Polyester and Cotton Fabrics*, PhD Dissertation, Heriot-Watt University, Edinburgh, pp. 152–177.

119. Aldib, M., (2015). Photochromic ink formulation for digital inkjet printing and color measurement of printed polyester fabrics, *Color. Technol., 131*(2), 172.

120. Fu, Z., (2006). In: *Digital Printing of Textiles*, Ujiie, H., ed., Woodhead Publishing Series in Textiles, Woodhead Publishing, pp. 218–232.

121. Feczko, T., Varga, O., Kovacs, M., Vidoczy, T., & Voncina, B., (2011). Preparation and characterization of photochromic poly(methyl methacrylate) and ethyl cellulose nano-capsules containing a spirooxazine dye, *J. Photochem. Photobiol. A Chem., 222*(1), 293.

122. Zhou, Y., Yan, Y., Du, Y., Chen, J., Hou, X., & Meng, J., (2013). Preparation and application of melamine-formaldehyde photochromic microcapsules, *Sensors Actuators B Chem., 188*, 502.

123. Aitken, D., Burkinshaw, S. M., Griffiths, J., & Towns, A. D., (1996). Textile applications of thennochromic systems, *Rev. Prog. Color. Relat. Top., 26*(1), 1.

CHAPTER 4

CIE COLORIMETRY

MICHAL VIK

CONTENTS

Abstract 155
4.1   Light Sources 156
4.2   CIE Standard Observer 170
4.3   CIE XYZ 175
4.4   Approximately Uniform Color Spaces.....................................182
4.5   CIE CAM 195
4.6   Color Difference Formulas 207
4.7   CIE 2015 Cone-Fundamental-Based CIE Colorimetry 214
Keywords 217
References..217

ABSTRACT

Since the beginning of the 20th century, several organizations have been working hard to define and propose the adoption of standard light sources to be used in color measurement. At the meeting of the International Gas Congress in 1900, several participants agreed that one of the main problems facing the industry was the photometry of incandescent gas mantles, and Commission Internationale de Photometrie (CIP) was founded. World War I forced a break in the activities of the Commission Internationale de L'Eclairage, designated as CIE, which had been formed in 1913, by a proposal presented in the CIP Congress in Berlin, Germany. In 1924, in Geneva, the first standard of light was adopted, based on a full radiator at the melting point of platinum, and the working group of colorimetry was established [1, 2]. To quantify and model the human color perception, colorimetry was introduced, which is the science of measuring color. This considers the physical characteristics of the light source, the object's spectral properties, and the physiological aspects of human vision [3].

The color of a physical sample depends primarily on the composition of the light reflected from it, which enters the observer's eyes. This is governed by two factors: the reflectance characteristics of the sample and the composition of the light falling on it. In the textile industry, for instance, a dyer determines the reflectance characteristics by his choice of dyes, but he has no control over the composition of lights under which the textile will be viewed. Natural daylight itself varies in spectral power distribution, from the reddish light of sunrise and sunset in the bluish light of cloudless northern sky. Furthermore, natural daylight is not available either at night or in many interior rooms. These limitations have motivated a long search for ways to simulate natural daylight, usually by choosing lamps and developing filters to modify their spectral power distributions (frequently blue glass filters). Before daylight could be simulated, it was necessary to characterize it by measurement to decide which phase to use and to standardize it.

4.1 LIGHT SOURCES

The perceived color of an illuminated object depends not only on its intrinsic color or spectral reflectivity, but also on the power spectrum of the light source. In this chapter, we will study some of the most important light sources used in colorimetric studies. Most sources of light emit a range of wavelengths. A standard light source is best described in a number of ways. The indication of its chromaticity coordinates is not enough, as they do not give any indication of the color rendering properties, which are extremely important to define a light source quality. The definition of color rendering indexes became very usual, especially after the introduction of discharge and solid-state light sources, which are differ in spectral power distribution (SPD) to thermal radiators.

To describe the light that is emitted from a source, it is necessary to express the radiance of the source at each wavelength, yielding the spectral radiance distribution or, more succinctly, the spectrum of the source. More specifically, we are concerned with the spectral radiant flux incident per unit area of our object that is we are concerned with the spectral irradiance provided by our illuminant or source.

In case of visual triplet, three subjects are necessary: light, object (sample), and observer. If we assume that the light source emits approximately equal amounts of all wavelengths, the light striking the sample will appear white. If the sample reflects equal amounts of all wavelengths, the color seen by the observer will be either white or some shade of gray. If, on the other hand, the sample absorbs some wavelengths of light and reflects others, the observer will see a colored sample. For example, if the sample absorbs blue part of the visual spectrum and reflects other wavelengths, the sample will appear to be yellow to the observer. When on the same sample will turn out only blue part of visual spectra, the sample will appear green, because sample will absorb a huge part of short wavelength blue light and due to absence of long wavelength in spectrum of blue light no red light is reflected. Therefore, the only wavelengths, which are both available from the light source and reflected by the sample, are mid-range green wavelengths. Because the color content of the light source has such a pronounced effect upon the colors being observed, it is necessary to define and specify a standard light source to be used visual assessment and in instruments for the determination of optical properties. The current recommendation of the CIE is that the visible spectrum be sampled at 1-nm increments between 380 and 780 nm. Illuminants are normally specified in terms of the relative energy tabulated for each wavelength or wavelength band. Another possibility is based on color temperature. The color temperature of a light source is the temperature of an ideal blackbody radiator that radiates light of comparable hue to that of the light source.

Color temperature T_C is used for a light source with the spectral distribution of energy identical with that given by the blackbody radiator (Planck's law).

Equivalent color temperature T_{CE} is used for a light source with the spectral distribution of energy not identical with that given by the Planck's law, but still is of such a form that the quality of the color evoked is the same as would be evoked by the energy from a blackbody radiator at the given color temperature.

Correlated color temperature T_{CP} **(CCT)** is used for a light source with the spectral distribution of energy such that the color can be matched *only approximately* by a stimulus of the blackbody radiator of spectral distribution.

Color temperature has been described most simply as a method of describing the color characteristics of light, usually either warm (yellowish) or cool (bluish), and measuring it in degrees of Kelvin (°K).

In the above variants of Planck's law, the *Wavelength* and *Wavenumber* variants use the terms $2hc^2$ and hc/k_B, which comprise physical constants only. Consequently, these terms can be considered as physical constants themselves, and are therefore referred to as the **first radiation constant C_{1L}** and the **second radiation constant C_2** with $C_{1E} = 2hc^2$ and $C_2 = hc/k_B$. E (*L*) is used here instead of B because it is the SI symbol for spectral radiance. The E in C_{1E} refers to that. This reference is necessary because Planck's law can be reformulated to give spectral radiant exitance $M(\lambda,T)$ rather than spectral radiance $E(\lambda,T)$, in which case C_1 replaces C_{1E}, with $C_1 = 2\pi hc^2$, so that Planck's law for *spectral radiant exitance* can be written as

$$M_\lambda = C_1 \cdot \lambda^{-5} \left(e^{\left(C_2/T\lambda\right)} - 1 \right)^{-1} \tag{4.1}$$

where $C_1 = 3.7415 \times 10^{-16}$ W.m², $C_2 = 1.43878 \times 10^{-2}$ m.K, are mentioned constants [4].

The CIE system considers three essential components, namely light sources, objects, and observers, which determine the color perception. It is a well-known phenomenon that different light sources may affect the color appearance of an object quite significantly. To standardize the light sources in practice, the CIE defined standard illuminants and artificial light sources representative of these illuminants.

4.1.1 WHAT IS THE DIFFERENCE BETWEEN A LIGHT SOURCE AND AN ILLUMINANT?

The terms light source and illuminant have precise and different meanings. A light source is a physical emitter of radiation such as a candle, a tungsten bulb, and natural daylight. An illuminant is the specification for a potential light source. All light sources can be specified as an illuminant, but not all illuminants can be physically realized as a light source.

In 1931, CIE recommended three standard illuminants, **A, B,** and **C.** Later, **D** and **F** were added, and in some cases, **E** is used illuminant, which has been added to represent an equal power distribution that is, a distribution in which the relative power does not vary with wavelength [5].

Illuminant **A** is intended to represent typical, domestic, tungsten-filament lighting. Its relative spectral power distribution is that of a blackbody radiator at a temperature of approximately 2856 K. CIE standard illuminant A is used in many calibration procedures, and it is the only illuminant for which the corresponding source can also be realized with reasonable accuracy; its central position has remained intact even after the ban of incandescent lamps for general use. The relative spectral power distribution $S_A(\lambda)$ is defined by Eq. (4.2):

$$S_A(\lambda) = 100\left(\frac{560}{\lambda}\right) \cdot \frac{e^{\left(\frac{1.435\times10^7}{2848\times560}\right)} - 1}{e^{\left(\frac{1.435\cdot10^7}{2848\cdot\lambda}\right)} - 1} \tag{4.2}$$

where λ is the wavelength in nanometers and the numerical values in the two exponential terms are, definitive constants originating from the first definition of Illuminant A in 1931. This spectral power distribution is normalized to the value 100 at the wavelength 560 nm.

Remark: Equivalent color temperature of 100 W incandescent lamp is approximately 2870 K, 40 W incandescent lamp ≈ 2500 K, and 100 W tungsten-halogen lamp ≈ 3000 K

- Standard illuminant **B** has a correlated color temperature of about 4874 K (205,17 MK⁻¹) and was intended to represent sunlight. Standard illuminant B is now obsolete.
- Standard illuminant **C** old standard for average daylight from the northern sky, defined in 1931 by the CIE with a T_{cp} = 6 774 K (147.62 MK⁻¹); has a significant lower UV content than the D-types; does not have CIE status of "standard illuminant" anymore (Figure 4.1).
- CIE Illuminant Series **D**. These series of illuminants that has been statistically defined in 1964 upon numerous measurements of real daylight [5]. Although mathematically described, they can hardly be realized artificially. The correlated color temperatures of the commonly used

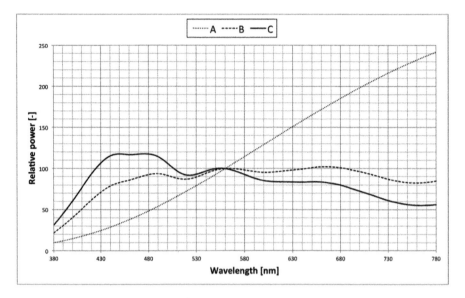

FIGURE 4.1 Spectral power distribution (SPD) of CIE light sources from 1931.

illuminants D50, D55 and D65 are slightly different to the values suggested by their names. Due to the revision of an estimate of one of the constant factors in Planck's law after the standards were defined, the correlated color temperature was shifted a little. For example, the CCT of D50 is 5003K and that of D65 is 6504 K. A whole range of series D is starting from 4000 K till 25000 K (250–40 MK^{-1}) [6].

Daylight illuminant D at a nominal correlated color temperature (T_{cp}) can be calculated using the following equations. These equations will give an illuminant whose correlated color temperature is approximately equal to the nominal value.

4.1.1.1 Chromaticity

The 1931 (x, y) chromaticity coordinates of the daylight (D) to be defined must satisfy the following relation:

$$y_D = -3.000x_D^2 + 2.870x_D - 0.275y_D \qquad (4.3)$$

with x_D being within the range of 0.250 to 0.380.

The correlated color temperature T_{cp} of daylight D is related to x_D by the following formulae based on normal (rectangular line) to the Planckian locus (or blackbody locus is the path or locus that the color of a blackbody would take in a particular chromaticity space as the blackbody temperature changes) on the u, v uniform chromaticity diagram.

1. for correlated color temperatures from approximately 4000 K to 7000 K:

$$x_D = \frac{-4.6070 \cdot 10^9}{\left(T_{cp}\right)^3} + \frac{2.9678 \cdot 10^6}{\left(T_{cp}\right)^2} + \frac{0.09911 \cdot 10^3}{\left(T_{cp}\right)} + 0.244063 \quad (4.4)$$

2. for correlated color temperatures from greater than 7000 K to approximately 25000 K:

$$x_D = \frac{-2.0064 \cdot 10^9}{\left(T_{cp}\right)^3} + \frac{1.9018 \cdot 10^6}{\left(T_{cp}\right)^2} + \frac{0.24748 \cdot 10^3}{\left(T_{cp}\right)} + 0.237040 \quad (4.5)$$

4.1.1.2 Relative Spectral Power Distribution

Relative spectral power distribution is then calculated based on the x_D, y_D chromaticity coordinates of selected daylight:

$$M_1 = \frac{-1.3515 - 1.7703x_D + 5.9114y_D}{0.0241 + 0.2562x_D - 0.7341y_D} \quad (4.6)$$

$$M_2 = \frac{0.0300 - 31.4424x_D + 30.0717y_D}{0.0241 + 0.2562x_D - 0.7341y_D}$$

where M_1 and M_2 are scalar multiples independent on wavelength [8]. The values of M_1 and M_2 have to be rounded to three decimal places, and with the rounded values, the spectral distribution of the daylight phase is calculated using the following equation for every 10 nm between 300 nm and 830 nm:

$$S(\lambda) = S_0(\lambda) + M_1 \cdot S_1(\lambda) + M_2 \cdot S_2(\lambda) \qquad (4.7)$$

where $S_0(\lambda)$, $S_1(\lambda)$, and $S_2(\lambda)$ are functions of wavelength λ, given the different sources [3, 4, 7].

The daylight SPD for other wavelengths between 300 nm and 830 nm can be found by linear interpolation or by an alternative method, which is described in Appendix C of the CIE Publication 15:2004 [4].

The relative spectral power distribution representing a phase of daylight with a correlated color temperature of approximately 6500 K (called also nominal correlated color temperature of the daylight illuminant), symbol: $S_{D65}(\lambda)$. CIE standard illuminant D65 contains power in the UV range and so does natural outdoor daylight. It is recommended that in the interest of standardization, D65 was used whenever possible. When D65 cannot be used, it is recommended that one of the daylight illuminants D50, D55, or D75 [7].

Differences between spectral power distribution of CIE illuminant D calculated following Eqs. (4.3) till (4.7) and Planckian radiator based on Eq. (1.6) and (4.1) respectively, are shown in the graph in Figures 4.2 and 4.3.

The CIE Publication 15:2004 [4] states: "At present no artificial source is recommended to realize CIE standard illuminant D65 or any other illuminant D of different correlated color temperature. It is hoped that new developments in light sources and filters will eventually offer sufficient basis for a CIE recommendation." The CIE has agreed on a formula to describe the quality of a daylight simulator for colorimetry, which was accepted also by ISO [9]. Presently, three major categories of daylight simulators are used: Xe-sources, T-sources, and F-sources. The Xe-sources are high-pressure short-arc xenon lamps. The agreement between the SPDs of the Xe-sources and that of the CIE D65 illuminant appears quite good throughout the spectral range from 300 to 800 nm, except for a considerable excess of irradiance around 470 nm and above 700 nm as visible in graph on Figure 4.4. The T-sources are either tungsten halogen lamps combined with one or more colored glass filters used as spectrum modulators. The agreement between the SPDs of the T-sources and D65 is quite good in the visible region of the spectrum. However, in the ultraviolet region (λ<380 nm), the T-sources exhibit serious deficiencies and the use of the spectrum modulators consist of conventional absorption filters, is insufficient (see SPL III D65 in Figure 4.4). The F-sources are fluorescent lamps. The discrepancies between SPDs

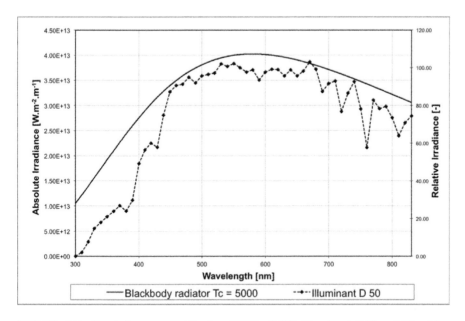

FIGURE 4.2 Spectral power distribution (SPD) of CIE illuminant D50 and Planckian radiator at 5000 K.

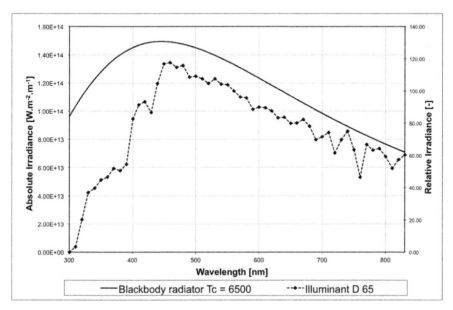

FIGURE 4.3 Spectral power distribution (SPD) of CIE standard illuminant D65 and Planckian radiator at 6500 K.

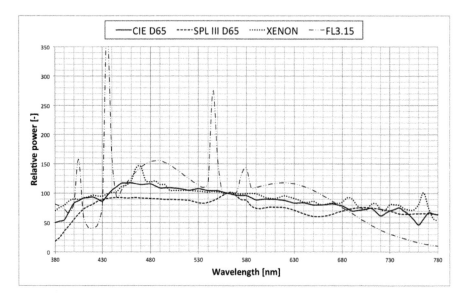

FIGURE 4.4 Spectral power distribution of D65 and its simulators – SPL III D65 – Macbeth type of T – source. Xenon is filtered by Oriel 150W xenon discharge lamp as Xe – source, FL3.15 is F – source.

of the F-sources and CIE D65 illuminant are generally quite severe at the wavelength of the emission lines as visible in graph of FL3.15 in Figure 4.4. Another discrepancy that is characteristic of F-sources occurs in the wavelength region above 650 nm, where these sources show a rapid decrease in spectral irradiance to the corresponding CIE daylight illuminant.

- Illuminant **E**. This is a hypothetical reference radiator. All wavelengths in CIE illuminant E are weighted equally with a relative spectral power of 100. Because it is not a Planckian radiator, no color temperature is given; however, it can be approximated by a CIE D illuminant with a correlated color temperature of 5455 K. The CIE standard illuminant D55 is the closest to match its color temperature.

Fluorescent lamps are widely used for store, office, and works lighting. A wide range of fluorescent lamps are available due to the use of different phosphor types and formulations. Ultraviolet light, mainly at 253.7 nm, emitted by mercury vapor excites the fluorescent coating on the inside of the glass tube. The phosphors are inorganic compounds of high chemical purity, and sometimes, some metals are added as activators to increase their efficiency. Although the following illuminants are not included in CIE standard illuminant category, they are frequently used in many standards [6].

- Illuminant **F1** to **F12** (in CIE Colorimetry standard 15, 2004 [4] are assigned as FL1 to FL12, and a new set of fluorescent illuminants is included with marking FL3.1 to FL3.15) are representative of fluorescent lamps of different types differ widely in spectral power distribution. The commonly used illuminants are only F2, F7, and F11. F2 represents a cool white fluorescent lamp (4230 K), F7 represents a broadband daylight fluorescent lamp (6500 K), and F11 represents a narrowband white fluorescent lamp (4000 K). The former code of the narrowband fluorescent lamp, which was replaced by F11, is Philips TL84. Relative SPDs of 12 CIE fluorescent illuminants are shown in Figures 4.5–4.7.

The HP-illuminants are typical high-pressure lamp spectra used at the time of publishing this book. The operation principle of high-pressure discharge lamps differs considerably from that of standard incandescent lamps. Light is produced by gas discharge that occurs in an arc tube between two electrodes after ignition. Electrical conductivity is established by ionized

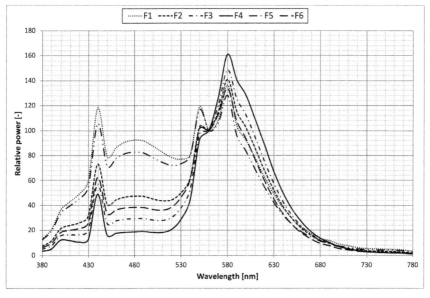

FIGURE 4.5   Spectral power distribution of CIE illuminants FL1 – FL6.

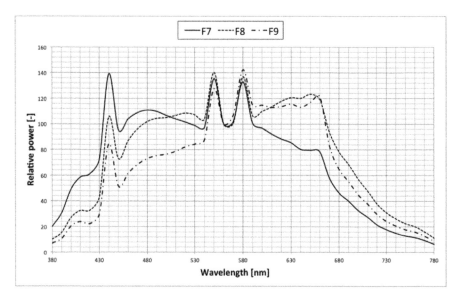

FIGURE 4.6   Spectral power distribution of CIE illuminants FL7 – FL9.

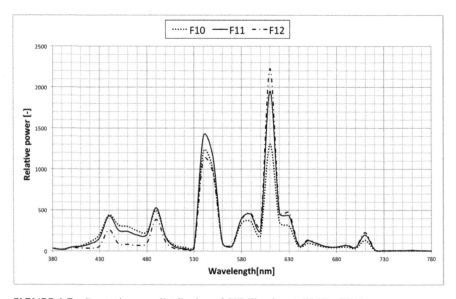

FIGURE 4.7   Spectral power distribution of CIE illuminants FL10 – FL12.

filler components. The electrodes are fed into a completely sealed discharge vessel. During gas discharge, the additives (metal halides) and mercury are excited by the current flow and emit the excitation energy in the form of their characteristic radiation. The mixture of different radiation components produces the desired color temperature and color reproduction properties. HP1 is standard high-pressure sodium lamp; HP2 is color enhanced high-pressure sodium lamp; and HP3 to HP5 are three types of high-pressure metal halide lamps, which SPDs are shown in the graph in Figure 4.8.

Light sources of principle importance in the near future are expected to be light emitting diodes (LEDs) and lasers. The central component part of an LED consists of a p–n semiconductor junction. The so-called white LED can be prepared by two ways:

- Combining multiple differently colored LEDs in a single package, but this is complicated and inefficient, and the color consistency is not good. The simplest multiple differently colored LEDs are based on combination of red, green and blue LED. The RGB mixing ratio required to reproduce a target white point varies according to the chromaticity coordinates of the primary color emitters being used. For example, the two sets of RGB LEDs have identical red and green emitter characteristics, but their blue emitters have different dominant wavelengths (465 nm and 476 nm, respectively). It becomes evident that changing even one LED's dominant wavelength results in dra-

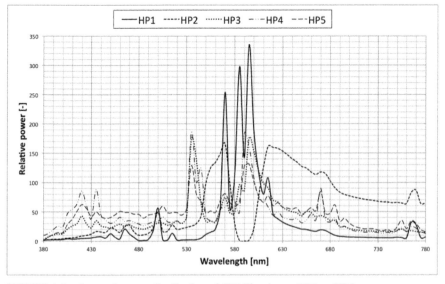

FIGURE 4.8 Spectral power distribution of CIE illuminants HP1 to HP5.

matic changes in the value of the RGB mixing ratio required to match the D65 target white point. The LED set with the 476 nm blue emitter will require a 2.1:4.3:1.0 RGB mixing ratio but simply changing to a 465-nm emitter shifts it to 4.1:10.6:1. SPD of typical RGB white LED is shown in graph in Figure 3.10.

- A single monochrome LED emitter, usually an InGaN blue LED, and phosphors. The technique involves coating a blue LED (commonly called the pump, as it pumps light into the phosphor) with a phosphor or mix of phosphors. The phosphor produces a broadband yellow light that combines with the blue to create a spectrum we perceive as white light. Based on the ratio between intensities of these two lights, it is possible to prepare different appearance of white light from warm white, through neutral to cool white; Figure 4.9 shows examples of three different blue LED-based white light sources (B-LED). As alternative presently available, "full spectrum" white LEDs that utilize a violet LED (V-LED on Figure 4.9) element achieves full-spectrum real white color thanks to the use of R/G/B phosphors.

Correlated color temperature (CCT) and spectral power distribution (SPD) are two common characteristics. Beside these two, we can characterize light sources by its CIE special metameric index, which are mainly used

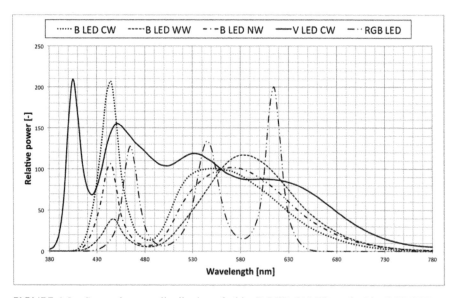

FIGURE 4.9 Spectral power distribution of white B-LED, V-LED, and white RGB LED.

in daylight simulators. Light sources that have different SPDs but produce identical relative absorptions by the three cone types will have the same chromaticity. At the same luminance, these lights will also appear to be identical under the same viewing conditions. Light sources of this type are known as metamers.

The last parameter, which is commonly used for rating the color property, is color rendering index (CRI). R_a is the ability of a light source to properly reproduce the colors of the illuminated objects, and it is derived from CIE 1960 (u, v) chromaticity coordinate and correlated color temperature by using the procedure CIE 13.3-1994. R_a is the average values of the eight special indices R_{is}. They are obtained by estimating the shift of coordinate (u, v) with respect to the reference light source for the specified reflectivity spectra of the test samples. Unless otherwise specified, the reference illuminant for light sources with correlated color temperature below 5000K should be a Planckian radiator and from 5000 K, one of a series of spectral power distributions of phases of daylight [10].

Different light sources have different color rendering properties. For instance, low-pressure sodium lamps, which emit light at virtually only one wavelength, result in the appearance of color being very different from that in daylight. The advent of fluorescent lamps enabled to vary spectral power distribution at will over quite a wide range, and it became desirable to have some means of expressing the degree to which any illuminant gave satisfactory color rendering. The CIE method to calculate CRI is based on resultant color shifts for 14 test-color samples (the reflectance values of these samples are supplied in the CIE publication [10]). Eight of the fourteen test-color samples were chosen from the Munsell color order system, covering the hue circle with moderate chroma and approximately equal lightness. The other six test-color samples, representing four highly saturated primary colors (R, B, G, and Y) as well as complexion and foliage colors were added in this method to indicate the color rendering property of a tested light source under extreme conditions. The CIE Color Rendering Index is given by:

$$R_a = 100 - 4.6\Delta E_i \qquad (4.8)$$

where ΔE_1 to ΔE_{14} are the values of ΔE_i for each number of CIE test color samples specified by the spectral radiance factors in CIE 1960 color space (CIE U*V*W*).

A reference source, such as blackbody radiator, is defined as having a CRI of 100. Hence, incandescent lamps have that rating, as they are, in effect, almost near to blackbody radiators. The best possible faithfulness to a reference is specified by a CRI of 100, while the very poorest is specified by a CRI below 0. A high CRI by itself does not imply a good rendition of color, because the reference itself may have an imbalanced SPD if it has an extreme color temperature.

4.2   CIE STANDARD OBSERVER

To describe colors, we need to know how people respond to them. Human perception of color is a complex function of context; illumination, memory, object identity, and emotion can all play a part. The simplest question is to understand which spectral radiances produce the same response from people under simple viewing conditions. This yields a simple, linear theory of color matching, which is accurate and extremely useful for describing colors.

As color sensations cannot be measured (not yet), for the measurement of color, one only asks whether two stimuli can be discriminated or not and does not consider the sensations at all. To eliminate any influence of the object's shape and surroundings, and to make comparison easy, the colors to be compared are presented as structure less colors (aperture colors), each filling half of the viewing field, which appears like a shining opening.

Particularly useful for the investigation of surface colors is the *Maxwellian mode of observation*: looking at an object with a magnifying glass, the distances can be chosen such that the whole aperture of the glass shows the same color. For a color, which does not appear to belong to a specific object, the separation of object and illuminant color, which in most cases is performed inadvertently and with remarkable precision, is not possible anymore. Comparing colors presented in that way, the experiment is reduced to the comparison of fluxes of light and of color stimuli.

If two stimuli agree on their spectral composition, $\varphi_{1,\lambda} = \varphi_{2,\lambda}$, they will appear equal to all observers under any conditions; this is called an **isomeric** match. Color stimuli may look equal also if their spectral functions are different from each other; such a match is called **metameric**. Different observers may disagree in their determination of metameric matches.

To account for metamerism, i.e., the fact that different stimuli can produce the same color sensation, one introduces the **valence V** of the stimulus with respect to the specific observer. If an observer perceives to stimuli as a match, we write:

$$V_1 = V_2 \tag{4.9}$$

meaning that the two valences are equal, and nothing is implied for the stimuli [11]. While normal color vision observers agree fairly well in their judgment of metameric equality, the equation above strictly holds only for the specific observer [13].

Obviously, in a useful color space, lights that appear the same (visually match) must occupy essentially the same point in the color space, i.e., the values of R, G, and B computed via the color-matching functions for the two visually matching lights must be essentially the same: $R_1 = R_2$, $G_1 = G_2$, and $B_1 = B_2$ [14].

4.2.1   COLOR MATCHING FUNCTIONS

In a typical experiment, a subject sees a colored light — the test light — in one half of a split field. The subject can then adjust a mixture of lights in the other half to get it to match. The adjustments involve changing the intensity of some fixed number of primaries in the mixture. In this form, a large number of lights may be required to obtain a match, but many different adjustments may yield a match.

The drawing in Figure 4.10 shows the outline of such an experiment. The observer sees a test light **M** and can adjust the amount of each of three primaries in a mixture that is displayed next to the test light. The observer is asked to adjust the amounts so that the mixture looks the same as the test light. The mixture of primaries can be written as $R_i(\mathbf{R}) + G_i(\mathbf{G}) + B_i(\mathbf{B})$; if the mixture matches the test light, then we write:

$$\mathbf{M} = R_M(R) + G_M(G) + B_M(B) \tag{4.10}$$

It is a remarkable fact that for most people, three primaries are sufficient to achieve a match for many colors, and for all colors if we allow subtractive

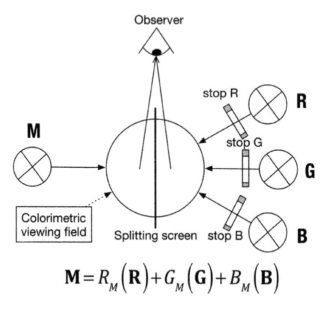

$$\mathbf{M} = R_M(\mathbf{R}) + G_M(\mathbf{G}) + B_M(\mathbf{B})$$

FIGURE 4.10   Visual colorimeter – setup a).

matching (i.e., some amount of some of the primaries is mixed with the test light to achieve a match).

Some people will require fewer primaries. Furthermore, most people will choose the same mixture weights to match a given test light. Wright and Guild have determinate them accurately for a number of normal observers using specified experiment [15]. These determinations have been collated and the mean results, transformed to a system having monochromatic stimuli of wavelengths 435.8 nm, 546.1 nm and 700 nm as reference stimuli and an equal-energy stimulus as basic stimulus, adopted as standard by the CIE. The relative amounts of the three reference stimuli required to match every wavelength of an equal-energy spectrum are known as distribution coefficients and are given the symbols $\bar{r}, \bar{g}, \bar{b}$ for red, green, and blue reference stimuli, respectively. It was noted in many cases, one of the tristimulus values $\bar{r}(\lambda), \bar{g}(\lambda), \bar{b}(\lambda)$ is negative, indicating that the color match was actually obtained using one or two primaries to desaturate the test stimulus. In the point of view of simplified scheme of visual colorimeter in Figure 4.11, one or two primaries were moved on the left side, where test light M is placed.

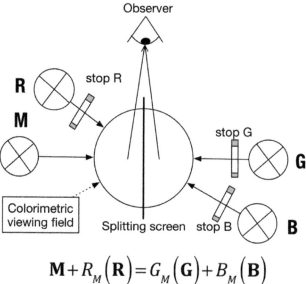

$$M+R_M(\mathbf{R})=G_M(\mathbf{G})+B_M(\mathbf{B})$$

FIGURE 4.11 Visual colorimeter – setup b).

In other cases, one or two tristimulus values are zero, indicating that color match was obtained, respectively, by the use of two primaries or one primary only. The spectral stimulus values $\bar{r}(\lambda),\bar{g}(\lambda),\bar{b}(\lambda)$ of the monochromatic stimuli of different wavelength, but constant radiance is appropriately called color-matching functions with respect to the given primaries **R, G,** and **B**. Figure 4.12 illustrates these functions.

The simple possibility to transform tristimulus values induced the CIE in 1931 to introduce special virtual primaries X, Y, and Z and to define the tables of the corresponding color matching functions as standard for colorimetry.

The new primaries X, Y, and Z has been chosen such that no negative tristimulus values X, Y, and Z occur in any color. In the unit plane, the spectral locus is then entirely within the triangle (XYZ), which naturally implies that the valences X, Y, and Z cannot be realized; they only supply a convenient coordinate system for measurements and graphical representation. The primaries **X** and **Z** have been assumed to lie in that plane; thus, $L_X = L_Z = 0$, and the color matching function of the valence **Y** then must be proportional to the photopic luminous efficiency function V_λ. Thus, the coordinate Y proportional to the luminosity.

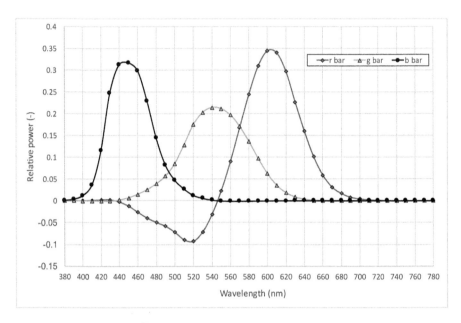

FIGURE 4.12   The $\bar{r},\bar{g},\bar{b}$ color-matching functions – 2° observer.

The requirements that $\mathbf{E} = 1/3(\mathbf{X} + \mathbf{Y} + \mathbf{Z})$ and that the *XYZ*-color space encompasses the gamut of real colors as closely as possible have been used to fix the valences $\mathbf{X}$ and $\mathbf{Z}$ and thus to get the color matching functions $\bar{x}(\lambda),\bar{y}(\lambda),\bar{z}(\lambda)$ as transforming of $\bar{r}(\lambda),\bar{g}(\lambda),\bar{b}(\lambda)$.

As the functions $\bar{x}(\lambda),\bar{y}(\lambda),\bar{z}(\lambda)$ are arbitrarily normalized (such that the maximum of $\bar{y}(\lambda)$ is equal to 1, as shown in Figure 4.13), we have written a proportionality factor before the summation sign. The hypothetical observer whom the CIE 1931 color-matching functions describe is often referred to as the CIE 1931 standard observer.

In 1964, the CIE standardized a second set of color-matching functions appropriate for larger field sizes. These color-matching functions consider the fact that human color matches depend on the size of the matching fields. The CIE 1964 10° color-matching functions are an attempt to provide a standard observer for these larger fields. The use of 10° color-matching functions is recommended by the CIE when the sizes of the regions under consideration are larger than 4° [11, 12].

The difference between a large and a small field can be clearly observed in a very large uniform field, where the appearance to the observer clearly

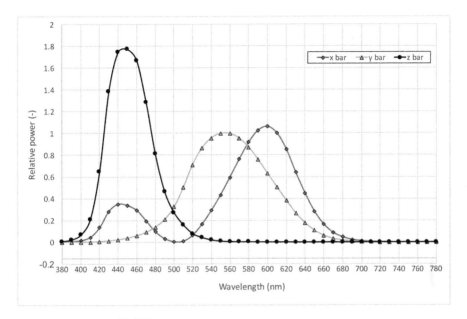

FIGURE 4.13   The $\bar{x}, \bar{y}, \bar{z}$ color-matching functions – 2° observer.

shows a central spot with a slightly different color. This central zone is known as the Maxwell spot, which has a diameter of about four degrees; when the eye looks for a different point, the Maxwell spot moves with the eye. This spot is most clearly seen with a moderate luminance in the field. This is due to the presence of the macular pigment in the central portion of the retina.

The plot shows a set of color-matching functions for normal human vision standardized by the CIE in 1931 – 2° and CIE in 1964 – 10°. It is interesting to notice that the spectral relative luminous efficiency of the eye $V(\lambda)$ is not defined for large fields, but only for a 2° field. Then, $y_{10}(\lambda)$ is slightly different from $V(\lambda)$, with the most noticeable difference being in the blue region, as visible in the graph on Figure 4.14.

4.3   CIE XYZ

The CIE XYZ color space encompasses all color sensations that an average person can experience. Hence, CIE XYZ (Tristimulus values) is a device invariant color representation. Metameric lights are lights that though of

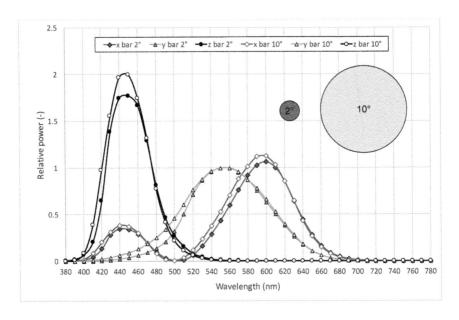

FIGURE 4.14 CIE color-matching functions $\bar{x}, \bar{y}, \bar{z}$

dissimilar spectral radiation are seen as the same by the observer. In a pro-
totypical color-matching experiment using additive lights, the metamers
are presented in a bipartite field. For 2° foveal fields, metamers have three
important properties that allow treatment of color mixture as a linear system
[16, 17]:

1. The additive property. When a radiation is identically added to both
   sides of a color mixture field, the metamerism is unchanged.
2. The scalar property. When both sides of the color mixture field are
   changed in radiance by the same proportion, the metamerism is
   unchanged.
3. The associative property. A metameric mixture may be substituted for
   a light without changing the metameric property of the color fields.

Grassmann's assumption of additivity and its quantitative formulation,
Trichromatic Generalization, allow handling quantities of color stimuli as
ordinary algebraic values, and are the basis of colorimetry. According to
Grassmann's laws, a color match is invariant under a variety of experimental
conditions that may alter the appearance of the matching fields. Metameric
matches will hold with the addition of a chromatic surround or following

pre-exposure to a moderately bright chromatic field. This implies that the background light differentially adapted some sites controlling the appearance of lights. A pair of metameric lights placed on an adapting field of moderate luminance (insufficient to bleach a significant proportion of the cone photo pigments) continues to be metameric even though the appearance of the lights may change. This phenomenon is referred to as the *persistence of color matches.*

The CIE Standard Colorimetric Observer CMF data are used to calculate the CIE tristimulus values of a stimulus with known spectral power distribution. It is important that the CIE system of color measurement is an internationally adopted method to specify the characteristics of a light source or an object color in terms of its tristimulus values. For a desired light source specified by its spectral power distribution S_λ in the visible range, the tristimulus values XYZ are given by [18]:

$$X = k \int_{380}^{780} S_\lambda \rho_\lambda \bar{x}_\lambda d\lambda \tag{4.11}$$

$$Y = k \int_{380}^{780} S_\lambda \rho_\lambda \bar{y}_\lambda d\lambda \tag{4.12}$$

$$Z = k \int_{380}^{780} S_\lambda \rho_\lambda \bar{z}_\lambda d\lambda \tag{4.13}$$

where S_λ is spectral power distribution of used light source; ρ_λ is spectral reflectance factor; $\bar{x}, \bar{y}, \bar{z}$ are CIE color-matching functions.

The normalizing constant k usually is chosen such that for an ideal white surface one would obtain the luminance Y of 100:

$$k = 100 \Big/ \int_{380}^{780} S_\lambda \bar{y}_\lambda d\lambda \tag{4.14}$$

The integration is approximated by summation, thus [12]:

$$X = \sum_{380}^{780} S_\lambda \rho_\lambda \bar{x}_\lambda \Delta\lambda \tag{4.15}$$

$$Y = \sum_{380}^{780} S_\lambda \rho_\lambda \bar{y}_\lambda \Delta\lambda \qquad\qquad (4.16)$$

$$Z = \sum_{380}^{780} S_\lambda \rho_\lambda \bar{z}_\lambda \Delta\lambda \qquad\qquad (4.17)$$

One particular feature of the *XYZ* system is that the $\bar{y}$ function was chosen to be equal to the photopic luminosity curve, V_λ, defined by the CIE in 1924. This choice leads to the Y tristimulus value being proportional to luminance. It is common practice to specify a light in terms of its chromaticity coordinates and luminance, (x, y, Y), from which the three tristimulus values are easily recalculated.

Tristimulus values *XYZ* are defined in the *RGB* color space by the following coordinates:

$r_X = 1.2750$ $g_X = 0.2778$ $b_X = 0.0028$
$r_Y = -1.7394$ $g_Y = 2.7674$ $b_Y = -0.0280$
$r_Z = -0.7429$ $g_Z = 0.1409$ $b_Z = 1.6020$

According to the vector theory, the correlation between primary color valence systems can be obtained simply from a linear transformation with 3 x 3 matrices. Then, for conversion of *RGB* to *XYZ,* we have:

$$\begin{bmatrix} X \\ Y \\ Z \end{bmatrix} = \begin{bmatrix} 2.770 & 1.750 & 1.130 \\ 1.000 & 4.590 & 0.006 \\ 0.000 & 0.056 & 5.590 \end{bmatrix} \cdot \begin{bmatrix} R \\ G \\ B \end{bmatrix} \qquad (4.18)$$

The abovementioned transformation was used during re-computing of standard observer functions $\bar{x}, \bar{y}, \bar{z}$ from measured matching functions of average Wright-Guild observer $\bar{r}, \bar{g}, \bar{b}$. It is important to remember the CIE RGB color space is based three monochromatic light stimuli at 700 nm (red), 546.1 nm (green), and 435.8 nm (blue) as the primaries to represent colors. In case of other light stimuli, it is necessary to calculate new transformation matrix to XYZ system Thus, the sRGB color space, for example, is defined with different transformation matrix as matrix in Eq. (4.18).

The fact that new color spaces can be constructed by applying linear transformations has an important implication for the study of color. If we restrict attention to what we may conclude from the color-matching experiment, we can determine the psychological representation of color only up to a free linear transformation. There are two attitudes one can take toward this fact. The conservative attitude is to refrain from making any statements about the nature of color vision that depend on a particular choice of color space. The other is to appeal to experiments other than the color-matching experiment to choose a privileged representation. At present, there is not universal agreement about how to choose such a representation, and we therefore advocate the conservative approach.

Because the visual representation of light is three-dimensional (see Figure 4.15), it is difficult to plot this representation on a two-dimensional page. Even more difficult is to represent a dependent measure of visual performance as a function of color coordinates. A strategy for plotting color data

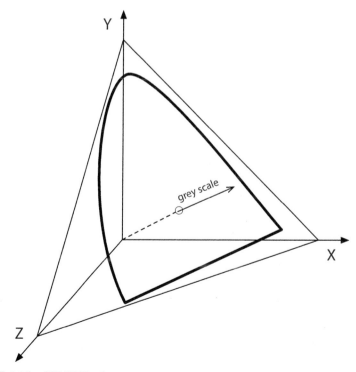

FIGURE 4.15   CIE XYZ color space.

is to reduce the dimensionality of the data representation. One common approach is through the use of chromaticity coordinates.

Although the *XYZ* tristimulus values do not correlate to color attributes, important color attributes are related to the relative magnitudes of the tristimulus values, called the chromaticity coordinates. The chromaticity coordinates are derived from the *XYZ* tristimulus values via a projective transformation and defined as:

$$x = \frac{X}{X + Y + Z} \tag{4.19}$$

$$y = \frac{Y}{X + Y + Z} \tag{4.20}$$

This color representation system is referred to as *xyY*. A third coordinate, z, can also be defined in a manner similar to the definition of *y* in Equation 4.20, but it is redundant, because:

$$x + y + z = 1 \tag{4.21}$$

With the two variables *x* and *y*, it is possible to represent colors in a 2D color diagram; this diagram is known as the *CIE chromaticity diagram* in the sense that the third variable, *Y*, has only the brightness content. Note that the *Y* component in the *XYZ* space represents luminance; brightness is a perceptual attribute associated with luminance. The CIE *x, y* chromaticity diagram is shown in Figure 4.16.

The horseshoe-shaped curve is sometimes called the *locus of spectral colors,* or *locus of monochromatic colors*. Any point on the locus corresponds to the chromaticity of monochromatic light of a particular wavelength. A scale of wavelength in that connection is shown on the curve. Chromaticities that lie along the straight line joining the open ends of the locus of spectral chromaticities (they are various purple hues) do not appear in the light spectrum, and accordingly are not monochromatic. They are called *the non-spectral purples* and are arbitrarily considered to have 100% saturation. Remember that in any color space, if we start with a particular color, with certain values of three primaries, and increase its luminance (without changing its chromaticity), the proportions of its three primary components

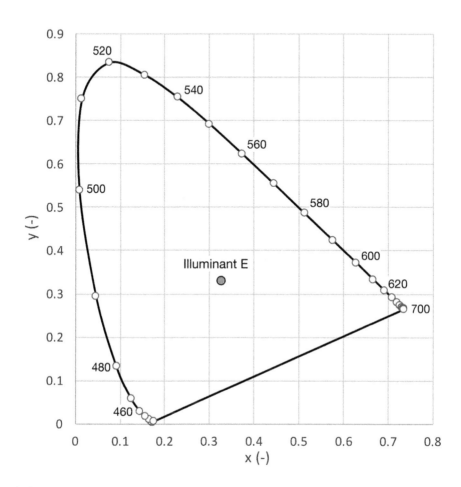

FIGURE 4.16 The CIE 1931 chromaticity diagram with the locus of spectral colors and the purple line. The wavelengths of spectral colors are given in nm.

remain unchanged. Thus, in the CIE system, for an unchanging chromaticity, X and Y increase by the same ratio that $(X+Y+Z)$ does, and so, x and y remain unchanged. Therefore, the point x, y in fact represents the chromaticity of a color. Keeping in mind that the variable Y indicates the luminance of the color of interest, the entire 1931 color space is often designated the xyY (or sometimes Yxy) space.

Beside the color coordinates x, y, it is possible to calculate two other coordinates, which are called excitation purity and dominant wavelength, or Helmholtz numbers [19]. The wavelength associated with the point on the horseshoe-shaped curve at which the extrapolated line intersects is the dominant wavelength. When the line is extrapolated in the other direction (from the end that terminates at the reference white point), the wavelength associated with the point on the horseshoe-shaped curve at which this line intersects is called the "complementary wavelength." When the line used to locate, the dominant wavelength does not intersect the horseshoe-shaped curve at all (i.e., it intersects the straight line that contains the purples near the bottom of the diagram), then the complementary wavelength instead of the dominant wavelength is used to describe the color most accurately, and in this case, it is called the "complimentary dominant wavelength." The excitation purity is defined to be the ratio of the length of the line segment that connects the chromaticity coordinates of the reference white point and the color of interest to the length of the line segment that connects the reference white point to the dominant wavelength. The larger the excitation purity, the more saturated the color appears or the more similar the color is to its spectrally pure color at the dominant wavelength. The smaller the excitation purity, the less saturated the color appears or the whiter it is. Pastel colors are very poorly saturated, for example.

The chromaticity coordinates x, y depends on the primaries X, Y, and Z, which have been chosen more or less arbitrarily. The chromaticity diagram depends on this choice; however, another choice will only lead to a *perspectival* distortion of the diagram, accompanied by a redefinition of units such that the simple rules for computing additive mixing remain valid. With respect to color differences, the chromaticity chart is not quite satisfactory: the distances on the chart are not nearly proportional to perceived differences. No wonder, any perceptions beyond color matching have until now been deliberately excluded. If now color differences are compared, this is beyond the initial restriction. A measure for the perceived distance may be obtained from the scattering of data points when color matching is repeated many times.

4.4 APPROXIMATELY UNIFORM COLOR SPACES

For purposes of routine colorimetry, it is not essential that all colors be represented in a uniform tridimensional system. Nevertheless, the most pressing problem has been to find simple equations, which represent the coordinates of color, or simple projective transformations that will yield high fidelity color coordinates. One group of uniform color scales is mostly based on Mac-Adam measurement of color discrimination [7, 18]. The second major group of uniform color scales was designed to simulate visually adjusted spacing of surface colors based on Munsell System. Conceived by Munsell, this system was developed not as an experiment in psychophysics, but to meet his needs as an artist for a means to identify and interrelate colors of surfaces [20].

The Munsell system consists of three independent dimensions, which can be represented cylindrically in three dimensions as an irregular color solid: hue, measured by "degrees" around horizontal circles; chroma, measured radially outward from the neutral (gray) vertical axis; and value, measured vertically from 0 (black) to 10 (white). Munsell determined the spacing of colors along these dimensions by taking measurements of human visual responses based on logarithmic relation of the Weber-Fechner law [21]. Former visual spacing was re-measured and re-assessed based on visual experiments with a gray background of 20% reflectance (approx. neutral grey of mid-level of lightness). The derived relationships between instrumentally computed Munsell colors and visual estimates are called the Munsell Renotation System [22, 23].

The Munsell system is almost visually uniform [24], but the problem is, that this system is not based on CIE colorimetry and relationship is based on converting tables [25, 26] or nonlinear computing. Studies by an OSA subcommittee [27] based on a more precise measurement of the luminance factors of the samples of the 1929 Munsell Book of Color indicated the definition of Munsell value recommended by this subcommittee is given by the fifth-order polynomial function:

$$\frac{100Y}{Y_{MgO}} = 1.2219V - 0.23111V^2 + 0.23951V^3 - 0.021009V^4 + 0.0008404V^5 \quad (4.22)$$

Equation (4.22) is frequently called as Judd polynomial function [28], which is rather written in following form:

$$\frac{100Y}{Y_{MgO}} = 1.2219V - 0.23111V^2 + 0.23951V^3 - 0.021009V^4 + 0.0008404V^5 \quad (4.23)$$

Due to change in reflectance standard from MgO to absolute white in 1959, it is necessary to correct coefficients in Eq. (4.22) as shown in the following equation [29]:

$$Y = 1.1913V - 0.22533V^2 + 0.23352V^3 - 0.020484V^4 + 0.0008194V^5$$
$$(4.24)$$

4.4.1 ANLAB

In 1944, Nickerson and Stultz proposed replacing the Weber derivations of V_x, V_y and V_z of the Adams chromatic value space by the Judd polynomial function [30, 31]. They also suggested scaling the resulting lightness coordinate by a factor of 0.23 to make unit steps in lightness more consistent with those of the opponent chromatic scales [1, 3, 18]. The resulting formula was widely used, largely due to publicity given it by Nickerson, and as a result became known as the ANLAB (Adams–Nickerson *LAB)* formula, given in following equations:

$$\Delta E_{ANLAB} = \sqrt{\left(\Delta L_{AN}\right)^2 + \left(\Delta a_{AN}\right)^2 + \left(\Delta b_{AN}\right)^2} \qquad (4.25)$$

where

$$L_{AN} = S\left(0.23V_Y\right) \approx 40\left(0.23V_Y\right) = 9.2V_Y \qquad (4.26)$$

$$a_{AN} = S\left(V_X - V_Y\right) \approx 40\left(V_X - V_Y\right) \qquad (4.27)$$

$$b_{AN} = S\left[0.4\left(V_Y - V_Z\right)\right] \approx 40\left[0.4\left(V_Y - V_Z\right)\right] = 16\left(V_Y - V_Z\right) \qquad (4.28)$$

Number 40 is scaling factor S, which became the factor most widely used, although other factors have been applied (42, 43.909, 50...). Results from the ANLAB formula are therefore ambiguous unless S is clearly stated, mentioned the usual notation being (for $S=40$) ANLAB (40) or one of the abbreviations, AN(40) and AN40.

At this time, several investigations came to consider the Judd polynomial function (Eq. (4.23), its modified form respectively) unnecessarily complicated, and suggested cube-root approximations of it. The other possibility was focused on a simplified form of iterative calculations [32] or look up method based on tabulated values [33, 34]. The approximation, which was

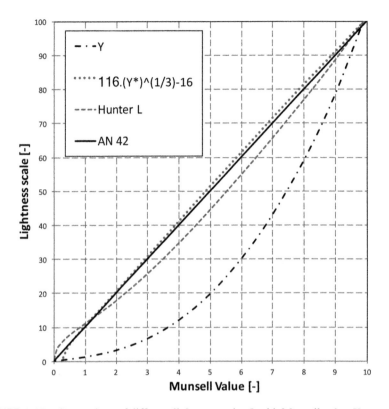

FIGURE 4.17 Comparison of different lightness scales L with Munsell value V.

finely used by CIE together with Hunter square-root approximation and Judd polynomial function as part of ANLAB with scaling factor 42, is shown on Figure 4.17

4.4.2 HUNTER LAB

Around 1958, Hunter developed a new formula, which was originally designed for use with illuminant C, and the 1931 2° standard observer. A general form of the equation suitable for use under other conditions is shown in equation [35]:

$$L_H = 100\sqrt{\frac{Y}{Y_0}} \tag{4.29}$$

$$a_H = \frac{K_a \left(\dfrac{X}{X_0} - \dfrac{Y}{Y_0} \right)}{\sqrt{\dfrac{Y}{Y_0}}} \tag{4.30}$$

$$b_H = \frac{K_b \left(\dfrac{Y}{Y_0} - \dfrac{Z}{Z_0} \right)}{\sqrt{\dfrac{Y}{Y_0}}} \tag{4.31}$$

where $X_{0,i}$ and $Z_{0,I}$ are co-ordinates of ideal white for the used illuminant, $X_{0,i}$ and $Z_{0,I}$ are co-ordinates of ideal white for the illuminant C and 2° observer, K_a and K_b are illuminant chromaticity coefficients given by Eqs. (4.32) and (4.33)

$$K_a = 175 \sqrt{\frac{X_{0i}}{X_{0C2}}} \tag{4.32}$$

$$K_b = 70 \sqrt{\frac{Z_{0i}}{Z_{0C2}}} \tag{4.33}$$

Whole color difference is calculated by the obvious form:

$$\Delta E_H = \sqrt{\left(\Delta L_H \right)^2 + \left(\Delta a_H \right)^2 + \left(\Delta b_H \right)^2} \tag{4.34}$$

4.4.3   GLASSER CUBE-ROOT

Glasser, Reilly, and others developed scales that are designed to provide an approximation to the Adams chromatic value scales [36]. The formulas apply a lightness-response function to X, Y, and Z separately before taking differences:

$$L_{GCR} = 25.29 G^{\frac{1}{3}} - 18.38 \qquad (4.35)$$

$$a_{GCR} = K_{a(GCR)} \left(R^{\frac{1}{3}} - G^{\frac{1}{3}} \right) \qquad (4.36)$$

$$b_{GCR} = K_{b(GCR)} \left(G^{\frac{1}{3}} - B^{\frac{1}{3}} \right) \qquad (4.37)$$

where

$$R = 1.1084 X + 0.0852 Y - 0.1454 Z \qquad (4.38)$$

$$G = -0.0010 X + 1.0005 Y + 0.0004 Z \qquad (4.39)$$

$$B = -0.0062 X + 0.0394 Y + 0.8192 Z \qquad (4.40)$$

$K_{a(GCR)} = 105$ for $R < G$ and $K_{b(GCR)} = 30.5$ for $B < G$

$K_{a(GCR)} = 125$ for $R > G$ and $K_{b(GCR)} = 53.6$ for $B > G$

4.4.4   CIELAB

CIELAB is now almost universally the most popular approximately uniform color space. Coordinates of a color in LAB are obtained as a non-linear mapping of the XYZ coordinates; the nonlinearity is introduced to emulate the logarithmic response of the human eye. The lightness component $L*$ is calculated as:

$$L^* = 116 f \left(\frac{Y}{Y_0} \right) - 16 \qquad (4.41)$$

$$a^* = 500 \left[f \left(\frac{X}{X_0} \right) - f \left(\frac{Y}{Y_0} \right) \right] \qquad (4.42)$$

$$b^* = 200 \left[f\left(\frac{Y}{Y_0}\right) - f\left(\frac{Z}{Z_0}\right) \right] \tag{4.43}$$

where X, Y, and Z are the CIE XYZ tristimulus values, X_0, Y_0, and Z_0 are the tristimulus values of the reference white, and the function $f(\omega)$, where ω is (X/X_0), (Y/Y_0), or (Z/Z_0), is given by Eq. (4.44):

$$f(\omega) = \begin{cases} \omega^{1/3} & \omega > (24/116)^3 \\ (841/108)\omega + 16/116 & \omega \le (24/116)^3 \end{cases} \tag{4.44}$$

Although this new color space was intended to deal with reflective "color," it was soon adapted for use in describing "luminous" colors as well. It has come into widespread use in various graphics software packages and for other color management purposes. It is a member of the luminance-chrominance family. The traditional designation of the space is $L^*a^*b^*$, where the asterisks remind us of the nonlinear nature of its three variables.

The luminance-like aspect of the reflective color is actually described by the variable *lightness*, indicated as L. It is nonlinearly related to the reflectance of the surface so as to more closely follow the human eye's perception. In CIELAB computing of such nonlinearity is the exponent 1/3 (root-square) as replacement of fifth-order polynomial function in ANLAB color space. For all the above expressions, if any of the ratios X/X_0, Y/Y_0, and Z/Z_0 is less than or equal to 0.008856, then that ratio is replaced by a linear relation as shown in Figure 4.18. The purpose of the linear piece of the nonlinear function is that with the traditional function (stated just above), when the inverse function is used (to convert from Lab to XYZ), the slope of the function becomes infinite at the origin, thus leading to implementation difficulties. The linear piece avoids this complication.

As was noted at the beginning, a coordinate system, which most intuitively relates to the human perception of color, defines a color in terms of luminance, hue, and saturation. In the a-b plane of the CIE $L^*a^*b^*$ color space, the angular position of the chrominance point (with respect to the set of axes) in fact corresponds to hue, and the radius from the origin to the point, divided by L^*, is indicative of saturation.

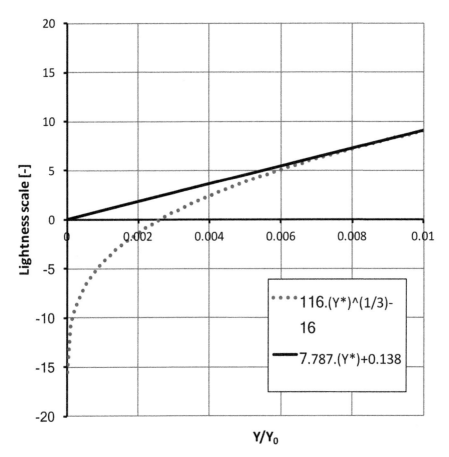

FIGURE 4.18 Derivation of the L^* curve.

To produce a more intuitive set of coordinates, we can recast the a-b chromaticity plane from Cartesian (rectangular) coordinates to polar coordinates, using the variable C to represent the radius to the chromaticity point and the variable h to represent the angle in degrees to the point, measured counterclockwise from the positive a axis (see Figure 4.19). Chroma value C^* essentially indicates the product of saturation and lightness, and h essentially indicates hue. The resulting space is called the CIE LCH space [37]. The perceptual correlates of chroma and hue are given by Eqs. (4.45) and (4.46).

$$C_{ab}^* = \sqrt{\left(a^*\right)^2 + \left(b^*\right)^2} \tag{4.45}$$

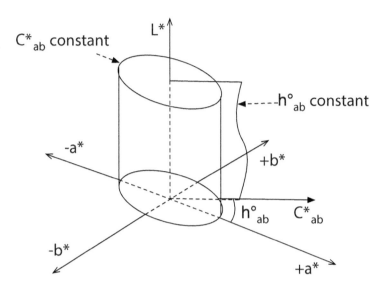

FIGURE 4.19 A three-dimensional representation of the CIELAB color space showing rectangular system of axis LAB and cylindrical system LCh.

$$h_{ab} = \arctan\left(b^*/a^*\right) \tag{4.46}$$

from interval 0–360°.

Because of its approximate perceptional uniformity, the distance in the CIE $L^*a^*b^*$ space is used to specify color tolerances.

$$\Delta E_{ab}^* = \sqrt{\left(\Delta L^*\right)^2 + \left(\Delta a^*\right)^2 + \left(\Delta b^*\right)^2} \tag{4.47}$$

Approximate correlates of certain perceived attributes of color and color difference (lightness, chroma, hue angle, and hue difference) could be calculated in the CIELAB or CIELCH color space in the following manner:

$$\Delta L^* = L^*_{sample} - L^*_{targed} \tag{4.48}$$

$$\Delta a^* = a^*_{sample} - a^*_{targed} \tag{4.49}$$

$$\Delta b^* = b^*_{sample} - b^*_{targed} \qquad (4.50)$$

It is convenient to be able to express the same color difference in terms of differences in L^*, C^*, and a measure that correlates with hue difference [38]. The sign convention for ΔH^* ensures that ΔH^* and Δh_{ab} will have the same sign. When these are compared with the other differences, it must not be neglected that Δh_{ab} is measured in degrees rather than CIELAB units, and thus has a different metric from these other differences. That means Δh_{ab}, being an angular measure, it cannot be combined with L^* and C^* easily [39]. Instead, the following expression is used:

$$\Delta C^* = C^*_{sample} - C^*_{targed} \qquad (4.51)$$

$$\Delta H^* = \sqrt{\left(\Delta E^*_{ab}\right)^2 - \left(\Delta L^*\right)^2 - \left(\Delta C^*\right)^2} \qquad (4.52)$$

As can be seen, ΔH^* requires an indirect calculation method. This method shown above does not produce a negative sign and therefore is not generally recommended. Nevertheless, sign of ΔH^* can be added based on Cooper's method determining appropriate hue-difference ΔH^* descriptor, which is shown in Figure 4.20.

Generally speaking, counterclockwise direction of hue-difference ΔH^* is positive and clockwise direction is negative sign. Based on geometric meaning of ΔH^* from Figure 4.21 an alternative calculation, direct definitions of ΔH^* were suggested by Huntsman [40], Sève [41], Stokes and Brill [42], and can be used to replace the original indirect definitions.

$$\text{Huntsman } \Delta H^* = \sqrt{2C_1^* C_2^* \left(1 - \cos \Delta h^\circ\right)} \qquad (4.53)$$

$$\text{Sève } \Delta H^* = 2\sqrt{C_1^* C_2^*} \cdot \sin\left(\Delta h^\circ\!\big/2\right) \qquad (4.54)$$

$$\text{Stokes-Brill } \Delta H^* = s\sqrt{2\left(C_1^* C_2^* - a_1^* a_2^* - b_1^* b_2^*\right)} \qquad (4.55)$$

where $s = +1$, if $a^*_1 b^*_2 > a^*_2 b^*_1$ and $s = -1$, if $a^*_1 b^*_2 \leq a^*_2 b^*_1$.

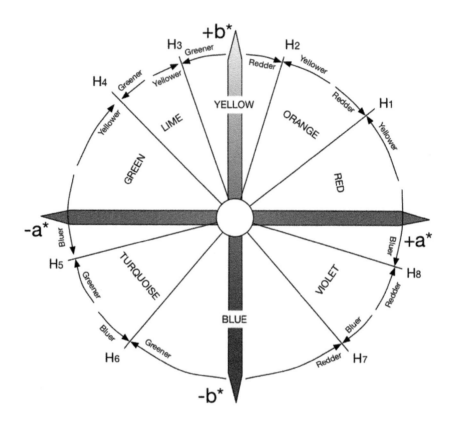

FIGURE 4.20 Cooper's method of determining appropriate hue-difference ΔH^* descriptor in the CIELAB.

4.4.5 CIELUV

In 1976, the CIE was unable to choose only one single color space as representative approximately uniform color space and agreed to a second one as well. This space is called as CIELUV and could be regarded as an improvement of CIE 1964 UCS [43]. The L^* function of the CIELUV space is the same as that of the CIELAB space. It was intended to provide perceptually more uniform color spacing for colors at approximately the same luminance. The 1976 CIE-UCS chart uses u' and v' coordinates. The symbols u' and v' were chosen to differentiate from the u and v coordinates of the similar but short lived 1960

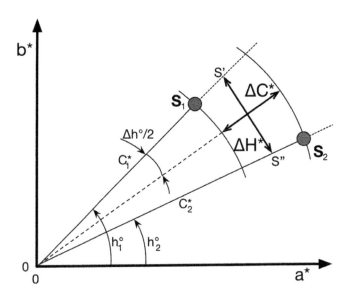

FIGURE 4.21   Geometric meaning of ΔH^*.

CIE-UCS system. The u' and v' chromaticity coordinates are also calculated from the XYZ tristimulus values according to the following formulae [4]:

$$u^* = 13L^* \left(u' - u'_0\right) \tag{4.56}$$

$$v^* = 13L^* \left(v' - v'_0\right) \tag{4.57}$$

$$u' = \frac{4X}{X + 15Y + 3Z} \tag{4.58}$$

$$v' = \frac{9Y}{X + 15Y + 3Z} \tag{4.59}$$

Values with $_0$ as subscript are related to ideal white.

In the CIELUV color space, not only correlates of chroma and hue can be defined, but in the u', v' diagram (see Figure 4.22), a correlate of saturation can be defined too.

$$s_{uv} = 13\sqrt{\left(u' - u'_0\right)^2 + \left(v' - v'_0\right)^2} \tag{4.60}$$

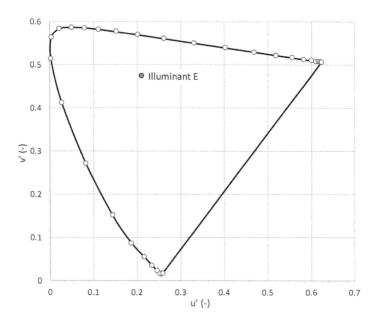

FIGURE 4.22   CIELUV – u'v' diagram.

$$C_{uv}^* = \sqrt{\left(u^*\right)^2 + \left(v^*\right)^2} = L^* s_{uv} \tag{4.61}$$

$$h_{uv} = \tan^{-1}\left(v^* \middle/ u^*\right) \tag{4.62}$$

CIELUV hue angle, h_{uv} shall lie between 0° and 360°. CIELUV color difference ΔE^*_{uv} between two color stimuli is calculated as the Euclidian distance between the points representing them in the space, similarly as in CIELAB:

$$\Delta E_{uv}^* = \sqrt{\left(\Delta L^*\right)^2 + \left(\Delta u^*\right)^2 + \left(\Delta v^*\right)^2} \tag{4.63}$$

After the CIELAB and CIELUV color difference formulas were introduced by the CIE 1976, it quickly became apparent that, despite being as good or better than any other formulas then available, they could be further improved by adjusting the ΔL^*, ΔC^* and ΔH^* components, depending on their position in the color space. However, the CIELAB color space is not quite perceptually uniform, particularly in the blue region of the color space. As a result, the Euclidean distance in the CIELAB space between two colors does not always correspond to the perceived color difference. This non-uniformity issue was addressed by the CIE by establishing an advanced color difference equation, as described below.

4.5   CIE CAM

Color appearance models predict how a patch of color is perceived under specific illumination. Generally, color appearance models (CAMs) consist of three steps. In the first step, chromatic adaptation is modeled with a von-Kries-style transform, carried out in a sharpened color space. The second step involves non-linear response compression. Performed in a cone response space, this step models photoreceptor behavior. The third step computes color opponent values and uses this to derive measures of lightness, hue, and chroma (as well as others), which describe the human visual response to a colored patch in the context of an environment illuminated by a given light source.

The analysis of colors based on their matching tristimulus values has a constitutive limitation. The color matching can reveal those colors that appear to match, but it does not reveal how the colors appear. In other words, the color matching can estimate what colors look similar but not what color look like [44]. The CIE XYZ or LUV or LAB systems can only be used under limited viewing conditions. Based on these models, two stimuli having identical CIE XYZ tristimulus values appear to match in color for average observer only if the requirements that were defined in section 4.2.1 are met. In addition, the two stimuli must be viewed with identical backgrounds, surrounds, illumination geometry, and surface characteristics, and their shape and size must also be equal. However, if any of these requirements are not fulfilled, the color match may no longer hold. Hence, many color matching applications cannot fulfill these requirements and because of that, the enhanced colorimetry must be implemented to include these requirements as color appearance model [44].

The human visual system is able to adapt to different lighting conditions. Even if there are great differences in light intensity, as for example, daylight compared to artificial lighting in a room, the human eye is able to adapt to perceived intensity and color. The result of this adaption is that objects are perceived in the same color, although the illumination is different. For example, a green apple is perceived in the same color in daylight at noon and in a room in the evening [45]. This adaption is called color constancy and is an important topic in color science. It is assumed that three different types of cones with different spectral sensitivities exist. They have their maximum spectral sensitivities at the colors of yellow-orange, green and blue-violet and are frequently captioned R, G, B [45]. Because their caption is misleading they are called ρ, γ, and β. Their sensitivity behavior is depicted in Figure 4.23.

Color perception works by processing the response of the cones (see Figure 4.24). It is assumed that the connection of the output results in an achromatic signal A and three color difference signals C1, C2, and C3. The achromatic signal is composed of the output of all rods and cones. The different occurrence of cones is simply represented by the term $2\rho + \gamma + \beta/20 + S$, where S denotes the scotopic input of the rods. Physiological studies

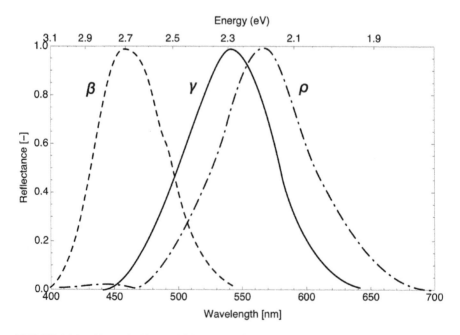

FIGURE 4.23   The probable sensitivity curves β, γ, and ρ of the three types of light receptor believed to be responsible for color vision as determined by indirect methods [44].

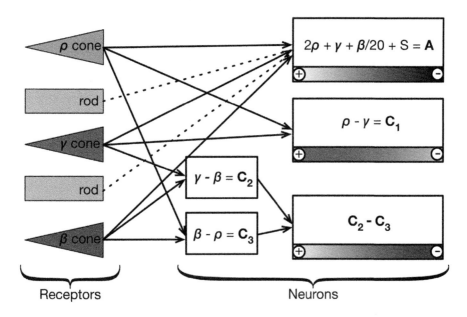

FIGURE 4.24 This image shows a simplified representation of the commonly assumed connection between cone and rod output, which results in human perception.

indicate that the output of color difference signals is further processed as two signals C1 and C2 − C3 because of the fact C1 + C2 + C3 = 0 [45].

It may result in the same color perception, although different spectral distributions are involved in the process, which is related to the different spectral sensitivity distributions of the cones. The effect of equal color perception of different colors is well known as metamerism and is an important challenge for many color industries [45].

One purpose of color vision models is to predict perceptual descriptors of colors under different viewing conditions. There are two main types of color vision models: color appearance models and neural models. Both predict perceptual descriptors, but the difference between them is that neural models also try to follow the stages of the visual system. However, quite often, evaluating color differences between two samples is more important than obtaining numerical values of perceptual descriptors. Initially, separate models were developed to fit color appearance and color difference data. However, the perfect solution would be to obtain a single model, which could reproduce both perceptual descriptors and color differences.

Color appearance model predicts the change in color appearance under different viewing conditions such as illuminant, luminance level, background color and surround. CIE Technical Committee 1-34 (TC1-34) describes color appearance model as "A color appearance model is any model that includes predictors of at least the relative color-appearance attributes of lightness, chroma and hue" [44]. Figure 4.25 shows the input and output parameters of the CIE color appearance model.

The summary of the inputs into CAM are given below:

1. Trichromatic values of color stimuli – XYZ
2. White point X_W, Y_W, Z_W, which are the tristimulus of the reference white under the test illuminant.
3. L_A, which specifies the luminance of the adapting field.
4. Y_b, which defines the luminance factor of the background field.
5. Surround conditions: average, dim, and dark.

There are many output parameters from the model, which typically are:

1. Lightness (J)
2. Brightness (Q)
3. Redness-greenness (a)
4. Yellowness-blueness (b)
5. Colorfulness (M)
6. Chroma (C)
7. Saturation (s)

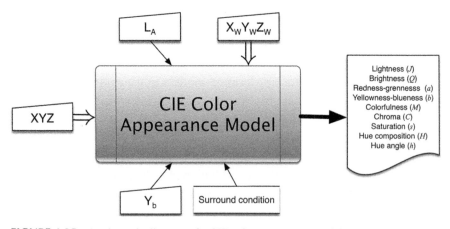

FIGURE 4.25 A schematic diagram of a CIE color appearance model.

8. Hue composition (H)
9. Hue angle (h)

These output parameters can be combined to form various spaces according to the different applications.

4.5.1 CIE CAM97S

Common structure of the color appearance model is made up of three components: a chromatic adaptation transform, a dynamic response function, and a color space for representing the correlates of the perceptual values. Chromatic adaptation transform in its simplest form is a multiplicative normalization of CIE XYZ tristimulus values to make them constant for whites in varying conditions. This approach is used in the CIELAB system, but most of the models use or base on the more fundamental approach where normalization is done on cone responses. This is also known as von Kries transform [46].

Color appearance models are not that a new idea in color science. Earliest versions of the color appearance models were published over 20 years ago by Hunt and Nayatani, and they were suggested only about 5 years after the CIE recommendations of the CIELAB and CIELUV for color difference specification. Subsequently, there has been a lot of scientific research covering color appearance and different phenomenon including it [47].

Nonetheless, the CIE LAB model can also be classified as a color appearance model. This stems from the fact that the CIE LAB model has also predictors of lightness, chroma, and hue. In addition, the CIE LAB model use a von Kries type of chromatic adaptation transform for the white point normalization. Although the transformation is done in CIE XYZ tristimulus space instead of physiological cone space, resulting to inaccurate hue shifts in the hue predictions [48].

The first CIE color appearance model CIECAM97s that was recommended in year 1997 was meant as a temporal model with the expectation that it would be revised as more data and theoretical basis became available. Overall, CIECAM97s was a success in providing a turning point that all the researchers agreed upon and started their work together to enhance the model. This was especially notable because members of group that recommended CIECAM97s model had personally created eight different color appearance models. CIECAM97s color appearance model's shortcomings were identified, and a new model, CIECAM02, was recommended in 2002 [4, 47, 49].

4.5.2 CIE CAM02

The CIECAM02 color appearance model is built on the basic structure of the CIECAM97s model, although it has many revisions. These revisions include a linear, von Kries-type chromatic adaptation transform, and a new non-linear response compression function and modified formulas for the perceptual color appearance correlates. The chromatic adaptation transform and the equations for the perceptual correlates are the main parts of the CIECAM02 model (Figure 4.26) [49].

The chromatic adaptation transform is used for modeling the adaption into changes in the chromaticity of the adopted white point. In CIECAM02, due to different viewing conditions, the CIE considered the differences in color perception by transforming to and from a single reference white. The forward model transforms tristimulus values viewed under a wide range of viewing conditions to the corresponding perceptual attribute correlates as viewed under a reference white, in this case the equal-energy illuminant (Illuminant **E**) with the perfect reflecting diffuser as the reference white. The inverse model transforms from this reference white to some other viewing condition. Besides the chromaticity of the white point, also its luminance effects on the degree to which an observer adapts to the white point. The

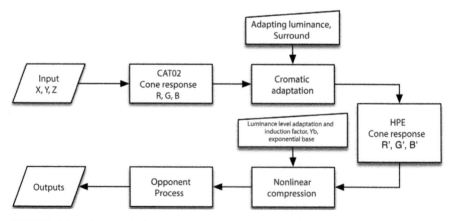

FIGURE 4.26   Schematic diagram of the CIECAM02 computation process.

transform is done based on the viewing condition parameters that have been defined for three types of surrounds: average, dim, and dark. The viewing condition parameters are also used in the calculation of perceptual attributes, which are brightness, lightness, chroma, saturation, colorfulness, and hue (Table 4.1) [49].

For intermediate surrounds, these values can be linearly interpolated. Then, the CAT02 forward matrix converts the sample CIE 1931 tristimulus values to a long, medium, and short wavelength-sensitive space.

$$
\begin{bmatrix} R \\ G \\ B \end{bmatrix} = M_{CAT02} \begin{bmatrix} X \\ Y \\ Z \end{bmatrix}
\tag{4.64}
$$

$$
M_{CAT02} = \begin{bmatrix} 0.7328 & 0.4296 & -0.1624 \\ -0.7036 & 1.6975 & 0.0061 \\ 0.0030 & 0.0136 & 0.9834 \end{bmatrix}
\tag{4.65}
$$

The degree of adaptation to the white point, the D factor, is computed. The range of D value will be from 1 for complete adaptation to 0 for no adaptation. In practice, the minimum D value will not be less than 0.659 (computed from Eq. 4.66) for a dark surround and will exponentially converge to 1 for average surrounds with increasingly large values of L_A.

$$
D = F \left[1 - \left(\frac{1}{3.6} \right) e^{\left(\frac{-(L_A+42)}{92} \right)} \right]
\tag{4.66}
$$

The D factor is then applied to weight chromatic adaptation:

TABLE 4.1   Viewing Condition Parameters for Different Surrounds

Viewing conditions	c	N_C	F
Average surround	0.690	1.0	1.0
Dim surround	0.5901.0	0.9	0.9
Dark surround	0.525	0.8	0.8

$$R_C = \left[\left(Y_W \frac{D}{R_W}\right) + (1-D)\right] R$$

$$G_C = \left[\left(Y_W \frac{D}{G_W}\right) + (1-D)\right] G \qquad (4.67)$$

$$B_C = \left[\left(Y_W \frac{D}{B_W}\right) + (1-D)\right] B$$

where R_W, G_W, and B_W are RGB values computed for the white point (perfect reflecting diffuser) using Eqs. (4.64) and (4.65).

The next step consists of computing constants, which are dependent on viewing conditions used during observation:

$$k = \frac{1}{5L_A + 1} \qquad (4.68)$$

$$F_L = 0.2k^4\left(5L_A\right) + 0.1\left(1 - k^4\right)^2 \left(5L_A\right)^{1/3} \qquad (4.69)$$

$$n = \frac{Y_b}{Y_W} \qquad (4.70)$$

The value n is a function of the luminance factor of the background. Its range is from 0 for a background luminance factor of zero to 1 for a background luminance factor equal to the luminance factor of the adapted white point. The n value can then be used to compute N_{bb}, N_{cb}, and z, which are then used during the computation of several of the perceptual attribute correlates such as achromatic response, lightness, chroma, etc.

$$N_{bb} = N_{cb} = 0.725\left(\frac{1}{n}\right)^{0.2} \qquad (4.71)$$

$$z = 1.48 + \sqrt{n} \qquad (4.72)$$

The post-adaptation signals for both the sample and the source white are then transformed from the sharpened cone responses to the Hunt-Pointer-Estevez cone responses as shown in Eqs. (4.73) and (4.74).

$$
\begin{bmatrix} R' \\ G' \\ B' \end{bmatrix} = M_{HPE} M_{CAT02}^{-1} \begin{bmatrix} R_C \\ G_C \\ B_C \end{bmatrix}
\tag{4.73}
$$

$$
M_{HPE} = \begin{bmatrix} 0.3897 & 0.6889 & -0.0787 \\ -0.2298 & 1.1834 & 0.0464 \\ 0.0000 & 0.0000 & 1.0000 \end{bmatrix}
$$

$$
M_{CAT02}^{-1} = \begin{bmatrix} 1.0961 & -0.2789 & 0.1827 \\ 0.4544 & 0.4735 & 0.0721 \\ -0.0096 & -0.0057 & 1.0153 \end{bmatrix}
\tag{4.74}
$$

It was found that Hunt-Pointer-Estevez sharpening provides better prediction of perceptual attribute correlates [50]. CIECAM02, like CIECAM97s, makes use of one space for the chromatic adaptation transform and another for computing perceptual attribute correlates. This adds complexity to the model, but pending additional research, it would appear that chromatic adaptation results are best predicted in a space that has some degree of sharpening. The nature and degree of this sharpening is still subject to debate, but it should be noted that CAT02 does incorporate some degree of sharpening. In comparison, the use of a space closer to the cone fundamentals, such as the Hunt-Pointer-Estevez or Stockman-Sharpe fundamentals, provides better predictions of perceptual attribute correlates. Blue constancy, a significant shortcoming for CIELAB, is considerably improved by using a space closer to a cone fundamental space [49].

The post-adaptation non-linear response compression is then applied to the output:

$$
R'_a = \frac{400\left(F_L R' \big/ 100\right)^{0.42}}{27.13 + \left(F_L R' \big/ 100\right)^{0.42}} + 0.1
$$

$$G'_a = \frac{400\left(F_L G'\!/\!100\right)^{0.42}}{27.13+\left(F_L G'\!/\!100\right)^{0.42}}+0.1 \qquad (4.75)$$

$$B'_a = \frac{400\left(F_L B'\!/\!100\right)^{0.42}}{27.13+\left(F_L B'\!/\!100\right)^{0.42}}+0.1$$

If any of the values of R', G' or B' are negative, their absolute values must be used, and the quotient term in Eq. (4.75) must be multiplied by a negative 1 before adding the value 0.1.

Temporary Cartesian representation and hue are calculated before computing eccentricity factor and perceptual attributes. These values are used to compute a preliminary magnitude t.

$$a = R'_a - \frac{12G'_a}{11} + \frac{B'_a}{11} \qquad (4.76)$$

$$b = \left(\tfrac{1}{9}\right)\cdot\left(R'_a + G'_a - 2B'_a\right) \qquad (4.77)$$

$$h = \tan^{-1}\left(b/a\right) \qquad (4.78)$$

$$e_t = \frac{\cos\left(h\pi/180 + 2\right) + 3.8}{4} \qquad (4.79)$$

Hue quadrature or H can be computed from linear interpolation of the data shown in Table 4.2

$$H = H_i + \frac{100\,(h' - h_i)/e_i}{(h' - h_i)/e_i + (h_{i+1} - h')/e_{i+1}} \qquad (4.80)$$

The chroma scale now requires the calculation of a new factor, t,

TABLE 4.2 Data for Conversion from Hue Angle to Hue Quadrature [44]

	Red	Yellow	Green	Blue	Red
i	1	2	3	4	5
h_i	20.14	90.00	164.25	237.53	380.14
e_i	0.8	0.7	1.0	1.2	0.8
H_i	0	100	200	300	400

$$t = \frac{(50\,000/13)\,N_c N_{cb}\, e_t \sqrt{a^2 + b^2}}{R'_a + G'_a + (21/20) B'_a} \tag{4.81}$$

Now, it is possible to calculate various appearance attributes as follows: The achromatic response, A:

$$A = \left[2R'_a + G'_a + \left(\frac{1}{20}\right) B'_a - 0.305 \right] N_{bb} \tag{4.82}$$

Lightness, J:

$$J = 100 \left(\frac{A}{A_W}\right)^{cz} \tag{4.83}$$

Brightness, Q:

$$Q = \left(\frac{4}{c}\right) \cdot \left(\frac{J}{100}\right)^{0.5} (A_W + 4) \cdot (F_L)^{0.25} \tag{4.84}$$

Chroma, C:

$$C = t^{0.9} \sqrt{J/100}\, (1.64 - 0.29^n)^{0.73} \tag{4.85}$$

Colorfulness, M:

$$M = C \cdot F_L^{0.25} \tag{4.86}$$

Saturation, s:

$$s = 100 \left(\frac{M}{Q} \right)^{0.5} \tag{4.87}$$

CIECAM02 includes three color attributes chroma (C), colorfulness (M), and saturation (s). These attributes together with lightness (J) and hue angle (h) form three color spaces:

a. J, a_c, b_c
b. J, a_M, b_M
c. J, a_s, b_s.

where coordinates for different color opponents are as follows:

$$a_C = C \cdot \cos(h) \qquad a_M = M \cdot \cos(h) \qquad a_s = s \cdot \cos(h)$$

$$b_C = C \cdot \sin(h) \qquad b_M = M \cdot \sin(h) \qquad b_s = s \cdot \sin(h) \tag{4.88}$$

While CIECAM02 does not provide an official color difference formula, CIECAM02 coordinates (J, a_c, b_c) correlate well with CIELAB ones (L*, a*, b*) and color differences can be computed as:

$$\Delta E_{02}^C = \sqrt{(\Delta J)^2 + (\Delta a_C)^2 + (\Delta b_C)^2} \tag{4.89}$$

Alternatively, color differences can be computed based on color opponents calculated by using colorfulness or saturation.

CIECAM02 can predict all the phenomena that can be predicted by CIECAM97s. It includes correlates of all the typical appearance attributes (relative and absolute) and can be applied over a large range of luminance levels and states of chromatic adaptation. This model has attempted to balance backwards compatibility with CIECAM97s, prediction accuracy for a range of data sets, complexity, current understanding of certain aspects of the human visual system, and invertibility. The model also describes the inverse equations for computing XYZ values given perceptual attribute correlates and viewing condition parameters. CIECAM02 has been defined for individual stimuli (i.e., color patches) presented in a particular environment. Trying to use CIECAM02 is not as simple as switching from XYZ to CIELAB. It requires careful thought about the goals that one hopes to achieve, careful selection of the CIECAM02 input parameters and management of the expectations. It is not the answer to all the problems, and it also produces new problems. Currently, there is no CIE recommendation for calculating color differences under non-daylight illuminants. The CIE has not officially withdrawn CIELAB and CIELUV color difference formulae, but for small color difference methods based on empirical color difference formulae, and for large color differences, the methods based on color appearance models, provide much better agreement with visual data.

4.6 COLOR DIFFERENCE FORMULAS

A color-difference formula is an important quality control tool in the production of colored goods. The unsatisfactory uniformity of the CIELAB space prompted researchers to investigate other color-difference data and develop better color-difference formulas [51]. It was soon discovered that this equation had its shortcomings. These shortcomings were that it was not considered that the human eye is more sensitive to small color differences in some regions of the color wheel and less sensitive in others. This means that a ΔE of 1.0 could be a small visible difference in one area of the visible spectrum (i.e., dark blue colors) and a large visible difference in other area (i.e., light pastel type colors). After 1976, many attempts to develop more accurate color-difference formulas for evaluating small-to-medium color differences were made based on modifications to the CIELAB distance of ΔE^* [52, 53, 54].

4.6.1 CMC(L:C)

The CMC (*l:c*) formula is based on the CIELAB color space and was recommended by the Color Measurement Committee of the Society of Dyers and Colourists [55]. This committee proposed the metric called the CMC color difference. The CMC color-difference formula, based on the CIELAB system, is defined as:

$$\Delta E_{CMC(l:c)} = \sqrt{\left(\frac{\Delta L^*}{l\Delta S_L}\right)^2 + \left(\frac{\Delta C^*}{c\Delta S_C}\right)^2 + \left(\frac{\Delta H^*}{S_H}\right)^2} \qquad (4.90)$$

$$L^*16 \qquad \text{then } S_L = \frac{0.040975 \cdot L^*}{1+0.01765 \cdot L^*},$$

$$L^* \le 16 \qquad \text{then } S_L = 0.511,$$

$$S_C = \frac{0.0638 \cdot C^*}{1+0.01131 \cdot C^*} + 0.638,$$

$$S_H = S_C \cdot (T \cdot f + 1 - f),$$

$$f = \sqrt{\frac{\left(C^*\right)^4}{\left(C^*\right)^4 + 1900}},$$

$$T = 0.56 + [0.2 \cdot \cos(h+168°)] \text{for} \quad 164° \le h < 345°,$$

$$T = 0.36 + [0.4 \cdot \cos(h+35°)] \text{for} \quad 345° \le h < 164°$$

The ratio $l{:}c$, typically fixed at 2:1, controls the lengths of the axes of the ellipsoids representing color tolerance. Some idea of the changing shape and size of the tolerance ellipsoid for different regions of CIELAB color space is shown in Figure 4.27. The CMC formula is now the standard formula used in industrial color control [56, 57].

4.6.2 *CIE1994*

The complexity of the CMC formula has been criticized and after analyzing a large set of psychophysical data it was suggested that simple weighting functions, S_L, S_C, and S_H, would be sufficient to improve the perceived color difference. Hence, CIE proposed a new formula, which is known as CIE94 [58–61], and the color difference $\Delta E_{CIE1994}$ is calculated by:

$$\Delta E_{CIE94} = \sqrt{\left(\frac{\Delta L^*}{k_L \Delta S_L}\right)^2 + \left(\frac{\Delta C^*}{k_C \Delta S_C}\right)^2 + \left(\frac{\Delta H^*}{k_H \Delta S_H}\right)^2} \qquad (4.91)$$

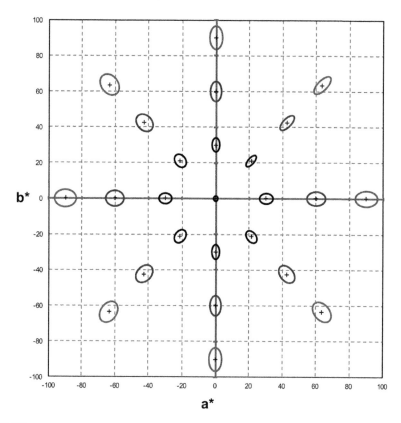

FIGURE 4.27 Cross-section through their centroids of CMC ellipsoids in CIELAB chromaticity plane ($\Delta E_{CMC} = 2$).

where $S_L = 1$; $S_C = 1 + 0.045 \cdot C^*$; $S_H = 1 + 0.015 \cdot C^*$.

The variables k_L, k_H, and k_C are parametric factors to be adjusted according to different viewing conditions such as texture, background, and separation. In typical applications, the parameters k_L, k_H, and k_C are set to unity. Presently, $\Delta E_{CIE1994}$ is abandoned [4].

4.6.3 DIN99

The DIN99 color space, which was developed in 1999 and has been adopted as the German standard, is other representative of UCS based on nonlinear transformation of the CIELAB color space [62]. It originated from an idea by Rohner and Rich [63], who presented it as the DCI-95 formula. In general, the color differences in the DIN 99 color space will be smaller, e.g., for higher chroma color pairs and therefore correlate better to the visual impression.

Temporary variable for redness: $\quad e = a^* \cdot \cos(16°) + b^* \cdot \sin(16°)$

Temporary variable for yellowness: $\quad f = 0.7 \cdot [a^* \cdot \sin(16°) + b^* \cdot \cos(16°)]$

$$
h_{ef} = \begin{cases}
\tan^{-1}(f/e) & \text{for} \quad e > 0 \quad \text{and} \quad f \geq 0 \\
\pi/2 & \text{for} \quad e = 0 \quad \text{and} \quad f > 0 \\
\pi + \tan^{-1}(f/e) & \text{for} \quad e < 0 \\
3\pi/2 & \text{for} \quad e = 0 \quad \text{and} \quad f < 0 \\
2\pi + \tan^{-1}(f/e) & \text{for} \quad e > 0 \quad \text{and} \quad f \leq 0 \\
0 & \text{for} \quad e = 0 \quad \text{and} \quad f = 0
\end{cases}
$$

Then, co-ordinates of DIN99 color space are computed:

$L_{99} = (1/k_E) \cdot 105.51 \cdot \text{In}(1 + 0.0158 \cdot L^*)$,

$a_{99} = C_{99} \cdot \cos(h_{ef})$,

$b_{99} = C_{99} \cdot \sin(h_{ef})$

where $C_{99} = \text{In}(1 + 0.045 \cdot G)/(0.045 \cdot k_{CH} \cdot k_E)$

$$h_{99} = h_{ef} \frac{180}{\pi}$$

The size of the perceived color differences can be influenced by external factors. The k factors can be used for this purpose (generally, it is not recommended to use factors different from 1) [59].

$$\Delta E_{00} = \sqrt{\left(\frac{\Delta L'}{K_L \cdot S_L}\right) +} \tag{4.92}$$

4.6.4 CIE2000

Additional attempts have been made to adjust the color difference equation to further improve the uniformity, and in 2001 a new color difference formula, CIEDE2000 (ΔE_{00}), was proposed and adopted by the CIE [64]. CIEDE2000 is the latest CIE recommended color-difference formula based on CIELAB [65]. This color difference formula is based on the modeling of five aspects: chroma dependency, hue dependency, and a rotation term in the blue area, lightness transformation and neutral gray modeling. Among the new parameters, a term for improving the performance in the blue colors and a rescaling factor of the CIELAB $a*$ axis improving the performance for colors close to the $L*$ axis, can be mentioned. It is important to notice that even though small noticeable improvements compared to $\Delta E_{CIE1994}$ could be made in some parts of the CIELAB color space the complexity of the ΔE_{00} formula may reduce the performance of the color difference prediction in other parts of the color space.

$$\Delta E_{00} = \sqrt{\left(\frac{\Delta L'}{K_L \cdot S_L}\right)^2 + \left(\frac{\Delta C'}{K_C \cdot S_C}\right)^2 + \left(\frac{\Delta H'}{K_H \cdot S_H}\right)^2 + R_T \left(\frac{\Delta C'}{K_C \cdot S_C}\right) \cdot \left(\frac{\Delta H'}{K_H \cdot S_H}\right)} \quad (4.93)$$

$$\bar{L}' = \left(L_1^* + L_2^*\right)/2$$
$$\bar{C} = \left(C_1^* + C_2^*\right)/2$$

$$G = \left(1 - \sqrt{\frac{\bar{C}^7}{\bar{C}^7 + 25^7}}\right)/2$$

$$a_1' = a_1^*\left(1 + G\right)$$
$$a_2' = a_2^*\left(1 + G\right)$$

$$C_1' = \sqrt{\left(a_1'\right)^2 + \left(b_1^*\right)^2}$$

$$C_2' = \sqrt{\left(a_2'\right)^2 + \left(b_2^*\right)^2}$$

$$\bar{C}' = \left(C_1' + C_2'\right)/2$$

$$h_1' = \begin{cases} \tan^{-1}(b_1/a_1') & \tan^{-1}(b_1/a_1') \geq 0 \\ \tan^{-1}(b_1/a_1') + 360° & \tan^{-1}(b_1/a_1') < 0 \end{cases}$$

$$h_2' = \begin{cases} \tan^{-1}(b_2/a_2') & \tan^{-1}(b_2/a_2') \geq 0 \\ \tan^{-1}(b_2/a_2')+360° & \tan^{-1}(b_2/a_2') < 0 \end{cases}$$

$$\bar{H}' = \begin{cases} (h_1'+h_2'+360°)/2 & |h_1'-h_2'| > 180° \\ (h_1'+h_2')/2 & |h_1'-h_2'| \leq 180° \end{cases}$$

$$T = 1 - 0.17\cos\cos\left(\bar{H}'-30°\right)+0.24\cos\cos\left(2\bar{H}'\right)$$
$$+0.32\cos\cos\left(3\bar{H}'+6°\right)-0.20\cos\cos\left(4\bar{H}'-63°\right)$$

$$\Delta h' = \begin{cases} h_2'-h_1' & |h_1'-h_2'| \leq 180° \\ h_2'-h_1'+360° & |h_1'-h_2'| > 180°; h_2' \leq h_1' \\ h_2'-h_1'-360° & |h_1'-h_2'| > 180°; h_2' > h_1' \end{cases}$$

$$\Delta L' = L_2^* - L_1^*$$
$$\Delta C' = C_2' - C_1'$$
$$\Delta H' = 2\sqrt{C_1'C_2'}\cdot\sin\left(\Delta h'\!/_2\right)$$
$$S_L = 1 + \frac{0.015\left(\bar{L}'-50\right)^2}{\sqrt{20+\left(\bar{L}'-50\right)^2}}$$
$$S_C = 1 + 0.045\cdot\bar{C}'$$
$$S_H = 1 + 0.015\cdot\bar{C}'T$$
$$\Delta q = 30e^{\left\{-\left(\frac{\Delta H'-275°}{25}\right)^2\right\}}$$
$$R_C = \sqrt{\frac{\bar{C}'^7}{\bar{C}'^7 + 25^7}}$$
$$R_T = -2R_C\sin\left(2\Delta\theta\right)$$

The primes in the color difference terms $(\Delta L')$, $(\Delta C')$, and $(\Delta H')$ denote corrections for neutral colors in lightness, chroma, and hue differences, respectively. The weighting factors, K_L, K_C, and K_H can be fitted to existing data sets, if they exist. For most applications, these weights are unknown and should be all set to 1.0. For the textile industry, the value of K_L may be considered as 2. Note however that this metric is only valid for 2° or 10° standard colorimetric observer (Figure 4.28).

It should be noted that any such color difference formula or color space (even where it is perfectly uniform) could only be strictly correct for one condition of viewing. Thus, as visual adaptation effects occur (in which the visual mechanism 'adjusts' as the level or color of the of illumination changes), or simultaneous contrast effects are introduced (in which the sample is seen in context with other colors surrounding it) the uniformity of the space must deteriorate. Visual adaptation may reduce the significance of

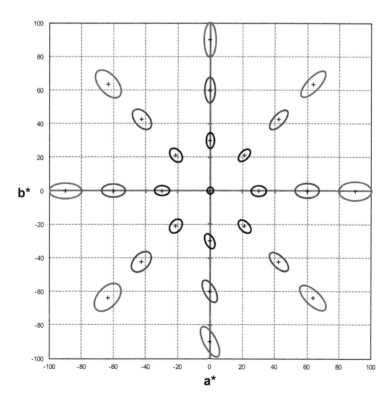

FIGURE 4.28 Cross-section through their centroids of CIE2000 ellipsoids in CIELAB chromaticity plane ($\Delta E_{CMC} = 2$).

such differences, but they cannot be totally ignored when comparing colors on different media or when viewing conditions change radically (both of which are likely to happen with the reproduction of complex color images and hence when using color management). However, for many practical purposes, such as quality control, these problems are not relevant [66]; the changes due to adaptation in controlled viewing conditions are small compared to the non-uniformity of the spaces themselves.

Nevertheless, it should be noted that the uniform color spaces are normally assumed to be "uniform" for illuminant D65 with an illuminance level of 1000 lux. It is important to ensure that the "uniformity" of this space also applies to any differences specified for graphic arts viewing conditions, using D50 with an illuminance level of 2000 lux. In future, it seems likely that color differences will be based on color appearance measurements (rather than simple measures of the stimulus), but there is much work to be done to get to that point. In the meantime, data are required to establish whether the existing formulae are satisfactory for our needs, and the extent to which the weighting of the various parametric constants in the models will improve them for graphic arts applications. That same data can also be used as a means of evaluating any future models, possibly including those based on appearance modeling if the experiments are sufficiently comprehensive.

4.7   CIE 2015 CONE-FUNDAMENTAL-BASED CIE COLORIMETRY

Since the establishment of the 1931 standard colorimetric observer, considerable knowledge on color vision mechanisms has been acquired. Establishing colorimetry directly on physiology is a new CIE approach to color specification [67]. The most fundamental aspect of applied colorimetry is the trichromacy of our visual system, which allows us to represent any color in terms of its tristimulus values. Computing tristimulus values for any object color requires the use of the spectral reflectance of the object color, the spectral power distribution of the scene illuminant, and the spectral characteristics of a colorimetric observer.

In 1931, the CIE recommended a first model of a standard observer with a $2°$ viewing angle. Color matching functions (CMF) related to this model were based on Wright and Guild measurements. Section 4.2 described transformation of $\bar{r}(\lambda),\bar{g}(\lambda),\bar{b}(\lambda)$ into $\bar{x}(\lambda),\bar{y}(\lambda),\bar{z}(\lambda)$ to prevent negative values in the CMFs and to make the $\bar{y}(\lambda)$ the same as CIE 1924 spectral luminous function $V(\lambda)$. Since 1924 and until recently, the photopic luminous efficiency function, $V(\lambda)$, has been the de facto sole spectral weighting function for quantifying light or, more formally, for quantifying luminous intensity, in candelas. $V(\lambda)$, which has a peak efficiency at approximately 555nm, does not, however, accurately represent what it was intended to do in 1924, namely to characterize the "human eye's visual sensitivity" to electromagnetic radiation. Based upon a subsequent, large body of neuroscience research, we now know that $V(\lambda)$ only approximates the spectral sensitivity of just two of the five photoreceptor types in the retina and, thus, does not fully characterize the spectral range of human visual (and non-visual) sensitivity to electromagnetic radiation. For this reason, it will be necessary to discuss universal luminous efficiency function "$U(\lambda)$" based on upon the spectral sensitivity of all five photoreceptors in the human retina, which is proposed here to be used for the quantification of light [68].

Each CMF defines the amount of that primary required to match monochromatic targets of equal energy. To measure CMFs, one needs to perform color matching many times. An observer makes a match using three monochromatic primaries (usually red, green, and blue light) against a reference spectrum, a monochromatic light at a certain wavelength. Using the same matching primaries, this process is repeated with a reference monochromatic light from short- wavelength region to long-wavelength region. In the end, $\bar{r}(\lambda),\bar{g}(\lambda),\bar{b}(\lambda)$ CMFs are obtained. This is called maximum saturation method [69].

CMFs can be determined without any knowledge of the underlying cone spectral sensitivities. The only restriction on the choice of primary lights is that they must be independent in the sense that no two will match the third. CMFs can be linearly transformed to any other set of real primary lights, and, as illustrated in Eq. (4.18).

In 2006, new physiologically-based color matching functions based on Stockman and Sharpe data were recommended by CIE [70]. Both researchers tried to find the real sensitivities of the human color receptors. To find the sensitivity functions, named cone fundamentals $\bar{l}(\lambda),\bar{m}(\lambda),\bar{s}(\lambda)$, for each cone type, was used measurements made in both with normal trichromats and color deficient observers [71]. To find $\bar{l}(\lambda)$ and $\bar{m}(\lambda)$ con fundamentals deuteranopes and protanopes were used as subject following König hypothesis [72]. With them the sensitivity toward a color attraction of a single cone type could be isolated and measured by overexciting the other type. To find $\bar{s}(\lambda)$

cone fundamentals measurement with normal trichromats under intense adaptation conditions and with monochromats who lack functioning M- and L-cones were combined with an analysis of the 10° color matching function data of Stiles and Burch made in 1958 [73].

The final cone fundamentals estimates resulted from a linear transformation of the 10° CMF guided mainly by the cone spectral sensitivity data. The latest publication "CIE 170-2:2015: Fundamental Chromaticity Diagram with Physiological Axes – Part 2: Spectral Luminous Efficiency Functions and Chromaticity Diagrams" provides a complete color space based on the new fundamental color matching functions from 2006. The comparison of resulting CIE x_F, y_F chromaticity diagram with CIE x, y chromaticity diagram is shown in Figure 4.29.

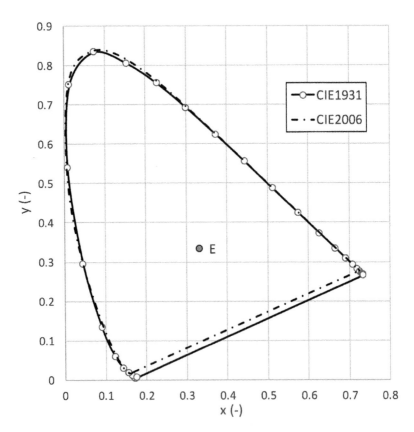

FIGURE 4.29 CIE 1931 chromaticity diagram and CIE 2006 chromaticity diagram based on cone fundamentals.

The CIE 2015 cone fundamental-based CIE Colorimetry probably brings improvement of CIE colorimetry based on CIE 1931 (2°) and CIE 1964 (10°) color matching functions. In near future, it will be necessary to answer two most important questions of CIE research strategy [74]:

- How accurate are cone-fundamental-based colorimetry results compared with those of 1931 and 1964 in predicting typical colorimetry applications such as color difference, color appearance, whiteness, color rendering, etc.?
- Can the cone-fundamental-based colorimetry be used to quantify the age metamerism effect and the size metamerism effect? There is an urgent need to quantify observer metamerism. Evidence suggests that the earlier CIE method underestimates these effects.

KEYWORDS

- **CIE CAM02**
- **CIE LAB**
- **CIE standard observer**
- **color difference formula**
- **colorimetry**
- **light source**

REFERENCES

1. McLaren, K., (1983). *The Color Science of Dyes and Pigments*, Adam Hilger Ltd., Bristol, pp. 10–198.
2. Wright, W. D., (1988). The 1931 and 1964 CIE systems of colorimetry: their significance to colorimetrists today, *Tex. Chem. Col.*, *20*(2), 19–22.
3. Hunt, R. W. G., & Pointer, R. M., (2011). *Measuring Color*, Wiley: Chichester (The Wiley-IS&T Series in Imaging Science and Technology), pp. 1–72.
4. CIE 15:2004 *Colorimetry*, 3rd edition, CIE Vienna, pp. 67.
5. Judd, D. B., & Wyszecki, G., (1963). *Color in Business, Science and Industry*, 2nd. ed., John Wiley & sons, New York.
6. DiLaura, D. L., Houser, K. W., Mistrick, R. G., & Steffy, G. R., (2011). *The Lighting Handbook,* Tenth Edn. Reference and Application, Illuminating Engineering Society of North America, New York, pp. 7-4–7-77.

7. Vik, M., (2017). *Colorimetry in Textile Industry*, VUTS, a.s. Liberec, pp. 26–33.
8. Judd, D. B., MacAdam, D. L., & Wyszecki, G., (1964). Spectral distribution of typical daylight as a function of correlated color temperature, *J. Opt. Soc. Am.*, *54*(8), 1031–1040.
9. ISO 23603:2005(E)/CIE S 012/E:2004 Standard method of assessing the spectral quality of daylight simulators for visual appraisal and measurement of color.
10. CIE 13-3-1995. *Method of Measuring and Specifying Color Rendering Properties of Light Sources*; CIE: Vienna, pp. 1–20.
11. Hunt, R. W. G., (1971). Color Measurement, *Rev. of Prog. in Coloration*, *2*, 11–19.
12. ASTM standard E 308–15, Standard practice for computing the colors of objects by using the CIE System, 2015.
13. Vik, M., (1995). *Fundamentals of Color Measurement* (in Czech), vol. 1, TUL: Liberec, pp. 15–34.
14. Vik, M., & Viková, M., (2000). Color-appearance phenomena – Metamerism, *Vlakna a Textil*, *7*(2), 126–127.
15. Judd, D. B., (1935). A Maxwell triangle yielding uniform chromaticity scales, *J. Opt. Soc. Am.*, *25*(1), 24–35.
16. Ohta, N., & Robertson, A. R., (2005). *Colorimetry, Fundamentals and Applications*, John Wiley & Sons: Chichester, pp. 57–61.
17. Klein, G. A., (2010). *Industrial Color Physics*, Springer New York, pp. 116–120.
18. Hunter, R. S., & Harold, R. W., (1987). *The Measurement of Appearance*, John Wiley & Sons, New York, pp. 101–106.
19. Schroeder, G., (1981). *Technical Optics (in Czech)*, SNTL Prague, pp. 142.
20. Nickerson, D., (1969). History of the munsell color system, *Color Engineering*, *7*(5), 42–51.
21. Hoffmann, Z., Krejci, A., & Wagner, J., (1985). *Psychosensorial Brightness Scale for Television* (in Czech), Academia Prague, pp. 5–49.
22. Agoston, G. A., (1982). *Color Theory and its Application in Art and Design*, (in Russian) MIR Moskau, pp. 120–129.
23. Nickerson, D., Tomaszewski, J. J., & Boyd, T. F., (1953). Colorimetric specifications of Munsell repaints, *J. Opt. Soc. Am.*, *43*(3), 163–171.
24. Schultze, W., (1966). *Education in Color and Colorimetry (in German)*, Springer Verlag, Berlin, pp. 30–32.
25. Kelly, K. L., Gibson, K. S., & Nickerson, D., (1943). Tristimulus specification of the munsell book of color from spectrophotometric measurements, *J. Opt. Soc. Am.*, *33*(7), 355–376.
26. Granville, W. C., Nickerson, D., & Foss, C. E., (1943). Trichromatic specifications for intermediate and special colors of the munsell system, *J. Opt. Soc. Am.*, *33*(7), 376–385.
27. Newhall, S. M., Nickerson, D., & Judd, D. B., (1943). Final report of the O. S. A. Subcommittee on spacing of the munsell colors, *J. Opt. Soc. Am.*, *33*(7), 385–418.
28. McLaren, K., (1981). Lab space: the key to successful instrumental shade passing, *Coloration Technology (J. Soc. D. Col.)*, *97*(12), 498–503.
29. Richter, M., (1976). *Introduction in Colorimetry (in German);* De Gruyter, W., Berlin, pp. 123–148.
30. Nickerson, D., & Stultz, K. F., (1944). Color tolerance specification, *J. Opt. Soc. Am.*, *34*(9), 550–570.
31. Nickerson, D., (1950). Tables for use in computing small color differences. *Am. Dyest. Rep.*, *39*(8), 139–173.

32. Gall, L., (1973). *Farbe und Lack*, *79*, 279–293.
33. Billmeyer, F. W., (1963). Table of adams chromatic value color coordinates, *J. Opt. Soc. Am.*, *53*(11), 1317–1317.
34. McLaren, K., (1970). The Adams-Nickerson color-difference formula, *Coloration Technology (J. Soc. D. Col.)*, *86*(8), 354–366.
35. Christie, J. S., & Hunter, R. S., (1982). *Scale* – Expansion coefficients for L, a, b opponent – colors scales, HunterLab. Application Note, HunterLab, Reston, pp. 1–3.
36. Glasser, L. G., McKinney, A. H., Reilly, C. D., & Schnelle, P. D., (1958). Cube-root color coordinate system, *J. Opt. Soc. Am.*, *48*(8), 736–740.
37. McLaren, K., (1972). Color-Difference measurement, *Rev. of Prog. in Coloration*, *3*, 3–9.
38. McLaren, K., & Rigg, B., (1976). XII-the SDC recommended color-difference formula: change to CIELAB, *Coloration Technology (J. Soc. D. Col.)*, *92*(9), 337–338.
39. McLaren, K., (1976). XIII—the development of the CIE 1976 (L*a*b*) Uniform color space and color-difference formula, *Coloration Technology (J. Soc. D. Col.)*, *92*(9), 338–341.
40. Huntsman, J. R., (1989). A fallacy in the definition of ΔH^*, *Color Res. Appl.*, *14*(1), 41–43.
41. Seve, R., (1991). New formula for the computation of CIE 1976 Hue difference, *Color Res. Appl.*, *16*(3), 217–218.
42. Stokes, M., & Brill, M. H., (1992). Efficient computation of ΔH^*ab, *Color Res. Appl.*, *17*(6), 410–411.
43. Schanda, J., (2007). *Colorimetry, Understanding the CIE System*, John Wiley & Sons: Hoboken, pp. 58–61.
44. Fairchild, M. D., (2013). *Color Appearance Models*, 3rd Edition, John Wiley & Sons Ltd, Chichester, pp. 115–301.
45. Hunt, R. W. G., (2004). *The Reproduction of Color*, 6th Edition; John Wiley & Sons Inc.: Hoboken, pp. 139–179.
46. Luo, R. M., & Hunt, R. W. G., (1998). The structure of the CIE 1997 color appearance model (CIECAM97s). *Color Res. Appl.*, *23*, 138–146.
47. Fairchild, M. D., (2002). Modeling color appearance, spatial vision, and image quality, in: *Color Image Science: Exploiting Digital Media,* MacDonald, L., & Luo, M. R., (ed.), Wiley: New York, 357–370.
48. Sharma, G., (2003). *Digital Color Imaging, Handbook*, CRC Press: Boca Raton.
49. Moroney, N., Fairchild, M. D., Hunt, R. W. G., Li, C. J., Luo, M. R., & Newman, T., (2002). The CIECAM02 color appearance model. In: *The 10th Color Imaging Conference, IS&T and SID*, Scottsdale, Arizona, 23–27.
50. Smet, K. A. G., Zhai, Q., Luo, M. R., & Hanselaer, P., (2017). Study of chromatic adaptation using memory color matches, Part I: neutral illuminants, *Optics Express*, *25*(7), 7732–7748.
51. Mahy, M., Van Eycken, L., & Oosterlinck, A., (1994). Evaluation of uniform color spaces developed after the adoption of CIELAB and CIELUV, *Color Res. Appl.*, *19*(2), 105–121.
52. Guthrie, J. G., & Oliver, P. H., (1957). Applications of color physics to textiles, *Coloration Technology (J. Soc. D. Col.)*, *73*(12), 533–542.
53. Kuehni, R. G., (1982). Advances in color-difference formulas, *Color Res. Appl.*, *7*(1), 19–23.

54. Marshall, W. J., & Tough, D., (1968). Color measurement and color tolerance in relation to automation and instrumentation in textile dyeing, *Coloration Technology (J. Soc. D. Col.)*, *84*(2), 108–119.

55. Clarke, F. J. J., McDonald, R., & Rigg, B., (1984). Modification to the JPC79 color–difference formula, *Coloration Technology (J. Soc. D. Col.)*, *100*(4), 128–132.

56. McDonald, R., (1990). European practices and philosophy in industrial color-difference evaluation, *Color Res. Appl.*, *15*(5), 249–260.

57. Harold, R. W., (1995). *"FOC" Meeting CMC Color Difference Formula*, Sirmione, 15–16. 10, Italy.

58. Alman, D. H., Berns, R. S., Snyder, G. D., & Larsen, W. A., (1989). Performance testing of color-difference metrics using a color tolerance dataset, *Color Res. Appl.*, *14*(3), 139–151.

59. Vik, M., (1995). Industrial evaluation of color differences – new approach (in Czech), *Book of Proceedings 27.* National conference of textile chemists and colorists of Czech Republic, Pardubice 14.16.11, Czech Republic, pp. 34–42.

60. Berns, R. S., Alman, D. H., Reniff, L., Snyder, G. D., & Balonon-Rosen, M. R., (1991). Visual determination of suprathreshold color-difference tolerances using probit analysis. *Color Res. Appl.*, *16*, 297–316.

61. Witt, K., (1995). *Farbe + Lack, 101*(11), 937–939.

62. Witt, K., (2005). Chroma effect in color space (in German), Farbe + Lack, *111*(11), 937–939.

63. Rohner, E., & Rich, D. C., (1995). An approximately uniform object color metric for industrial color tolerances, *AIC Interim Meeting on Colorimetry*, Berlin, Germany, 3–6.

64. Luo, M. R., Cui, G., & Rigg, B., (2001). The development of the CIE 2000 colour-difference formula: CIEDE2000, *Color Res. Appl.*, *26*(5), 340–350.

65. Berns, R. S., (2001). Derivation of a hue-angle dependent, hue-difference weighting function for CIEDE2000, *AIC Color 01*, Rochester 24–29. 06., USA.

66. Judd, D. B., (1939). Specification of uniform color tolerances for textiles, *Textile Res. J., 9*, 253–263.

67. Vienot, F., (2016). Cone fundamentals: A model for the future of colorimetry, *Lighting Res. Technol.*, *48*, 13–48.

68. Rea, M. S., (2015). The lumen seen in a new light: Making distinctions between light, lighting and neuroscience, *Lighting Res. Technol.*, *47*, 259–280.

69. Wyszecki, G., & Stiles, W. S., (1982). *Color Science: Concepts and Methods, Quantitative Data and Formulae,* 2nd ed.; Wiley: New York, pp. 293.

70. CIE 170–1:2006: *Fundamental Chromaticity Diagram with Physiological Axes* – Part 1.

71. Stockman, A., & Sharpe, L., (2006). Physiologically-based color matching functions, ISCC/CIE expert symposium '06 *"75 Years of the CIE Standard Colorimetric Observer"*, Ottawa, Canada, 16–17.

72. Stockman, A., & Sharpe, L. T., (2001). Cone spectral sensitivities and color matching, in: Gegenfurtner, K., & Sharpe, L. T., (eds.), *Color Vision: from Genes to Perception,* Cambridge, Cambridge University Press, UK, pp. 53–87.

73. CIE 170–2:2015: Fundamental chromaticity diagram with physiological axes – Part 2: spectral luminous efficiency functions and chromaticity diagrams.

74. 873_CIE Research Strategy (August 2016) – Topic 6, CIE, Vienna, pp. 1–20.

CHAPTER 5

INSTRUMENTATION

MICHAL VIK

CONTENTS

Abstract 221
5.1   Specular and Diffuse Reflection 221
5.2   Geometric Conditions for Colorimetry 225
5.3   Measuring Devices in Colorimetry 228
5.4   Measurement 249
Keywords 280
References 280

ABSTRACT

Light interacts with matter in many different ways. Smooth surfaces such as metals are shiny, but water is transparent. Stained glass, solutions of dyestuffs, filters transmit some colors, but absorb others. Some materials appear milky white because they scatter the incoming light in all directions, etc.

In this chapter, we will look at a whole host of measurement of these optical phenomena in a wide range of materials from the point of view of colorimetry, surface colorimetry mainly. Before we can begin to do this, we will first classify the different kinds of reflectance. We will then introduce the materials that we will study and clarify methodology on how to measure it. [1, 2].

5.1   SPECULAR AND DIFFUSE REFLECTION

The reflection pattern of a surface is a direct consequence of surface topography. Perfectly smooth surfaces reflect most of the light in a well-defined direction giving rise to the phenomenon of specular reflection. Surfaces that are irregular or rough scatter light in all directions, thereby producing a diffuse reflectance pattern. The specular and diffuse patterns are illustrated in Figure 5.1.

The specular reflection occurring at a smooth surface is formulated in terms of Fresnel's reflection law. By contrast, no general theory is completely valid for perfectly diffuse reflection from surfaces. However, phenomenologically formulated Lambert's cosine law tries to accommodate the fact that a diffusely reflecting surface appears to be uniformly bright in all directions. According to this law, the reflected intensity from such surface is given by Eq. (5.1):

$$I_r = \left(\frac{I_i}{\pi}\right)\cos\theta_r \cos\theta_i \qquad\qquad (5.1)$$

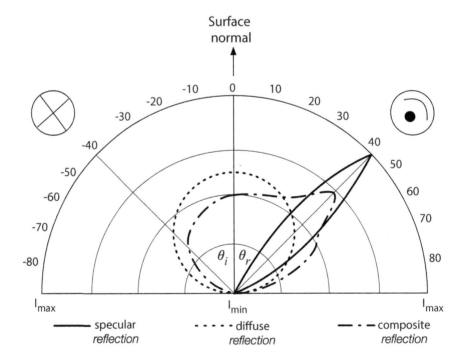

FIGURE 5.1 Polar reflectance curves.

where I_i is the intensity at perpendicular incidence.

Thus, in perfectly diffuse reflection, the reflected intensity is distributed symmetrically with respect to surface normal, without regard to the angle of incidence. Unfortunately, we do not have any ideal perfectly diffusing surface. Most surfaces encountered in practice are neither perfectly specular nor perfectly diffuse. The reflected intensity from such surfaces is a combination of both specular and diffuse reflection, i.e., the total reflected intensity I_r can be written as:

$$I_r = I_s + I_d \tag{5.2}$$

where I_s is the specular part, and I_d is the diffuse part of reflected intensity. A polar reflectance curve for such a surface is illustrated in Figure 5.1.

Depending on how light acts, products can be classified as opaque, translucent, or transparent. It is important to understand that the entire color impression is certainly composed of at least two components: the light interactions in the volume and at the boundary surface of a colored sample (see Figure 5.2). In other words, the reflection or transmission from the volume is superimposed on the boundary surface reflection. This reflection is caused by different refractive indices at the boundary surface. Such kinds of boundary surfaces are present in paints, coatings, plastic materials, emulsion paints, or

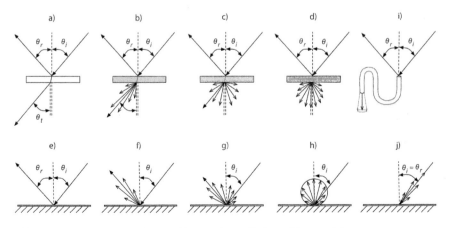

FIGURE 5.2 Simplified scheme of interaction radiation with materials.

ceramics. This distinction cannot be made for undefined surfaces such as of textiles, uncoated papers, plasters, or suede leather.

Transparent materials (Figure 5.2a): In the field of optics, transparency is the physical property of allowing light to pass through the material without being scattered. On a macroscopic scale (one where the dimensions investigated are much, much larger than the wavelength of the photons in question), the photons can be said to follow Snell's Law. Examples of transparent materials are glass, transparent solutions of dyestuffs, etc.

Semi-transparent materials (Figure 5.2b) and **Semi-translucent materials** (Figure 5.2c) are characterized by the mixed transfer of light – regular and diffuse. This category includes materials such as plastic foils with milk shade, colloidal solutions with low concentration, etc.

Translucent materials (Figure 5.2d): Translucent materials allow light to pass through them, but they diffuse the light in a way that makes objects on the opposite side appear blurred. Examples of translucent materials are frosted glass, oilpaper, some plastics, ice, and tissue paper.

Glossy surfaces, Figure 5.2e): At smooth interfaces between two materials with different dielectric properties, specular reflection occurs. The direction of incident ray, reflected ray, and the surface normal vector span the plane of incidence perpendicular to the surface of reflection. The angles of incidence and reflection are equal:

$$\theta_r = \theta_i \tag{5.3}$$

Semi-gloss, Figure 5.2f) and **Semi-matte surfaces**, Figure 5.2g): Very few materials have pure specular surface reflectivity. Most surfaces show a mixture of matte and specular reflection. As soon as surface micro roughness has the same scale as the wavelength of radiation, diffraction at the microstructures occurs. At larger scales, micro facets with randomly distributed slopes relative to the surface normal are reflecting incident light in various directions. Depending on the size and slope distribution of the micro roughness, these surfaces have a great variety of reflectivity distributions ranging from isotropic (Lambertian) to strong forward reflection, where the main direction is still the angle of specular reflection. A typical example of such materials is alumina, satin, etc.

Matte surfaces, Figure 5.2h): cause a strong diffuse reflection of light and weak specular reflection of light, resulting in a less saturated, duller color. A surface of perfectly matte properties is a **Lambertian surface**, which means that it adheres to *Lambert's cosine law* (5.1). As a consequence of this, the radiance of a Lambertian surface does not depend on the viewing angle. This can be understood by considering that while the radiant intensity decreases with cos θ, the area seen by the observer, within the same solid angle as before, increases with 1/ cos θ. An example of matte surface is barium sulfate pellet, chalk, etc.

Optical waveguide, Figure 5.2i) is a physical structure that guides electromagnetic waves in the optical spectrum. Common types of optical waveguides include optical fiber and rectangular waveguides. Optical fibers consist of a light-carrying core and a cladding surrounding the core. There are generally three types of construction: glass core/cladding, glass core with plastic cladding, or all-plastic fiber. Optical fibers typically have an additional outside coating, which surrounds and protects the fiber.

Retro reflective surfaces are typically used on warning clothing, traffic signs, etc. Retro reflection is defined as reflection in which radiation is returned in directions close to the direction of the incident radiation as visible on Figure 5.2j. Such materials are based on corner reflectors or glass spheres.

When one observes, for example, a colored paper or other object that has some gloss, the color of the object changes, depending on the viewing angle. This indicates that the reflection from a glossy object depends greatly on the angles at which the light is incident and viewed. Although usually less dramatic, the transmission of an object can also vary with incident and viewed angles. The degree of correlation between measured values and visual appraisals depends on the degree to which the geometric conditions of measurement simulate the geometric conditions of viewing. This is very important to describe fully condition used during measurement especially in case, when data are transferred to customer and opposite.

5.2 GEOMETRIC CONDITIONS FOR COLORIMETRY

There are many different types of devices. Colorimetric specifications are derived from spectral or tristimulus measurements. The measured values depend on the geometric relationships between the measuring instrument and the sample. Reflection and transmission are purely physical phenomena, which mean the related measurable quantities describe properties of how materials interact with radiation, as was mentioned before. The reflection or transmission spectrum gives the fraction of the incident light that an object reflects or transmits as a function of wavelength. Of the many factors that influence them, the ones of interest here are those to radiation: the direction of incidence, the direction of observation, and the aperture angles of the incident and observed beams [3]. Under uniform irradiance, the geometrical reflectance properties of a uniform, isotropic, flat surface are characterized in terms of the bidirectional reflectance distribution function (BRDF). The BRDF is defined by Eq. (5.4):

$$f_r\left(\theta_i,\phi_i,\theta_r,\phi_r\right)=\frac{dL_r\left(\theta_i,\phi_i,\theta_r,\phi_r,E_i\right)}{dE_i\left(\theta_i,\phi_i\right)}=\frac{dL_r\left(\theta_i,\phi_i,\theta_r,\phi_r,E_i\right)}{L_i\left(\theta_i,\phi_i\right)\cdot\cos\theta\cdot d\Omega_i} \qquad (5.4)$$

where dL_r is the differential radiance of the surface in direction (θ_r, Φ_r) when irradiated by irradiance dE_i from direction (θ_i, Φ_i), and Li is the total irradiating radiance over a differential incident solid angle $d\Omega_i$. The directions and solid angles are illustrated in Figure 5.3.

Reflectance standards are usually made to resemble Lambertian radiators as closely as possible. These kinds of standards are commonly characterized

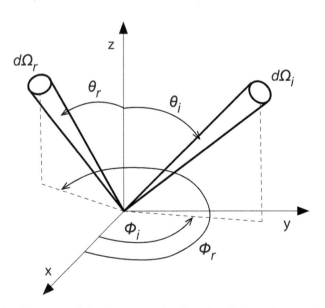

FIGURE 5.3 Geometry of the incident and reflected radiation. The sample surface is parallel to the x-y-plane.

by their reflectance factor. We can speak about three configurations of the observed radiation are consistent with the good technical realization:

- Directional (the beam consists of quasi-parallel rays)
- Conical (the beam has the shape of a narrow cone)
- Hemispherical (the beam is uniformly distributed over the half-space)

Based on that we can speak about:

- Radiance factor β (directional observation)
- Reflection factor R (conical observation)
- Reflectance factor ρ (hemispherical observation)

If the illumination condition is included in the classification, nine geometries must be listed because there are three possibilities for illumination and three for viewing (see Figure 5.4).

It should be noted that the nine standard reflectance terms defined by Nicodemus et al. [4] "are applicable only to situations with uniform and isotropic radiation throughout the incident beam of radiation." They then state that "If this is not true, then one must refer to the more general expressions." This

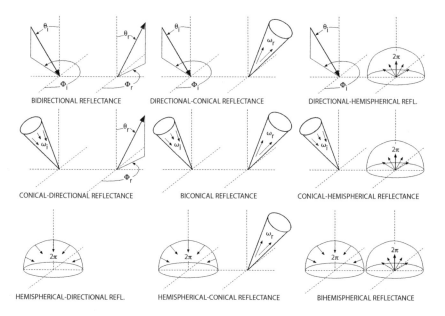

FIGURE 5.4 Relations of incoming and reflected radiance terminology used to describe reflectance quantities.

implies that any significant change to the nine reflectance concepts when the incident radiance is anisotropic lies in the mathematical expression used in their definition. It should be noted that the nine standard reflectance terms defined in [4] "are applicable only to situations with uniform and isotropic radiation throughout the incident beam of radiation." They then state that "If this is not true, then one must refer to the more general expressions." This implies that any significant change to the nine reflectance concepts when the incident radiance is anisotropic lies in the mathematical expression used in their definition [5].

To ease comparisons when reflection or transmission is measured, the CIE defined some standard terms and geometrical conditions [1]. The geometrical conditions denoted by symbols, such as d:8° or 45°:0°. The symbol "d" indicates the use of an integrating sphere, and the numbers before and after the colon indicates the angles of incidence and viewing, respectively. For instruments using an integrating sphere a glossy sample can be measured in two conditions either with the specular component included or excluded. Hence, the geometry is distinguished by adding the letter "e" for excluded or "i" for included. The current CIE recommendations for reflection measurements with diffuse geometries are shown in Figure 5.5.

In some national standards it is possible to find supportive recommendations for geometrical arrangement relating to transmission measurement τ [6–8].

a. Diffuse: eight-degree geometry, specular component included (di:8°)
b. Diffuse: eight-degree geometry, specular component excluded (de:8°)
c. Eight degree: diffuse geometry, specular component included (8°:di)
d. Eight degree: diffuse geometry, specular component included (8°:de)
e. Diffuse: diffuse geometry (d:d)
f. Alternative diffuse geometry (d:0°)
g. Forty-five degree annular geometry (45°a:0°)
h. Zero degree directional geometry and annular reading (0°:45°a)
i. Forty-five degree directional geometry (45°x:0°)
j. Zero degree directional geometry and directional reading (0°:45°x)

5.3 MEASURING DEVICES IN COLORIMETRY

The human eye is the oldest means of measuring color, but it has certain drawbacks related to h the analytic specification of color. These include individuality of spectral response and poor color memory. Instrumental methods

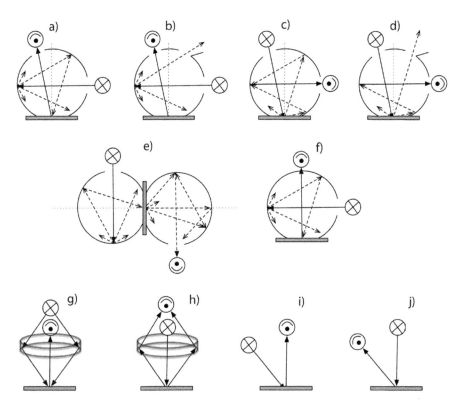

FIGURE 5.5 Measuring conditions for CIE colorimetry.

of measuring color are therefore required, and such instruments must evalu-ate the color by illuminating the material with light of a standard spectral composition and have sensors with standard spectral responses, correspond-ing to those of a human observer with normal color vision. This underlines the importance of the CIE's definition of both illuminants and the "standard observer."

While radiometry deals with electromagnetic radiation of all wave-lengths, photometry deals only with the visible portion of the electromag-netic spectrum. The human eye is sensitive to radiation between 380 and 780 nm and only radiation within this visible portion of the spectrum is called light.

The accurate measurement of color is a difficult task. While cameras and image scanners record color information, they are until now not suitable for the most accurate color measurements. For the accurate measurement of color, the instrument must have carefully controlled sensors (filters and detectors) and optical geometry. If the color of reflective material is to be measured, the illumination must be precisely known and matched with the sensors. There are physical problems with the color media that further complicate the task. Properties such as specular reflectance and translucence should be considered. These properties will be discussed in the following sections.

If only the tristimulus values or some derivative of them, such as L*a*b* values, are desired, a colorimeter can be used. This device uses filters to match the vector space defined by the eye. Thus, a measurement is limited to one set of viewing conditions. The colorimeter need measure only the quantities that can be transformed into tristimulus values. The minimal system consists of three filters and three detectors or changeable filters and a single detector. These devices are much less expensive than the devices designed to measure the entire spectrum.

The most complete information is obtained by measuring the entire visible spectrum. A spectroradiometer measures radiant spectra; a spectrophotometer measures reflective spectra. Having the spectrum allows the user to compute tristimulus values under any viewing condition. To measure the spectrum requires that the spectrum of the light be spread physically. This can be done with a prism, following Newton's famous early experiments. However, to make accurate measurements, the exact spreading of the spectrum must be known and controlled. This is done with optical gratings, which can be made more precisely and are much less bulky than prisms. A system of lenses is used to focus the spectrum onto the detector. The high-quality optics required for this task greatly increases the cost of these color measurement instruments.

After the spectrum is spread, the intensity of the light at each wavelength must be measured. This can be achieved in several ways. The most common method currently is to use a linear CCD array. A movable slit can also be used to limit the wavelength band being measured. The exact spread of the spectrum on the array is determined by measuring a known source. Interpolation methods are used to generate the data at the specified wavelengths.

The range of wavelengths and their sampling interval varies among the instruments. Higher precision instruments, record data over a larger range and at finer sample spacing. The CIE recommends a range of 360–830 nm with a sample spacing of 1 nm for computation of color. However, there are few instruments that actually take data at this specification. High-end instruments sample at 1 nm over a typical range of about 360 nm to 780 nm. The most popular instruments sample at 10 nm over 400 nm to 700 nm.

5.3.1 COLORIMETER

The simplest device to measure colors is a tristimulus colorimeter whose spectral responsivity mimics the color matching functions of the CIE standard observer. It measures, as suggested by its name, color tristimulus and reports these as color values in CIEXYZ, and CIELAB. The key elements include a light source (for measuring materials) usually illuminant C or D65, a set of color filters and commonly a silicon photodiode as a photo-detector. The spectral separation is obtained by either using color filters placed in front of the photodiode or by using spectrally different light sources to illuminate the sample, often using 45°:0° geometry, i.e. circumferential 45° illumination and normal viewing. To report CIE colorimetry, the sensitivities of the color filters are intended to have a close match to a set of color matching function.

For filter-based colorimeters, four independent filters are usually used, two approximate the $y(\lambda)$ and $z(\lambda)$ color matching functions and two approximate the two humps of the $x(\lambda)$ color matching functions. Alternatively, three filters are used as shown in Figure 5.6. Then, the first hump of $x(\lambda)$ color matching function is measured as 20% of signal from blue filter, which is corresponding with $z(\lambda)$ color matching function.

In general, the tristimulus colorimeter is easy to use, rapid, and inexpensive, but does not provide any detailed spectral information. Nevertheless, colorimeters are often used to calibrate and characterize self-luminous sources,

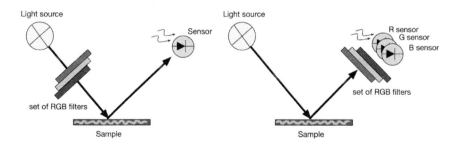

FIGURE 5.6 Basic configurations of a tristimulus colorimeter a) semi-monochromatic illumination, b) polychromatic illumination.

such as computer monitor displays, although there might be some performance restrictions in terms of accuracy due to the aging of the color filters and poor reproducibility to agree with the CIE color matching functions. It is also important to note that for surface colors, the colorimetric values measured are valid for one illuminant only, the instrument light source, but it is often desirable to know the tristimulus values or color difference for different illuminants. Thus, a colorimeter cannot give any indication of metamerism [9].

5.3.2 SPECTROPHOTOMETER

Presently, the most common color measurement instrument in the industry is the spectrophotometer, which measures the ratio of reflected to incident light (the reflectance) from a sample at many points across the visible spectrum. Why? Because for demanding color measurement, a spectral approach is definitely needed. Strictly it is impossible to define the absolute color values of a sample; hence, we always work with some kind of approximations, some of us closer than others. The main components of all spectrophotometers for color measurement include a light source, a wavelength selection device (monochromator), and a photo detector as shown in Figure 5.7. The reflected or transmitted light is passed on to the wavelength selection device or spectral analyzer, where the light is split into its spectral components to be measured. In a spectrophotometer, prisms, gratings, and interference filters are the technologies used to separate light into narrow bands. Both prisms and gratings separate the wavelength spatially.

The classical spectrometer consists of an input slit, a rotating dispersive element (prism or grating), an output slit and a single detector. This

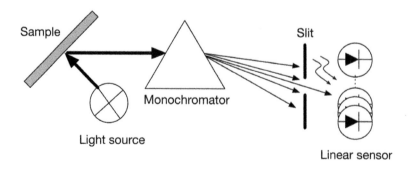

FIGURE 5.7   Basic features of a spectrophotometer.

arrangement allows the separation of polychromatic radiation into its spectral components. The main advantages are the high sensitivity and the low stray light. Several drawbacks, such as the non-parallel measurement, the moveable elements and the space consuming dimensions, were overcome by the development of array spectrometers. This kind of spectrometer uses a detector array instead of a single detector and therefore only needs fixed components.

Principles of Optical Wavelength Measurements
In this section, three common mechanisms used for the isolation of spectral information from optical signals are reviewed from [10–14]:

- **Interferometry:** The interferometric spectrometers modulate light in Fourier or modal space contrary to the spatial dispersion-based spectrometers that modulate light in the image plane. The Fourier transform spectrometers are particularly very useful when there is a need to measure a spectrum using one detector. Application area of this systems is mainly focused on FTIR spectroscopy.
- **Resonance:** Resonant effect can be created using optical cavities or by quantum mechanical material processes. The resonant effect is used to create spectroscopic filters such as the thin film filters, metal nanoparticles, organic dyes, etc. Using these devices, spectral analysis can be achieved using electronic detectors with intrinsically spectral sensitivity. Nevertheless, such kind of systems are out of obvious practice in colorimetry.
- **Spatial dispersion:** Employs the use of dispersive elements such as gratings and prisms (which redirects, refracts, and diffracts waves as a function of wavelengths) to spread light into a spectrum such that the spectral components can be measured.

Despite the difference in instrumentation, all spectroscopic techniques share several common features.

5.3.2.1 Bandpass and Spectral Resolution

Bandpass and spectral resolution both refer to the ability of the instrument to separate adjacent spectral lines. The bandpass is the spectral interval that may be isolated and depends on factors such as the entrance slit width, the width of a pixel on the detector and system aberrations.

Spectral sampling interval is the spacing between sample points in the measured spectrum. Sampling is independent of resolution and typically varies in dependence on used grating, slit and number of detector active (really used) pixels.

Spectral resolution is the smallest achievable spectral separation $d\lambda$ of two resolved spectral lines. This is a more subjective term, requiring a definition of when the lines are resolved. Figure 5.8a shows a spectrum of a monochromatic light source. If it were recorded by a "perfect" instrument, the output spectrum would look like:

In a), but a real instrument will give a line profile with a finite width as in b). This is called the instrumental line profile. The bandpass of the instrument is the measured Full Width at Half Maximum (FWHM) of this line profile and is wavelength dependent. If the entrance slit and pixel width are the only factors broadening the line, the instrumental profile is a convolution of the two rectangular slit functions $\Delta\lambda_1$ and $\Delta\lambda_2$, see Figure 5.9.

The width of $\Delta\lambda_1$ equals the width of the entrance slit image, and the width of $\Delta\lambda_2$ is the pixel width. In an optimized instrument, these two are equal and the line profile is triangular. The spectral bandpass is then the linear dispersion times the width of the entrance slit image, w_{ent}:

$$BP = FWHM \approx \frac{d\lambda}{dx} \cdot w_{ent} \qquad (5.5)$$

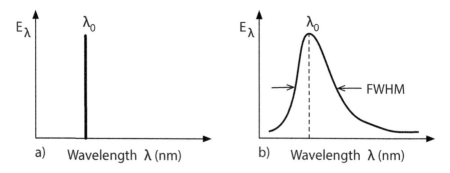

FIGURE 5.8 Schematic representation of spectral resolution: a) shows the spectrum of a monochromatic light source and b) illustrates the broadening of the spectrum when it is recorded by an instrument.

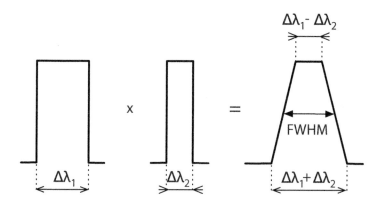

FIGURE 5.9   The instrumental profile is a convolution of the exit slit and the entrance slit image.

It is common to use at least three pixels per bandpass, also called pixels per bin, at the detector. Then, the maximum number of channels produced by the spectrometer depends on the bandpass of the instrument and not on the number of pixels along the wavelength axis of the CCD.

Most spectrophotometers used in colorimetry work in the range 400–700 nm with sampling intervals reporting reflectance at 5-, 10-, or 20-nm intervals. Typically, data are measured for 31 wavelength intervals centered at 400 nm, 410 nm, 420 nm, ..., 700 nm. If it has been checked that measurement data obtained only at 10 nm or 20 nm intervals satisfy the need of the observer, computation methods as described e.g. in ASTM E308 tables [15] might be used. This publication contains weighting factors for both the CIE 1931 standard colorimetric observer and the CIE 1964 standard colorimetric observer and a number of illuminants and practical light sources used in colorimetry. Table 5.5 of the ASTM publication has been developed for the case when the instrument manufacturer has built in a correction to zero bandwidth. Table 5.6 of the ASTM publication provides weighting factors for the case when a correction to a zero bandwidth is required.

Because we are measuring the ratio of incident to reflected light with a spectrophotometer, the type of light source illuminating the sample should not matter.

But simple reflectance measurements are insufficient when determining the color and appearance of luminescent samples. When light is incident on non-luminescent material part of the light is absorbed, part of it may be transmitted and the rest is reflected with the same wavelength. When illuminating a luminescent material, part of the absorbed light excites electrons in the sample from the ground state to higher energy states. The ground state and the other energy states are divided into vibrational states as depicted in Figure 2.3 by a Jablonski diagram (Chapter 2). Thus, the spectral power distribution (SPD) of the light source can have a dramatic effect on the measured color.

The reflectance of a fluorescent sample cannot be determined in the same way as the reflectance of a non-fluorescent material [14]. If a fluorescent sample is illuminated with monochromatic radiation and the reflectance is detected polychromatically, the reflectance values in the excitation range of the sample are increased because they include the emitted radiation occurring on longer wavelengths. In the opposite case of measurement fluorescent sample (polychromatic illumination and sensor array detection), the emitted light is detected as the reflected radiation with the same wavelength as the incident radiation and measured spectral reflectance is in some range of wavelengths frequently greater than the 1 unit (see Figure 5.10).

This creates erroneous data, and finally, the colorimetric parameters derived from these data do not agree with the visual perception of the color of the fluorescent sample. This would mean denial of the law of conservation

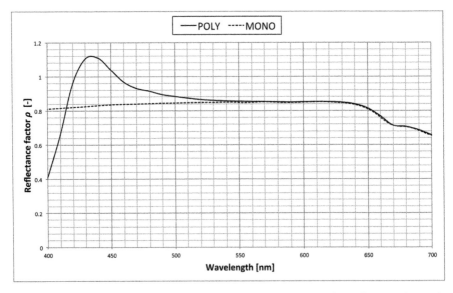

FIGURE 5.10 Reflectance curves measured by polychromatic and monochromatic arrangement of spectrophotometer.

of energy. Therefore, it is necessary to consider the fact that in the case of luminescent (fluorescent and phosphorescent) materials with total radiance factor β_T uneven reflectance ρ, respectively β_R, but is equal to the sum of the contributions of the spectral radiance factor withdrawn $\beta_R(\lambda)$ and the spectral luminescence factor $\beta_L(\lambda)$ according to Eq. (5.6).

$$\beta_T(\lambda) = \beta_R(\lambda) + \beta_L(\lambda) \tag{5.6}$$

Measurement of spectral luminescence factor $\beta_L(\lambda)$ is done using either of the two methods: two monochromators on device called spectrofluorimeters or by abridged fluorescence colorimetry.

5.3.2.2 Abridged Fluorescence Colorimetry

Figure 5.7 shows an optical system that illuminates the sample with the entire spectrum of light and the reflected light is passed through a monochromator. This system will properly measure both fluorescent and non-fluorescent samples, but usually suffers from lack of light energy. In order to obtain sufficient energy, it is necessary to use a high level of illumination in integrating sphere, which can result in heating of the sample or it's overexposure. The spectral content of the light source used in this *polychromatic light* instrument is very important when measuring fluorescent samples and simulator D65 is recommended for this purpose. Unfortunately, simulating daylight is not an easy task. The difficulty of simulating D65 is due to the discontinuities of the gradient in the spectral distribution of this illuminant, see Figure 4.3. The two major methods for separating the fluorescent and reflected components from a sample are the filter and two-mode methods. The methods proposed by Eitle and Ganz [16] and Allen [17] are categorized as the filter method.

The method of Eitle and Ganz (1968) was based on the measurement of the total radiance factor using a non-filtered source and a series of sharp short-wavelength cutoff filters. The wavelengths of the incident light that excite fluorescence would be excluded using the short-wavelength cutoff filters. The problem regarding the overlap region is still inherent in this method. These filters reduce the amount of fluorescence at this region, but the excitation is not completely eliminated. Hence, the estimated reflected radiance factor, $\beta_R(\lambda)$, would be higher in this region [14]. The other method for separating the spectral radiance factor of a fluorescent sample into reflected and fluorescent components was developed by Allen [17]. This method is referred to as the calculation method. The basic idea of this method is similar to the Eitle and Ganz method. Instead of using a series of sharp short wavelength cutoff filters, Allen proposed to use just two filters, one a fluorescence-killing filter and the other a fluorescence-weakening filter. Whole calculation of reflected and fluorescent component is described in Ref. [14].

5.3.2.3 Bispectral Measurement

Bispectral spectrophotometric instruments can make colorimetric measurements by considering the contribution of both the fluorescent and the reflected component to the total radiance of a sample. For the bispectral method, one monochromator is located between the instrument light source and the sample to be measured. The function of the monochromator is to separate the radiation from the instrument's light source into its spectral components before it reaches the sample. The second monochromator is located between the sample and the photodetector, which separates the radiation leaving the sample surface into its spectral components, see Figure 5.11. The $\beta_L(\lambda)$ is determined accurately by a monochromatic light hitting the sample form the first monochromator while the second monochromator transmits light at the same wavelength to the detector; this way, any fluorescence generated will be eliminated at any given wavelength. This is a simple method of getting the $\beta_L(\lambda)$ without measuring the complete spectrum of light emitted by the sample at each wavelength of irradiation.

5.3.2.4 Reference Standards

In most applications that utilize reflectance measurements, the reflectance of a sample is measured relative to a known reflectance standard. In CIE terminology, the "perfect reflecting diffuser" is the reference standard for reflectance factor. For this reason, it is important to have reliable calibration facilities capable of measuring these standards in absolute means. The absolute value of reflectance is usually very difficult to measure. During the 12th century, several methods were developed to perform this measurement,

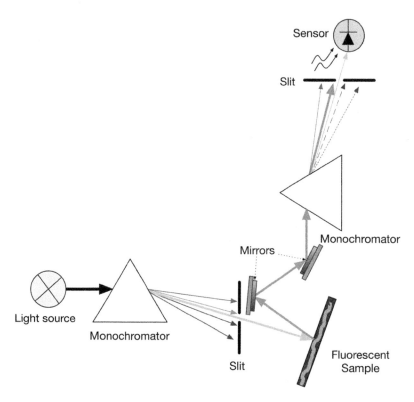

FIGURE 5.11 A simplistic design of the components of a spectrofluorimeter.

and number of materials were tested on their suitability as reflectance standards. The main limitation on measurement accuracy is the fact that there is no ideal perfectly diffusing comparison standard. The CIE publication 46 [18] summarized properties of materials that can be used as secondary standards of reflectance. Until the 1959 CIE Session in Bruxelles [19], the colorimetric measurement was referred not to the perfect reflecting diffuser, but to smoked MgO, and its reflectance factor was taken to be 1 for all wavelengths. Data of several secondary standards are still referred to this value, for example, the colorimetric values of samples of Munsell Book of Colors [20] or so-called Judd polynomial function, see Eq. (4.22).

The CIE report summarized properties and reflection values of:

- Smoked MgO, pressed powder of MgO, pressed powder of $BaSO_4$. These samples resemble more or less a Lambertian surface and thus

can be used to transfer radiance and reflectance factor data between different instruments. Their drawback is that they usually have a very fragile surface, limited stability and sensitivity on used pressure.
- Glasses, tiles and plastics. These materials are usually more stable, but their reflectance characteristics deviate more from the ideal Lambertian distribution [21].

Present day are used as reference reflectance standard based on pressed or sintered PTFE samples. The pressed samples are less stable, and sintered ones may have slightly lower reflectance, but are more stable.

5.3.2.5 Calibration

All measurement of surface colors in reflected light are referred to the ideal matte surface (ideal white diffusor), which reflects the incident radiation with a reflectance factor $\rho = 1$ independent of the angle. The principle of measurement of the spectral reflectance R is sketched below in Figure 5.12.

The first step is to measure the dark current (sometimes referred as dark counts, black level or the background intensity), which is the output signal from the sensor with zero illumination. This is typically due to leaking current, imperfect charge transport, or thermally generated electrons within the semiconductor material. The dark current will depend on the integration time used, as well as on the temperature at which the sensor is operated.

In the next step, the reference intensity is measured. This corresponds to the reflection of a perfect matte white sample. Numerical subtraction of the dark counts from the reference signal yields the intensity I_{REF}. In the third step, the intensity reflected from the sample is measured, and the dark counts get subtracted. This difference defines the intensity I_{SAM}. Finally, the ratio of I_{SAM} and I_{REF} defines the spectral reflectance R_{λ} as written in Eq. (5.7), where reflected light intensity (I_{λ}) is related to incident beam intensity ($I_{0\lambda}$):

$$\rho_{\lambda} = \rho(\lambda) = R_{\lambda} = \frac{I_{\lambda}}{I_{0\lambda}} = \frac{I_{SAM}(\lambda)}{I_{REF}(\lambda)} = \frac{S_{SAM}(\lambda) - S_{BS}(\lambda)}{S_{WS}(\lambda) - S_{BS}(\lambda)} \tag{5.7}$$

In most reflection spectrophotometers, the reflection of light by the sample is measured at each wavelength relative to that of a white standard such as a plate coated with MgO or $BaSO_4$. These standards give almost 100%

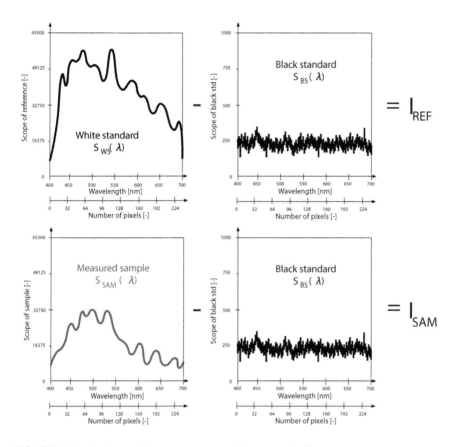

FIGURE 5.12   Scheme of the measurement of the spectral reflectance.

diffuse reflection (light rays reflected in all directions) between 380 and 750 nm (BaSO$_4$ ≈ 0.992; MgO ≈ 0,980 [22]). They are, however, rather fragile and often a working standard such as a ceramic tile of known reflectance is used, as was written previously. Calibration of the instrument with a white standard tile allows calculation of the reflectance of the sample relative to that of a perfect diffuse reflector having 100% reflection of the incident light at all wavelengths if is Eq. (5.7) corrected by reflectance factor of white standard:

$$\rho(\lambda) = \frac{S_{SAM}(\lambda) - S_{BS}(\lambda)}{S_{WS}(\lambda) - S_{BS}(\lambda)} \cdot \rho_{WS}(\lambda) \qquad (5.8)$$

5.3.2.6   Stray Light

Stray light constitutes a constant, false addition to the signal level at a given wavelength. Stray light is typically measured by blocking a portion of the incoming spectrum and measuring the residual light at those wavelengths. One method is to use a long pass filter to absorb shorter wavelengths, assigning the recorded intensity in this blocked wavelength region to the stray light generated by the remaining long-wavelength light, and then specifying the ratio of the two. Any instrument will reach a point where an increase in sample concentration will not result in an increase in the reported absorbance, because the detector is simply responding to the stray light. In practice the concentration of the sample or the optical path length must be adjusted to place the unknown absorbance within a range that is valid for the instrument. The effects of stray light can be – to some degree – numerically corrected after the measurement. The discussion of a full stray light correction on a given spectrometer is beyond the scope of this book. A simple recommendation is: if you have two light sources measuring device (deuterium and halogen lamp), it is better to switch off halogen lamp during UV range of measurement.

5.3.3   SPECTRORADIOMETER

Radiometry is the science of measurement of radiant energy, including light with respect to absolute power. Spectroradiometers are designed to measure radiometric quantities (irradiance and radiance) in a narrow spectral bandpass as a function of wavelength. They have the same principal components as the spectrophotometer, except for the light source. They are designed to measure the spectral properties of light sources, such as viewing booth, monitor displays, and projectors. The tele-spectroradiometer (TSR) is the most frequently used instrument in this category. The key components of the TSR are a telescope, a monochromator, and a detector. For most of the spectrophotometers, the measurement range interval of a TSR is between 380 nm and 780 nm. But the sampling interval of the TSR often has a spectral resolution ranging from 1 nm to 10 nm, which is sufficient for color work. The advantage of the TSR is the measurement of the object at a distance corresponding to its actual observing position including common viewing distance and viewing conditions. In other words, it can measure all forms of colors (surface and self- luminous), which are particularly important for cross-media reproduction.

Line-spectra have a potential for trouble in spectroradiometer measurement unless the relationship between monochromator spectral bandwidth and measurement wavelength interval is properly controlled. The rule is: the monochromator spectral bandwidth (width at 50% of peak response) should be an integer multiple of the wavelength interval between measurements. Thus, for example:

- 5 nm bandwidth with 5 nm steps – OK.
- 5 nm bandwidth with 3 nm steps – Errors.
- 5 nm bandwidth with 1 nm steps – OK.
- 5 nm bandwidth with 10 nm steps – Errors.

Failure to observe this relationship will result in failure to record the true power in any spectral lines present, resulting in colorimetric errors. Strictly speaking, the relationship holds for the case of a monochromator with a triangular bandpass function, but most monochromators do yield a good approximation to this when fitted with matching entrance and exit slits.

However, due to the high costs and complexity, the TCR is not a common measurement device used in the industry. Simplified alternative are tele-colorimeters, which has on the other side the same disadvantages as contact colorimeters in point of view of metamerism measurement.

5.3.4 SYSTEMS OF IMAGE ANALYSIS

The main purpose of the sensors in a camera is to convert the incoming light into electrical or electronic signals that represent the color image at each spatial position within the field of view (FOV). All practical DSCs use either charge-coupled devices (CCDs) or complementary metal-oxide semiconductor (CMOS) sensors in 2D arrays. Because all the sensors in a CCD or CMOS array have the same spectral response to the incident light, images detected by a sensor array are, in principle, gray scale images. To provide a color image, each cell of the sensor array is covered with an optical filter with a particular absorption spectrum.

Color images can be produced with a single image sensor in combination with a color filter array (CFA – Bayer grid, etc.) or color filter mosaic.

A color filter array is a mosaic of color filters placed over the sensors of an imaging device. In such a mosaic, at least three different color filters must be arranged with a particular spatial disposition so that each photodetector is sensitive to only one spectral band or color, Figure 5.13.

The Bayer grid arranges the G filter on a diagonal grid, and the R and B filter on a rectangular grid, as shown in Figure 5.13. Thus, the G image is measured at a higher sampling rate; this is based on the fact that the peak sensitivity lies in the medium wavelengths of light, corresponding to the G portion of the spectrum. Furthermore, the curve of absorption of green is similar to the luminosity function V_λ, and the luminance component has a higher bandwidth than the chrominance components.

Other methods for color separation include color sequential, where the image is composed of sequential exposures while switching the filters, and multi-sensor color, where the light is separated into red, green, and blue colors using a beam splitter and detected by three separate monochrome sensors, as shown in Figure 5.14.

Scanners are typically designed to scan images on paper or transparencies using its in-built light source. There is no need to capture the stationary object in a single exposure and typically linear sensor arrays are used to scan the image, moving along the direction perpendicular to the sensor array. Usually three (linear) sensor arrays are used, corresponding to the three channels, R, G, and B, but there are also solutions using three different lamps, obtaining the color image from three successive measurements with a single array [23].

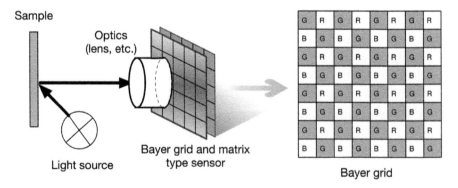

FIGURE 5.13 Optical analysis system and color filter array – Bayer grid.

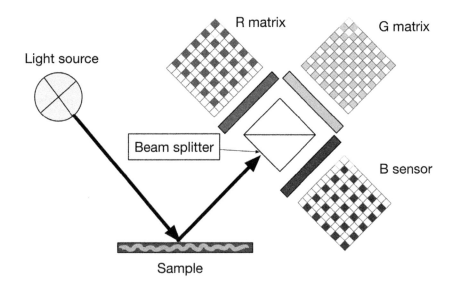

FIGURE 5.14   Scheme of a multisensor color capturing matrix system.

It is important to understand that each pixel receives only a specific range of wavelengths according to the spectral transmittance of the filter that is superposed to that specific pixel. Indeed, one pixel "sees" only one color channel. The color information is obtained by directly combining the responses of different channels, and then, it is essential that the convergence properties of light be perfectly controlled. Three-channel imaging is by its nature always metameric, i.e. based on metameric matching rather than spectral matching. However, the sensitivities of typical imaging devices differ from the CIE color matching functions, thus produces metameric matches that differs from those of a human observer. The limitations of metameric imaging are further expanded when the effects of illumination are considered. For example, it is possible for a metameric imaging system to be unable to distinguish a white object under a red light from a red object under a white light [24].

A multispectral imaging system is, by definition, an imaging system can provide complete spectral information instantaneously at each pixel of the captured image of a scene. Therefore, a multispectral image not only presents a high spatial resolution, but also provides the radiance spectrum of the illuminant, the reflectance spectrum of the object, or the combined color signal, at each pixel. Knowing the spectral information of the scene, the complete colorimetric information of the image is also known, as the tristimulus values in any color system can be calculated from the spectral information [25].

Spectral information at each pixel of the image is also useful to obtain printings or reproductions of a scene with high color accuracy, and even avoiding the metamerism due to changes in illuminant. Since the multispectral imaging systems appearance, spectral measurements were carried out exclusively using spectroradiometer or spectrophotometer, which are quite slow and only provide one spectral measurement each time.

The main component of a multispectral imaging system is a CCD or CMOS camera, whose response must be linear versus the received radiance and the exposure time. This property is named the reciprocity law [26]. Digital image acquisition systems are commonly classified in literature depending on the number of acquisition channels [27]: monochromatic with 1 acquisition channel, RGB or trichromatic with 3 acquisition channels, multispectral with 4–9 acquisition channels, hyperspectral with 10–100 acquisition channels, and ultra-spectral with more than 100 acquisition channels [28].

Multispectral imaging systems allow both, estimating the reflectance spectrum at each pixel by means of a previous spectral analysis carried out with direct measurements and images of color samples to establish a relationship between the digital levels of the system's response and the reflectance spectra [27], and improving the accuracy of the colorimetric characterization and subsequently improving the accuracy of the XYZ tristimulus values estimate.

The RGB image contains three gray scale channel images (see Figure 5.15a), which are acquired through 3 filters, as illustrated in Figure 5.15b.

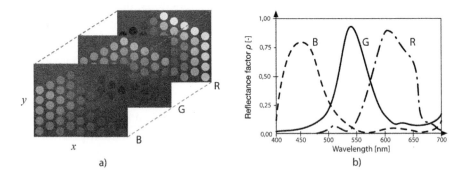

FIGURE 5.15   (a) Structure of RGB image (b) Typical spectral sensitivities of RGB camera.

The spectral image can contain multiple gray scale channel images, Figure 5.16a, which can be acquired through narrow band filters shown in Figure 5.16b). When the spectral image is captured by using the 400 to 700 nm region by 10 nm steps, the image consists of 31 different gray scale channel images. Each channel image contains information about one narrow spectral channel band. Because of the large amount of color information, the storage size of a spectral image can rise to hundreds of megabytes or even to gigabytes.

5.3.4.1 Dark Current and Electronic Gain

The dark current can be corrected by subtracting the dark current values from the captured images. However, this must be done for each element (pixel), because it will vary significantly over the CCD sensor matrix. Furthermore, it must be done for each individual exposure time used (i.e., for each color channel), since it will depend on the integration time. By capturing images without any illumination (i.e., with the lamp turned off and the lens cap on), using identical exposure times as the real images, these dark current images can be used as offset correction data. Since the dark current level will vary with the temperature at which the sensor is operated, and possibly over time, it is recommended to capture reference dark current images, prior to each image acquisition.

The electronic gain refers to the gain of the analog signal registered by each of the CCD elements. Electronic gain is characterized by capturing images of a diffuse white standard under uniform illumination. The camera should be slightly out of focus to eliminate the effect of any local

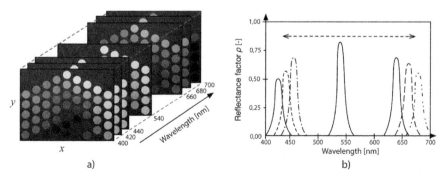

FIGURE 5.16   a) Structure of spectral image b) Typical spectral transmittance of interference filters.

inhomogeneity of the reflecting surface (it is mainly important in case of magnification of captured standard – microscopic level of reading). The electronic gain can then be measured for each individual CCD element, and of course compensated. Similarly, for the mentioned linear sensors, the correction for dark current and electronic gain is done for each individual element (pixel) of the matrix, and for each individual color channel because of the different exposure times used. Following this, Eq. (5.9) is rewritten in the form of corrected signal from every pixel and each channel $P_c(x,y)$:

$$P_C(x,y) = \xi_C \frac{S_{C\,SAM}(x,y) - S_{C\,BS}(x,y)}{S_{C\,WS}(x,y) - S_{C\,BS}(x,y)} \tag{5.9}$$

where $S_{C\,SAM}(x,y)$ is signal of acquired image for every pixel P and each channel c, $S_{C\,BS}(x,y)$ is signal for every pixel P and each channel c with the illumination turned off and the lens cap on, $S_{C\,WS}(x,y)$ is signal for a reference white reflection standard under uniform illumination and ξ_C is a channel dependent normalization factor, which can be chosen to give the corrected and normalized image desired maximal values [29].

Given that the noise power is independent of optical power, it is advantageous to make multiplexed measurements. In this case, one measures through a set of filters where each filter a transmits a selection of wavelengths covering approximately half the spectral region. Thus, the signal-to-noise ratio becomes higher for each measurement, when compared to narrowband measurements. Subsequently, the spectrum can be calculated as a linear combination of the measurements. The noise contributions are uncorrelated, and the overall gain in signal-to-noise ratio is proportional to the number of spectral channels N.

The past few decades have seen enormous progress in the development, production, and application of light sensing devices, many of which have resulted from advances in microelectronics, silicon chip fabrication technology mainly. In particular, these advances have major impact in many industries and in the introduction of increasingly sophisticated and versatile spectrophotometers.

5.4 MEASUREMENT

Color measurement is not only useful to check incoming goods but also during production. Thanks to modern opto-electronic components, color measurement is presently easy to handle and to understand and less expensive than before. Nevertheless, inexperienced users should read the following fundamental information to learn what color measuring technology is about and how to use it best in their specific field of application.

5.4.1 SAMPLE PREPARATION

One color measurement method for sliver or yarn wound on a cone or bobbin is to measure the sliver directly. In this case, no sample preparation is required. However, a positioning device is typically used at the instrument measuring port to consistently present the wound of sliver to the instrument. The way to choose the best method of measurement of a sample, including a selection of the instrument type, the appropriate illumination geometry and measurement, sample preparation method, etc., is a multistage process. This process is not always straightforward, because the choice of the device is largely dependent on laboratory equipment; moreover, a number of other factors may not be completely controlled. It is therefore important to ensure a maximum quality measurement in a given time and economic conditions. Hence, we always try to answer two basic questions:

a. Is a given measurement aspects of the use of the sample?
b. Measuring the repeatable within tolerances required for the job?

Correct answers to the above two questions need to be found before the foundation of the methodology of preparation and measurement of the sample. For example, the curtain can be measured on both the light transmission and light reflection depending on whether we are interested in the appearance or functionality in terms of shading.

For samples that are not opaque: prepare them by increasing sample thickness, so that they are opaque (or nearly so), unless one is intentionally measuring over white and/or black backgrounds.

Non-opaque samples such as textile substrates, most papers, and many printing ink substrates, usually must be measured as if they were opaque, to meet particular color application needs. In these cases, it is best to increase sample opacity by increasing sample thickness, so that the color measurement will not be influenced adversely by the "show-through" of a black (or white) background. Textile cloth can be folded multiple times, yarns and threads wound in multiple layers, and papers stacked, to meet the desired opaque (or most nearly so) requirement. If multiple sample folds' result in sample "billowing" into the sphere, then it is best to limit the thickness, and use a white sample background.

For samples with directional surface orientation: measure them always at the same single orientation, or measure them at the same four (4) orientations (90 degrees apart) with data averaged.

Optically directional samples (such as corduroy fabric, calendared vinyl, and card wound yarns and threads), typically measure differently in D/8° spectrophotometers, as they are rotated in the sample port. The directional geometric properties of these samples can result in measurement errors, whenever the standard and batch are measured differently. The measurement errors can be virtually eliminated by always measuring with the same sample orientation or by measuring the sample 4 times (at 90 degree intervals) and averaging the data.

For samples with inconsistent color and/or surface effects: measure them multiple times (moving the sample between reads to increase the effective area measured), with data averaged.

Inconsistent (or irregular) samples can yield different measurements, depending upon what part of the sample is measured. This measurement variability can be virtually eliminated by measuring the sample multiple times and averaging the results, moving the sample between measurements. This technique can also be successfully employed in cases where the sample may appear to be uniform but may not be.

As practical, utilize the largest sample measuring area available (LAV, SAV, USAV), to achieve averaging over a larger sample area. The largest viewing area (aperture value) available should generally be utilized, as it extends the measurement over a larger portion of the sample.

If possible, always measure standards and batches (of colors to be compared), using the same sample measuring area. Use the same viewing area for measuring the standard, as will be used in measuring production batches. This technique is desirable because different viewing areas (even of the same spectrophotometer) will typically not measure exactly the same.

If possible, always measure standards and batches (of colors to be compared) using the same spectrophotometer under the same conditions. If possible, use the same spectrophotometer under the same conditions to measure any samples to be compared, because the readings of no two spectrophotometers agree exactly.

As already mentioned, in the construction of measuring instruments, it is envisaged that the sample will be equal and will not penetrate through the metering orifice into the integrating sphere, because this measurement was incorrect. In cases where we need to measure subtle patterns (such as soft fiber flake, pile fabrics, etc.) leads to their penetration through the aperture, as can be seen in Figure 5.17.

In many cases, the sample may penetrate the interior of the integrating sphere even greater, because most of the apparatus is equipped with a specimen holder with thrust. Penetration of the sample can be avoided either by using the cover glass, as shown in Figure 5.17 or by using an aperture of smaller diameter as shown in Figure 5.18.

When measuring samples that may penetrate through the measuring aperture into the space in the integrating sphere, it is necessary increasingly to observe the number of measurements from which the corresponding colorimetric values are then calculated. As will be shown in Section 5.4.6, perform only one or two measurements, such as is happening in practice, often leads to high measurement uncertainty, particularly in Table 5.4, perform only one or two measurements, such as is happening in practice often leads to high measurement uncertainty. For this reason, the measuring aperture 30 mm recommended that at least four measurements, because three measurements are insufficient number of materials as documented in Table 5.1:

Table 5.1 also shows that the use of the metering orifice of smaller diameter, on the one hand, partially solves the problem of intrusion of the

FIGURE 5.17 Example of sample inherence through measuring aperture without cover glass (right side) and effect of cover glass (left side) on the sample.

FIGURE 5.18 Amount of inhered sample; it is possible to reduce by the application of measuring apertures with a small diameter.

TABLE 5.1 Dependence of Standard Deviation of Measured Values on the Size of Aperture and Presence of Cover Glass

Used aperture	30 mm with glass		30 mm without glass		20 mm without glass		5 mm without glass	
Number of readings	*4*	*3*	*4*	*3*	*4*	*3*	*4*	*3*
canvas	0.03	0.11	0.02	0.09	0.03	0.10	0.05	0.19
sanded terry cloth	0.08	0.15	0.04	0.07	0.04	0.07	0.18	0.27
corduroy	0.12	0.27	0.17	0.61	0.13	0.31	0.66	0.98
velour	0.15	0.34	0.13	0.52	0.20	0.59	0.73	1.19
fleece	0.09	0.17	0.21	0.71	0.25	0.83	0.37	1.04

sample into the space in the integrating sphere; on the other hand, it must be remembered that the measured sample area is significantly smaller (area of aperture of 5 mm diameter is 36 times smaller than the area of aperture of 30 mm diameter). The practice should therefore apply the rule that during the measurements we use the largest possible aperture, which allows the instrument and sample. Otherwise, we theoretically should increase the number of measurements so that the total measured area of the sample is matched.

As other factors during the measurement can increase the measurement error, the usual practice is such that for the comparison of samples, the following is true for a number of measurements per sample:

- 4 measurements using the aperture of diameter 30 mm
- 8 measurements using the aperture of diameter 20 mm
- 12 measurements using the aperture of diameter 10 mm
- 20 measurements using the aperture of diameter 5 mm

In case of translucent materials, it is important to remember the problem of *aperture vigneting*, especiallt if it is necessary to prepare temporary device for laboratory measurement. When reflecting samples are being measured with small apertures, errors can occur as result of light diffusing in the substrate of the samples, if the illuminating and measurement areas are co-incident. This kind of errors can be avoided by having either the illuminating is at least 2 mm larger in all directions than the viewing aperture, or the viewing aperture at least 2 mm larger in all directions than the illuminating aperture [30]. This is illustrated in Figure 5.19.

a) Correct setup in case of opaque materials – illuminating and viewing apertures of the same size. No light lost.
b) Incorrect setup in case of translucent materials – illuminating and viewing apertures of the same size. Some light lost.
c) Correct setup in case of translucent materials – edge of viewing aperture at least 2 mm beyond edge of illuminating aperture. No light lost.
d) Correct setup in case of translucent materials – edge of viewing aperture at least 2 mm inside edge of illuminating aperture. Light lost is equal to light gained.

Remember, a lens system is needed to restrict either the illuminating beam or viewing beam in all CIE recommended geometries.

If we look at the specific case of the three textile products listed in Figures 5.20 and 5.21, we can see two different approaches.

As previously mentioned, a variant of the cover glass over the measuring opening of the device is required to ensure correct measurement within the

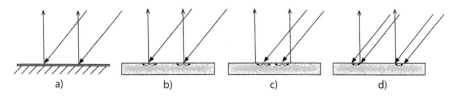

FIGURE 5.19   Simplified scheme of aperture vigneting.

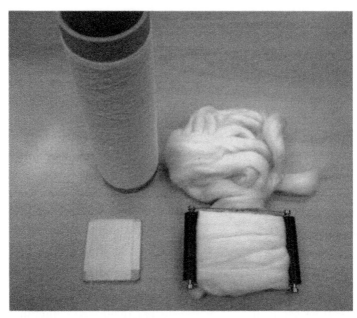

FIGURE 5.20 Yarn winded on a small cavity, card wound, free sliver and sliver inside of skein holder.

FIGURE 5.21 Accessory with cover glass for the measurement of carved carpets – sample is protected against penetration into the measuring device.

integrating sphere, but it is necessary to count the change in appearance of the measured material to compression. Therefore, preparation for the measurement of the notch carpet equipped with a spring with a relatively small pressure of about 10 cN during contact.

As mentioned earlier, since the yarn is being measured through a glass interface, the instrument will measure the sample as a duller color than it actually is. This is because glass causes low reflectance values to be higher, and high reflectance values to be lower. For quality control applications when the color difference is being measured and the "standard" was measured the same way, the error introduced by the glass is less significant. However, accuracy can be improved by calibrating the instrument through glass, or when a higher absolute accuracy of measurement is desired, a glass correction can be used Eq. (5.10):

$$R_{corr.}(\lambda) = \frac{R_g(\lambda) + T_c - 1}{R_g(\lambda) + T_c - 1 - T_d \cdot R_g(\lambda) + T_d} \tag{5.10}$$

where $R_{corr.}(\lambda)$ is the corrected value of spectral reflectance; $R_g(\lambda)$ is the value of reflectance behind glass; T_c = transmittance of glass to collimated light (based on Fresnel equation for glass with refractive index 1.5 is this value 0.92).

T_d = transmittance of glass to diffuse light (nominally equal to 0.87 (result from Fresnel's equations for diffuse illumination in case of common glass with refractive index 1.5)

The above calculation assumes a spectrophotometer with a diffuse measuring geometry, preferably di:0°, however, the variant di:8° is applicable as documented spectral reflectance curves in the graph of Figure 5.22; hence, colorimetric evaluation is shown in disjoint graph of Figure 5.23. Correlation with data measured on the template (sample without cover glass – R_{real}) is high, as documented in the graph in Figure 5.22. Error spectral reflectance in this case was 2.8% compared to 24% for uncorrected data. Further, the measured color difference between the corrected measurement and the draft was ΔE_{CMC} (2:1) = 0.3 versus ΔE_{CMC} (2: 1) = 5 in the uncorrected measurement.

The graph in Figure 5.23 shows that the main difference between the corrected and uncorrected measurement is specific purity ΔC^*. This is because the influence of the primary reflection (gloss) on the measured spectral data.

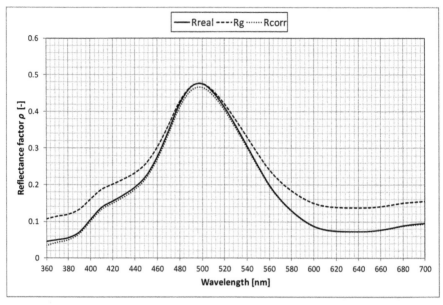

FIGURE 5.22 Reflectance chart of sample measured with *(Rg)* and without cover glass *(Rreal)* and corrected reflectance factor by using Eq. (7.11) – Rcorr.

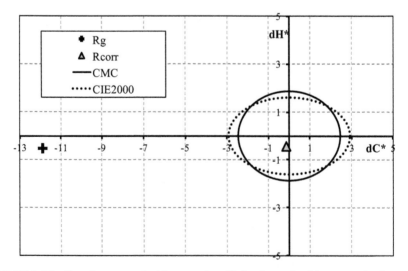

FIGURE 5.23 Sample measured with cover glass *(Rg)* and sample with corrected reflectance factor *(Rcorr)* projected on dC*dH* plane (as the target was used sample measured without cover glass *(Rreal)*.

As noted previously, the primary reflection is uncolored – it has the same spectral composition as the incident light, which in our case is white.

If we return to the samples shown at Figure 5.20, it is clear that in the case of a textile strand, which could be measured by using the cover glass, we can measure sample ingress to the interior of the integrating sphere measuring limit by the clamping of which makes it possible to pre-stress the spring. If yarns are usually used, two measurement techniques. One is a laboratory where yarn rewind to different paper cards (template), as shown in Figures 5.20. The advantage of this approach is reproducible measurements in addition to the possibility of archiving for future use. The second technique for measuring the yarn is typically used for operational controls when using fixation products (Figure 5.24) can be advantageously measured cross rewind packages, warp beams, etc.

Beside mentioned techniques it is possible to do direct measurement of samples with longitudinal shape (such as yarn, card sliver, etc.) without necessity to scroll it on paper card. The task of the positioning device, as its name implies, is fixing the mutual position of the device and the sample, which is not planar. According to the diameter of the winding, choosing a suitable preparation for the measurement is performed several times, if possible, around the perimeter of the sample. Although cross rewinding coils have a slightly conical shape, fixing the sample is sufficient and convenient measurement as shown in Figure 5.25. Most yarn samples will have some level of directionality. Directionality is when the sample has distinct lines

FIGURE 5.24 Set of Accessories (Positioning Device) for the Measurement of Tubular Samples with Different Diameters (HunterLab).

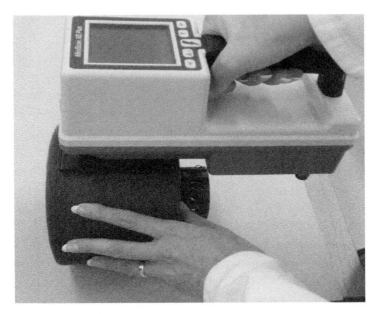

FIGURE 5.25   Measurement of X-bobbin by using Accessory from Figure 5.24.

or texture that when measured at different angles of rotation (yield varying results). Averaging multiple placements of the sample with a 90° rotation between readings will minimize the effect of directionality.

5.4.2   BACKGROUND EFFECT

Very often the differences between measurements taken at different sites are due to different background used in the measurement. As stated previously, a number of samples fall within the category of translucent objects where it is necessary to consider also the amount of light is able to pass through the material. So, if we measure one sheet of paper, one layer of fabric or some plastic moldings for various backgrounds will result in a difference in colorimetric data as shown in the graph in Figure 5.26.

Other graphs in Figures 5.27–5.29 are clearly seen to reduce differences between measurements on the black a white background with an increase in the number of layers.

It is therefore clear that, where we can ensure a sufficient number of layers of the sample, the measurement is not influenced by the substrate. This is especially important for portable devices because of desktop devices

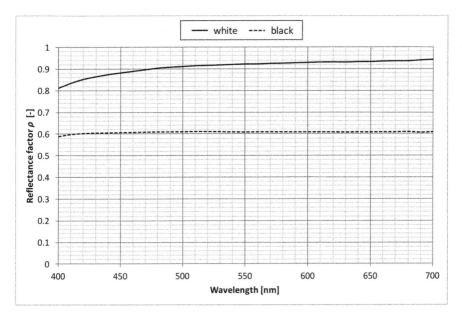

FIGURE 5.26 Plot of reflectance factor on black and white backgrounds for 1 layer of sample.

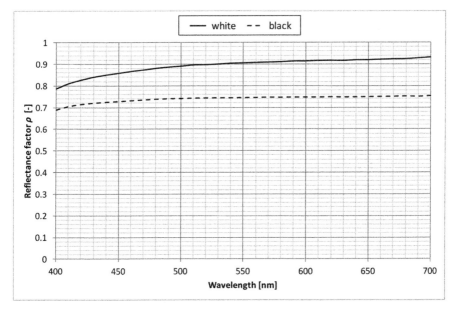

FIGURE 5.27 Plot of reflectance factor on black and white backgrounds for 2 layers of sample.

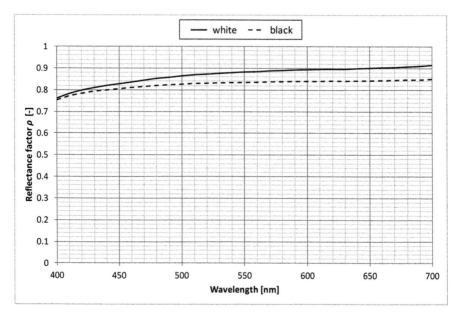

FIGURE 5.28 Plot of reflectance factor on black and white backgrounds for 4 layers of sample.

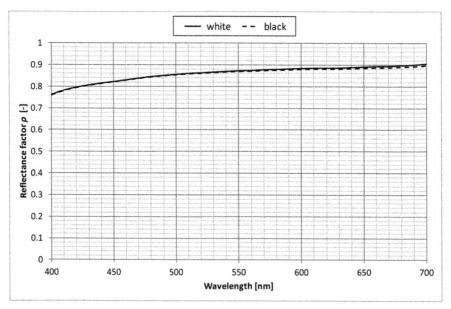

FIGURE 5.29 Plot of reflectance factor on black and white backgrounds for 8 layers of sample.

are used specimen holders that have a white or light gray contact area. If the measurement result compared with portable instruments by which we measure the same sample again on a light table, once on a dark table or tables with different designs, it is logical that depending on the degree of light transmission will be confronted with different spectral reflectance and colorimetric coordinates.

As already mentioned, this problem is usually solved so that measuring multiple layers of material, which is important to the supplier and the customer mutually agree on the number of measured layers. The easiest way to determine the number of layers based on coverage of the device measuring port.

Measuring hole (measuring aperture of device – aperture) it is necessary to cover with one layer of material that we hold hands so we do not cover the aperture and perform measurements. If during the measurement was seeing a flash of light (light pattern caused by instrument lamp), cover the aperture by another layer of material. This process is repeated until no light pattern visually fails. In the case of fabrics, it is usually 4–8 layers [31]. If for various reasons this procedure is unusable, it is necessary to use correction procedures to eliminate the effect of background on the measured values. For this purpose, it is used for measuring the black and white background. This method is referred in Chapter 6 as example of its usability in case of color-changeable materials. The monochrome background method is also often referred to as a method of contrast ratio (CR – Contrast Ratio method), typically, the method used to determine CR opacity.

5.4.3 ERRORS OF MEASUREMENT

In practice, none measurements, measuring methods or measuring instruments are absolutely accurate. To assess the results in physical and technical measurements can be chosen different approaches. When determining the measurement inaccuracies, there are two basic approaches: older and simpler fault, newer and more comprehensive evaluation process through the measurement uncertainties. Currently, the second method is preferred. During the measurements we can see the influences that affect the deviation between the actual measured value and actually measured value. The resulting difference between both values is sometimes formed by a complex combination of individual factors. Put simply, the difference between correct and the measured value depends on the accuracy of measuring instruments, precision measuring methods, but also the actual measured sample. The errors, which are the results of the colorimetric measurements always burdened with the most clearly manifested when we make a series of parallel measurements of the same perfectly homogeneous sample (e.g., color calibration standard), we get results that are exceptional cases (typically due to rounding) and somewhat different [32].

5.4.3.1   Important Points to Remember

1. A measurement without units is meaningless.
2. A measurement without an estimated range of uncertainty (or error) is also meaningless.
3. Always quote the final result of a measurement in the form **a** ± **b**, with **a** (measurement) and **b** (error) given to the same number of decimal places. For example, a measurement of the color difference: $\Delta E^* = 0.94 \pm 0.14$. This expression means that the true value is likely (the probability being approximately 68%, see below) to lie between 0.80 and 1.08, i.e., if you repeat the measurement a large number of times about 68% (i.e., two-thirds) of the results will lie between 0.80 and 1.08.
4. Errors are usually only known roughly, so quote your errors to no more than two significant figures. One significant figure is usually sufficient (i.e., 0.94 ± 0.14, or possibly 0.942 ± 0.137, but not 0.94217 ± 0.13794).

Errors are expressed in absolute or relative terms. Generally, measurement of any quantity is done by comparing it with derived standards with which they are not completely accurate. Thus, the errors in measurement are not only due to error in methods but are also due to derivation being not done perfectly well. An error may be defined as the difference between the measured value and the actual value. For example, if the two operators use the same device or instrument for finding the errors in measurement, it is not necessary that they may get the similar results. There may be a difference between both measurements. The difference that occurs between both the measurements is referred to as an error. The absolute error $\Delta(x)$ indicates the difference between the measured value x_y and real x_s.

Measurement uncertainty characterizes the range of readings about the result of the measurement, which can reasonably be assigned to the value of the measured variable. Uncertainty concerns not only the actual measurements, but also includes inaccuracy METRIC instruments, the value of the constants, corrections, etc., where the overall uncertainty of the result depends.

As mentioned earlier the errors are classified according to the nature and effect of the three categories [33]:

Systematic Errors, Bias (another sign the outliers and outlier data) are caused by an exceptional cause, typically incorrect typing result, the sudden failure of the meter, incorrect settings of measurement conditions, and the like. The measured value is significantly different from other values obtained in repeated measurements, as shown in the graph in Figure 5.30.

Such measurements are to be excluded from processing to falsification of the result. A number of control programs for spectrophotometers allows for instant comparison of measurements during the so-called. Repeated measurement. The operator thus has the possibility of deviating measurements exclude

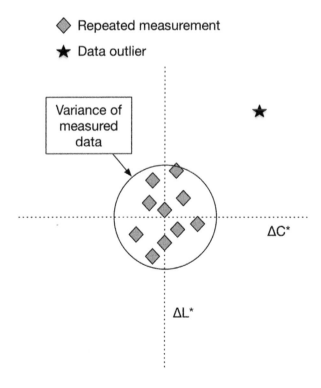

FIGURE 5.30 Example of an outlier in the measurement of color differences.

from the calculation of the arithmetical average estimate, typically, when a movement of the sample or instrument measuring aperture being outside the specific part of the sample, or a sample of significant inhomogeneity, etc.

Random (or indeterminate) errors are unpredictable and cannot be ruled out. When repeating measurements are changing their size and sign of how the law with the expected probability distribution. The determination of their size is based on repeated measurements using statistical methods corresponding to the respective probabilistic model, represented by the appropriate distribution law of random errors. In practice, it is very often a normal-Gaussian (Figure 5.31), which is used in most applications.

Let X be a randomly varying quantity, of which n independent measurements were obtained under the same conditions. In most of the practical cases, the best estimate of the value of X $(x_1, x_2,...x_p...x_n)$ is the *arithmetic mean* $\bar{x}$ of the results of all measurements:

$$\bar{x} = \frac{1}{n}\sum_{i=1}^{n} x_i \qquad (5.11)$$

The observations in the sample vary as the result of random fluctuation in the measurement instrument, the measured matter and others. The *standard deviation* $s(x)$ is calculated by taking the square root of the variance [33]:

$$s(x) = \sqrt{\frac{\sum_{i=1}^{n}\left(x_i - \bar{x}\right)^2}{n-1}} \qquad (5.12)$$

The value of the standard deviation, unlike variance, has the same units as the measured quantity, and therefore often is more convenient for use as the measure of dispersion of values about the mean. The *experimental standard deviation of the means* $s(\bar{x})$ is equal to square root of its variance [34]:

$$s(\bar{x}) = \frac{s(x)}{\sqrt{n}} = \sqrt{\frac{\sum_{i=1}^{n}\left(x_i - \bar{x}\right)^2}{n(n-1)}} \qquad (5.13)$$

As mentioned before, systematics errors are errors whose value under the same conditions of measurement does not change is constant in magnitude

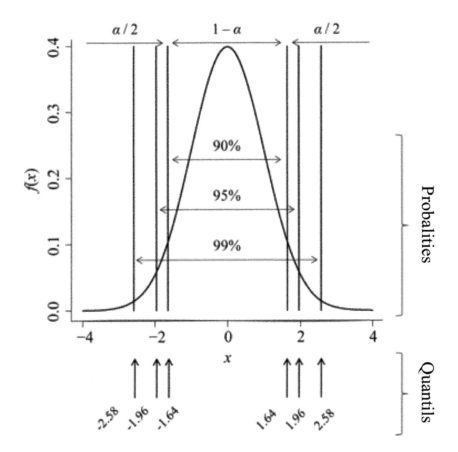

FIGURE 5.31 Key quantiles of standard normal density, bell-shaped probability density (Gaussian).

and sign, or who are changing the measurement conditions vary according to the particular addiction and a way "systematically" influence the measurement result. Random errors cannot be quantified exactly since the magnitude of the random errors and their effect on the experimental values is different for every repetition of the experiment. So statistical methods are usually used to obtain an estimate of the random errors in the experiment. One of possible estimation of random errors is based on formula for calculating the absolute error.

$$\Delta(x) = x_m - x_s \qquad (5.14)$$

From the perspective of the user measurement techniques, systematic errors are sympathetic that they can largely be identified and their effect can be reduced by appropriate compensation to eliminate the causes, which give rise, eventually a combination of these methods. Survey and removal of systematic errors is difficult and expensive, and therefore carried out only where it is inevitable. In any experiment, care should be taken to eliminate as many of the systematic and random errors as possible. Proper calibration and adjustment of the equipment will help reduce the systematic errors leaving only the accidental and human errors to cause any spread in the data. This part is called fixed systematic errors. The resulting measurement error is expressed as the sum of systematic and random components of ε, which can be written:

$$\Delta(x) = |e| + |\varepsilon| \qquad (5.15)$$

and its maximum value can be estimated as:

$$\Delta_{max} = (\overline{x} - x_s) + 2s \qquad (5.16)$$

where systematic component is $e = x - xs$ and the random component is $\varepsilon = s$, respectively $\varepsilon = 2s$

The coefficient of expansion of the standard deviation is related to the probability coverage interval and the type of distribution. That means, the value of 1.96 is based on the fact that 95% of the area of a normal distribution is within 1.96 standard deviations of the mean as shown in the graph in Figure 5.31.

5.4.4 SOURCE OF ERRORS

The entire measurement process is encumbered with a number of imperfections and problems that inevitably be reflected in the results of measurements and errors. According to the sources of imperfections, the errors can be divided into the following items:

- *Instrument error* – Conventional spectrophotometers are of the double-beam configuration, where the output is the ratio of the signal in the sample beam to the signal in the reference beam plotted as a function of wavelength. It is incumbent upon the experimenter to ensure that the only difference between the two beams is the unknown.

Therefore, if liquid or gas cells (cuvettes) are employed, one should be placed in each beam. For gas cells, an equal amount of carrier gas should be injected into each cell, with the unknown to be sample placed in only one cell, destined for the sample beam. For liquids, an equal amount of solute should be placed in each cell. A critical issue with liquid and solid samples is the beam geometry. Most spectrophotometers feature converging beams in the sample space. If the optical path (the product of index of refraction and actual distance) is not identical for each beam, a systematic difference is presented to either the entrance slit or the detector. In addition, some specimens (e.g., interference filters) are susceptible to errors when measured in a converging beam. Most instruments also have a single monochromator, which is susceptible to stray radiation, the limiting factor when trying to make measurements of samples that are highly absorbing in one spectral region and transmitting in another. Some recent instruments feature linear detector arrays along with single monochromator to allow the acquisition of the entire spectrum in several milliseconds; these are particularly applicable to reaction rate studies. Conventional double-beam instruments are limited by these factors to uncertainties on the order of 0.1%. For lower uncertainties, the performance deficiencies found in double-beam instruments can largely be overcome by the use of single-beam architecture. The mode of operation is sample-in–sample-out. If the source is sufficiently stable with time, the desired spectral range can be scanned without the sample, then rescanned with the sample in place. Otherwise, the spectrometer can be set at a fixed wavelength and alternate readings with and without the sample in place must be made. Care should be taken to ensure that the beam geometry is not altered between sequential readings.

- *Errors of installation* – such kind of errors are caused by wrong connections and setup of measuring device, by interferences between devices mainly if two devices are connected with one computer.
- *Errors of method* – errors that are caused by imperfection of method by using physical constant approximation and inadequate model of measured properties.
- *Errors of assessment* – source of these errors are limitation of human sense, current health condition of observer, low concentration, etc.

- *Errors of evaluation* – errors relating to data treatment (approximate relations, rounding, less number of used constant, errors of interpolation, extrapolation, linearization, etc.)
- *Influence of environment* – variation of temperature and humidity. Consider that almost all dyes and pigments change color as temperature changes. Measure samples at the same temperature each time, if possible.

The facts that pigments and dyes change their color properties with the change in temperature potentially serve a major source of errors in any color measurement operations. Almost all colorants (except white and black and some blues) change color as their temperature changes. The color changes can be significant, and it is typically in the range of about 0.6 to 1.1 ΔE_{CIELAB} for most high chroma colorants, for a 10°C temperature change. If production samples (batches) must be measured at a temperature different from that when the standard was measured, then (if possible) re-measure the standard at the new temperature.

The effects of humidity are less dramatic. Still, as a wet road appears darker than when dry, so too will physical standards alter their look if they are not maintained under standard condition.

From the point of view of colorimetry, all mentioned errors could be reduced into three main parts as shown in Figure 5.32.

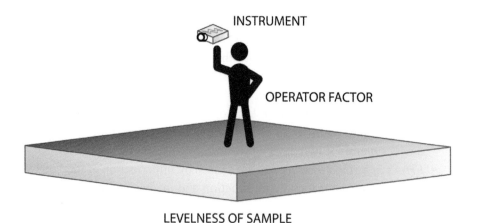

FIGURE 5.32   Simplified scheme of relationship of uncertainty parameters.

These three factors are most frequent in industry. In scheme in Figure 5.32, we can see obvious size of such errors. High negative effect on resulting data has measured sample, which can be uneven, sensitive on external stimuli as temperature and relative humidity.

The specimens must be visually checked for inhomogeneity of the grain and gloss level. The surface must be free of foreign matter. Areas showing damage not intentionally caused for testing purposes must be excluded from the measurement. It is important to remember:

If you are able to see inhomogeneity, device is able to read it also!

Textiles sensitive to stroking must be brushed prior to starting the measurement. Commercial standard VW 50190 gives the following recommendation: start with 1-time against the pile direction (specimen appears darker), then 10-times in the pile direction (specimens appear lighter and structure appears more homogenous.

Remember that some of samples have hidden unevenness due to continue change in color, which are frequently invisible, but will be measured as high standard deviation between individual readings. The best of materials will have point-to-point non-uniformities in the reflectance factor on the order of 0.1% and as high as 0.5%.

Precision measurements exist only where the individual instruments are carefully maintained and checked. As mentioned before, standard deviation expresses the degree of dispersion of values of one-dimensional variable about the mean. With two- of three-dimensional variable, equivalent expression takes form of covariance matrix rather than a single value. Color values in the CIELAB space have three-dimensional distribution; however, expressing their dispersion as covariance matrix is impractical and inconvenient for everyday use. Instead, it is possible to express it in familiar units of CIELAB color difference as Mean Color Difference from Mean (MCDM).

Let us have n sets of CIELAB vectors $\mathbf{C}_i$ representing a set of measurements of the same color stimulus. The dispersion of these values about the mean can be computed as follows:

$$MCDM = \frac{\sum_{i=1}^{n} \Delta E\left(C_i, \bar{C}\right)}{n} \tag{5.17}$$

where ΔE is the color difference calculated with any chosen CIELAB-based formulae (i.e., (4.47)), and $\bar{C}$ is the mean vector of CIELAB values.

The calculation of MCDM as an alternative to standard deviation assumes perceptual uniformity of the color space, i.e. that the distribution of color deviations about the mean is spherical. Naturally, the use of advanced color difference formulae improves the perceptual uniformity of color difference calculation. Thus, the use of CIEDE2000 formula in Eq. (4.93) is expected to lead to more reliable MCDM estimation then the use of standard CIELAB Euclidian distance formula.

The British Ceramic Research Association (present day LUCIDEON) produces a set of twelve colored tiles. These tiles are almost used by instrument manufacturers to establish traceability of colorimetric devices to national standards laboratories. These tiles may be used to verify instrument operation in a corporate color lab and for improvement of measured data traceability between different devices in company (supportive software such as Datacolor Guardian or X-Rite NetProfiler it is necessary to order). To profile a color-measuring or color-producing device means to quantify its input-output relationships.

Nevertheless, all highly colored materials are subject to thermochromism to varying degrees. This reversible change in color with temperature has been a problem with many standards but can be overcome in the CCSII by use of Thermo chromic Correction data determined by the National Physical Laboratory. The basic data is supplied free with sets of CCSII. In 1997, Malkin and others published article [37] entitled "Master spectral reflectance and thermochromism data." Approximate correction of measured colorimetric values can be calculated by the following equation:

$$C_c = \left[\frac{T_C - T_M}{10} \cdot \Delta TC \right] + C_M \qquad (5.18)$$

where C_C is the corrected color coordinate (L^*_C, a^*_C, b^*_C); C_M is the color coordinate as measured (L^*_M, a^*_M, b^*_M); ΔTC is the thermochromic change (ΔL^*, Δa^*, Δb^*) in Table 5.2; T_M is the temperature of the measurement; T_C is the temperature to which color coordinate is to be corrected.

It is important that the recommended range of the temperature is 25°C to 35°C. Outside of this range, the corrected values can be out of real thermochromic data. In relation to this item, it is advisable to check for temperature

TABLE 5.2   Thermochromism Measure ΔTC – The Changes in CIEL*a*b* for a 10°C Rise in Temperature for Specular Included, Illuminant D65/10° Observer

Tile color	ΔL^*	Δa^*	Δb^*	ΔE^*
Pale Grey	−0.03	−0.02	0.02	0.04
Mid Grey	−0.03	−0.02	0.04	0.05
Difference Grey	−0.04	0.04	0.03	0.06
Deep Grey	0	0.01	0	0.01
Deep Pink	−0.1	−0.44	−0.19	0.49
Red	−0.37	−0.71	−0.61	1.01
Orange	−0.45	0.56	−0.66	0.98
Yellow	−0.27	0.7	−0.11	0.76
Green	−0.18	0.66	−0.04	0.69
Difference Green	−0.18	0.69	−0.05	0.71
Cyan	−0.1	0.31	0.01	0.33
Deep Blue	0	−0.04	0.05	0.06

difference between the textile sample placement position and the room temperature. If is temperature difference between laboratory, where is instrument placed, and the room of sample production or storage up to 12°C, then unacceptable difference of color coordinates is caused. For precise measurement are generally acceptable difference of two or three centigrade.

5.4.5   MEASUREMENT PROCEDURE

A repeated reading only serves as a quality check of the sample or standard. The measurement opening must be matched to the specific specimen shape. In the case of textile specimens, as large a measuring aperture as possible must be selected (means in case of uniform color samples). Flat-plane surfaces with no pressure marks or damaged parts must be selected for the measurement areas. When selecting the measuring points, it must be ensured that they are distributed over the entire specimen and are at least 3 cm from the edge where possible. For translucent textiles, an underground that corresponds to the backing of finished part must be selected; in other cases, it is possible to follow the procedure of calculation of corrected reflectance factor for translucent materials [31].

The sample (target or batch) must be subjected to at least 4 individual measurements distributed over the entire area of the defined surface (see Figures 5.33 and 5.34).

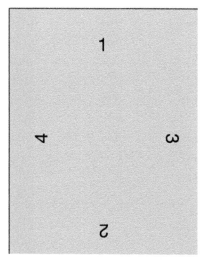

FIGURE 5.33 Methodology of sample rotation during repeatable measurement – 4 times multiple measurement typically by using devices equipped with diffuse measuring geometry.

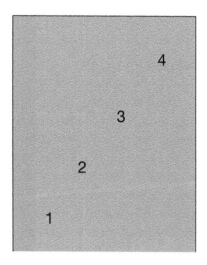

FIGURE 5.34 Methodology of sample positioning during measurement – 4 times multiple measurement typically by using devices equipped with measuring geometry 45°a:0°.

Annular 45°a:0° or circumferential 45°c:0° instrument illuminate the sample using many lights in a ring around the sample, 45° from it. Due to this construction it isn't necessary to average of readings with 90° rotation of sample between readings as obvious for directional (bi-directional typically) 45°x:0° or spherical geometry-based instruments.

Another recommendation is focused on measuring the aperture size. As mentioned before, large possible aperture (LAV) must be used; nevertheless, the diameter of aperture can be highly variable if we compare desktop and portable devices. If general recommendation is at least 4 individual readings, then in case of desktop device, the diameter of LAV can be 30 mm, and many of portable devices are equipped with fixed apertures of 5 mm diameter. This is the reason for reduction in traceability of the measured data. As a solution to this problem, it is possible to use a simple method of enhancement of number of reading, which is described in the following Table 5.3:

Recommendation of comparative size of measured area of sample by instruments with different size of aperture isn't absolute. Similar size of measured area as 4 readings with 30 mm aperture is related to 36 readings of instrument equipped by 10 mm aperture and device with only 5 mm aperture needs to measure 144 readings! Due to this reason is number of readings reduced, as shown in Table 5.3 and resulting size of measured area of sample is approximately seven times smaller in case of 5 mm aperture (20 readings) than 4 readings of 30 mm aperture. At each measuring point, we will do 4 reading with 90° rotation and as result we will obtain 20 individual readings. By this simple method, we will have reproducible measurement made by both devices and this measurement will be also traceable (not identical).

Specific problem is connected to data presentation. As was mentioned before that reflectance measurements are generally more accurate when the

TABLE 5.3 Relation between Aperture Diameter and Resulting Measured Area Following Obvious Number of Readings

Aperture diameter [mm]	Recommended number of reading (–)	Resulting measured area (mm²)
30	4	2827
20	8	2513
10	12	942
5	16–20	314–393

multiple readings are applied. Nevertheless, on the end is an average, mean value of readings, used as output value of whole measurement. Many of portable devices present resulting values with precision at one decimal place because then readings appear more stable. That implicate important question: how many decimal places are relevant in case of comparison of different results? As scientists or engineers, we get a large amount of the numbers we report and use in our calculations from measured observations. In this instance, a number is determined to be significant or not by the accuracy and precision of the measuring device. With a number derived from a measurement, the last digit to the right expresses the uncertainty. For example, if you are sure that your spectrophotometer can deliver an accurate measurement to a tenth of a colorimetric coordinate unit (L*, a*, b*), then you would be justified in reporting color to a tenth of a colorimetric coordinate unit. For example, if one measured a lightness value of 90.1 this number would contain three significant figures with the last digit expressing the uncertainty. The uncertainty would be plus or minus 0.05 u (unit). Even if the instrument could spit out 10 digits passed the decimal point one should only report the significant digits. Errors can arise in calculations if insignificant figures are used in a calculation. If a number resulting from a measurement is used in a calculation that involves multiplication or division all significant figures should be carried through the calculation, and then, the result should be rounded at the end of the calculation to reflect the term used in the calculation with the fewest significant figures. For example, 10.4 x 5.0 should be reported as 52 and not 52.0. If the calculation involves addition and subtraction a different rule applies, one should preserve common decimal places of the numbers involved. For example, if two numbers obtained from a measurement are used in an addition, 10.1 + 1000.234 the reported number should be 1010.3, not 1010.334!!! Notice that 10.1 has 3 significant figures and 1000.234 has 7 significant figures and the result of the addition has 5 significant figures.

General rules for determining the number of significant figures in a number:

- All non-zero numbers are significant.
- All zeros between significant numbers are significant, for example the number 1002 has 4 significant figures.
- A zero after the decimal point is significant when bounded by significant figures to the left, for example the number 1002.0 has 5 significant figures.

- Zeros to the left of a significant figure and not bounded to the left by another significant figure are not significant. For example, the number 0.01 only has one significant figure.
- Numbers ending with zero(s) written without decimal place possess an inherent ambiguity. To remove the ambiguity writes the number in scientific notation. For example, the number 1,600,000 is ambiguous with regard to the number of significant figures it contains; the same number written 1.600 x 106 obviously has four significant figures.

5.4.5.1   Several Notes

1. It is important to know the accuracy and precision of the measuring device one is using, and it is important to report only those digits that have significance.
2. It is generally accepted that the uncertainty is plus or minus 0.5 unit at the level of the uncertainty, for example, the "true value" for the number 0.003 can be described as being bounded by the numbers 0.0025 and 0.0035. It is important to note that in some instances, scientists or engineers will want to express an uncertainty that exceeds 1 at the level of the uncertainty and this should be noted explicitly in the following fashion, 0.003 ± 0.002.

Examples:

$L^* = 91.5 \pm 0.92$	$L^* = 92 \pm 1$
$L^* = 91.5 \pm 0.81$	$L^* = 91.5 \pm 0.9$
$\Delta L^* = 1.1 \pm 0.09$	$\Delta L^* = 1.10 \pm 0.09$
$\Delta L^* = 1.104 \pm 0.09$	$\Delta L^* = 1.10 \pm 0.09$
$\Delta L^* = 1.105 \pm 0.09$	$\Delta L^* = 1.11 \pm 0.09$

5.4.6   *UNCERTAINTY IN MEASUREMENT*

The objective of a measurement (or any other quantitative investigation) is to determine an estimate for the true value of the measurand. This estimate, i.e. the measurement result, may be an individual measured value. The term uncertainty represents a concept, according to which the true measurement value is unknowable because, after all the known sources of error are being corrected for, there will always be uncontrollable variations in measurement.

The value of uncertainty is "...associated with the result of the measurement, that characterizes the dispersion of the values that can be reasonably attributed to the measurand" [35].

Guide to the Expression of Uncertainty in Measurement (GUM) [36] classifies uncertainty components according to their method of determination into type A and type B:

- Type A: Evaluation using statistical analysis of measurement series,
- Type B: Evaluation using means other than statistical analysis of measurement series.

With respect to the suggested methodology, GUM does not differentiate between uncertainty components due to systematic effects and uncertainty components due to random effects. It is, however, assumed that, as far as possible, recognized systematic errors are either eliminated by technical means or corrected by calculation. For the uncertainty budget, a component remains that accounts for the uncertainty arising from any such action.

A typical example for a **type A evaluation** is the determination of an estimate of the standard deviation σ of an assumed normal distribution. If x1, x2, ..., xn are the results of repeated measurements of the quantity concerned, then the experimental standard deviation s of the measurement series {x1, x2, ..., xn} can be used as an estimate of the standard deviation σ of this normal distribution.

The Standard Uncertainty $u_A(y)$ is given by the experimental standard deviation of the mean:

$$u_A(y) = s(\bar{y}) = \frac{s(y)}{\sqrt{n}} = \sqrt{\frac{\sum_{i=1}^{n}(y_i - \bar{y})^2}{n(n-1)}} \qquad (5.19)$$

If we are doing less number of individual reading then ten (n<10), the standard uncertainty calculated as standard deviation of the mean (5.19) is less reliable, therefore we need to use *Expanded Uncertainty*, i.e., product of the standard uncertainty $u_A(y)$ and an appropriate coverage factor k_A from Table 5.4:

If we imagine obvious number of multiple readings, for example 4, which was mentioned in section 5.4.1 as measurement of color difference, when following Table 5.5 shows individual data:

TABLE 5.4 Size of Coverage Factor k_A for Different Count of Repetitive Readings

Number of reading n	10	9	8	7	6	5	4	3	2
Coverage factor k_A	1	1.2	1.2	1.3	1.3	1.3	1.7	2.3	7.0

TABLE 5.5 Example of Measured Color Difference

Number of reading n	1	2	3	4
ΔE^*	1.1	0.9	1.1	0.8

Then, the result of expanded uncertainty by using Eqs. (5.11) and (5.19) with coverage factor will as follows:

$$\bar{x} = \frac{1}{n}\sum_{i=1}^{n}x_i = \frac{1}{4}(1.1+0.9+1.1+0.8) = 0.975$$

$$u_A = k_A \cdot s(\bar{y}) = 1.7 \cdot$$

$$\sqrt{\frac{(1.1-0.975)^2 + (0.9-0.975)^2 + (1.1-0.975)^2 + (0.8-0.975)^2}{4(4-1)}} = 0.1275$$

$$x = \bar{x} \pm u_A = 0.98 \pm 0.13$$

Again, the more readings you use, the better the estimate will be. In this case, it is the estimate of uncertainty that improves with the number of readings (not the estimate of the mean or "end result"). In ordinary situations, 10 readings is enough. For a more thorough estimate, the results should be adjusted to take into account the number of readings.

A typical example of a **type B evaluation** is the transformation of a maximum/ minimum into a standard uncertainty. Suppose that only a minimum value x_{min} and a maximum value x_{max} are known for the characteristic value (reference value) attributed to a reference material. If all values in this interval are equally likely candidates of the true value, the mean and the standard deviation of a rectangular distribution with boundaries x_{min} and x_{max} can be used for the reference value x and its standard uncertainty $u(x)$.

$$x = \frac{\left(x_{max} + x_{min}\right)}{2} \qquad (5.20)$$

$$u(x) = \frac{\left(x_{max} + x_{min}\right)}{\sqrt{12}} \qquad (5.21)$$

However, if there is a reason to believe that values in the center of the interval are more likely than values at the boundaries, then, e.g., a symmetrical triangular distribution with boundaries x_{min} and x_{max} can be chosen instead of the rectangular distribution (uniform distribution). This gives:

$$u(x) = \frac{\left(x_{max} + x_{min}\right)}{\sqrt{24}} \qquad (5.22)$$

A Type B evaluation of standard uncertainty is usually based on scientific judgment using all the relevant information available, which may include:

- previous measurement data,
- experience with, or general knowledge of, the behavior and property of relevant materials and instruments,
- manufacturer's specifications,
- data provided in calibration and other reports, and
- uncertainties assigned to reference data taken from handbooks.

Generally, if the manufacturer provides the standard uncertainty, it is used directly. If very little information is available on an input quantity and its supposed variation interval comes under the form:

$$u_B\left(z_j\right) = \frac{z_{j\,max}}{k_s} \qquad (5.23)$$

where k_s is the factor depended on probability laws that are connected with source of uncertainty and z_j, which indicate the number of standard deviations that a particular value may be distant from the mean of the distribution. In case of normal distribution is $k_s = 2$ for 95% of the distribution, or 3 for 99,7% of the distribution. Where the information is insufficient (in some Type B estimates), you might only be able to estimate the upper and lower limits of uncertainty. You may then have to assume the value is equally likely to fall anywhere in between, i.e. a rectangular or uniform distribution. The standard uncertainty for a rectangular distribution is found from: $k_S = \sqrt{3} = 1.73$, for triangular $k_S = \sqrt{6} \doteq 2.45$ and for U-shaped $k_S = \sqrt{2} \doteq 1.41$.

If there are several sources of uncertainty, they should be combined to produce the combined standard uncertainty. This is done by taking square root of sum of squares of all the uncertainty values associated with the measurand, e.g.:

$$u_B(y) = \sqrt{\sum_{j=1}^{p} A_j^2 u_B^2(z_j)} \qquad (5.24)$$

where $u_B(z_j)$ are uncertainties of individual sources, A_j it's expansion factors.

Similarly, individual standard uncertainties calculated by Type A or Type B evaluations can be combined validly by "summation in quadrature" (also known as "root sum of the squares"). The result of this is called the combined standard uncertainty, shown by $u_C(y)$. Summation in quadrature is simplest where the result of a measurement is obtained by addition or subtraction.

5.4.6.1   How to Reduce Uncertainty in Measurement?

Always remember that it is usually as important to minimize uncertainties, as it is to quantify them. There are some good practices, which can help to reduce uncertainties in making measurements generally. A few recommendations are:

- Calibrate measuring instruments (or have them calibrated for you) and use the calibration corrections, which are given on the certificate.
- Make corrections to compensate for any (other) errors you know about.
- Make your measurements traceable to national standards – by using calibrations, which can be traced to national standards via an unbroken chain of measurements. You can place particular confidence in measurement traceability if the measurements are quality assured through a measurement accreditation.

- Choose the best measuring instruments and use calibration facilities with the smallest uncertainties.
- Check measurements by repeating them, or by getting someone else to repeat them from time to time or use other kinds of checks. Checking by a different method may be the best of all.
- Check calculations, and where numbers are copied from one place to another, check this too.
- Use an uncertainty budget to identify the worst uncertainties, and address these.
- Be aware that in a successive chain of calibrations, the uncertainty increases at every step of the chain.

Remember that there are many more sources of elemental uncertainty. If you need to classify them, apply the logic that if the analysis of the uncertainty can be done with statistics it is a Type A uncertainty. If statistics cannot be used (for example, if the uncertainty is systematic and repeatable) then it is to be treated as a Type B uncertainty.

If you wrongly classify a Type A elemental uncertainty as Type B, or vice versa, the consequences may not be significant. While the mistake will change the attribution of uncertainty to different components of the uncertainty analysis, the final estimated uncertainty would be the same, irrespective of wrong Type A/B classification!

KEYWORDS

- colorimeter
- colorimetry
- error
- image analysis
- measurement
- spectrophotometer
- spectroradiometer
- uncertainty

REFERENCES

1. CIE 15:2004 *Colorimetry*, 3rd edition, CIE, 2004.
2. JCGM 100:2008 Evaluation of measurement data – guide to the expression of uncertainty in measurement, *JCGM*, 2008.
3. Meyer, B., & Zollinger, H. R., (1989). *Colorimetry (in German)*, SANDOZ: Basel, pp. 5–34.
4. Nicodemus, F. E., et al., (1977). Geometrical considerations and nomenclature for reflectance. Washington, DC: National Bureau of Standards, US Department of Commerce.
5. Schaepman-Strub, G., Schaepman, M. E., Painter, T. H., Dangel, S., & Martonchik, J. V., (2006). Reflectance quantities in optical remote sensing—definitions and case studies, *Remote Sensing of Environment, 103*, 27–42.
6. Schroeder, G., (1981). *Technical Optics* (in Czech), SNTL Prague, pp. 131–143.
7. ASTM D 1003–00, (2000). Standard Test Method for Haze and Luminous Transmittance of Transparent Plastics.
8. ISO 14782 Plastics – Determination of Haze of Transparent Materials, 1999.
9. Hunt, R. W. G., & Pointer, R. M., (2011). *Measuring Colour*, 4th Edition; Wiley: Chichester, (The Wiley-IS&T Series in Imaging Science and Technology), pp. 114–215.
10. Hiltunen, J., (2002). *Accurate Color Measurement*, University of Joensuu, Department of Physics, Vaisala Laboratory, Dissertation 30.
11. Budde, W., Erb, W., & Hsia, J. J., (1982). International inter-comparison of absolute reflectance scales. *Color Research & Application, 7*, 24–27.
12. Zwinkels, J. C., & Erb, W., (1997). Comparison of absolute d/0 diffuse reflectance factor scales of the NRC and the PTB, *Metrologia, 34*, 357–363.
13. Nevas, S., Holopainen, S., Manoocheri, F., Ikonen, E., Liu, Y., Lang, T. H., & Xu, G., (2005). Comparison measurements of spectral diffuse reflectance. In: *Proceedings of the 9th International Conference on New Developments and Applications in Optical Radiometry*, Davos, Switzerland, pp. 17–19.
14. Germer, T. A., Zwinkels, J. C., & Tsai, B. K., (2014). Spectrophotometry: accurate measurement of optical properties of materials (*Experimental Methods in Physical Sciences, vol. 46*), Elsevier Amsterdam, pp. 271.
15. ASTM standard E 308–15, Standard practice for computing the colors of objects by using the CIE system, 2015.
16. Eitle, D., & Ganz, E., (1968). Method allowing estimation of color coordinates of fluorescent samples (in German), *Textilveredlung, 3*(8), 389–392.
17. Allen, E., (1973). Separation of the spectral radiance factor curve of fluorescent substances into reflected and fluoresced components, *Applied Optics, 12*, 289–293.
18. CIE 046–1979, A review of publications on properties and reflection values of material reflection standards, 1979.
19. Schanda, J., (2007). *Colorimetry, Understanding the CIE system*, John Wiley & Sons: Hoboken, pp. 57.
20. Newhall, S. M., Nickerson, D., & Judd, D. B., (1943). Final report of the OSA. subcommittee on spacing of the munsell colors, *J. Opt. Soc. Am., 33*(7), 385–418.
21. Verril, J. F., (1987). Physical standards in absorption and reflection spectrophotometry, in: *Advances in Standards and Methodology in Spectrophotometry*, Burgess, C., Mielenz, K. D., eds., Elsevier Amsterdam, 111–124.

22. Budde, W., (1976). Calibration of reflectance standards, *Journal of Research of the National Bureau of Standards – A. Physics and Chemistry*, *80A*(4), 585–595.

23. Sharma, G., (2003). *Digital Color Imaging, Handbook*, CRC Press: Boca Raton, pp. 92–98.

24. Johnson, G. M., & Fairchild, M. D., (1999). Full-spectral color calculations in realistic image synthesis, *IEEE Computer Graphics and Applications*, *19*(4), 47–53.

25. Wyszecki, G., & Stiles, W. S., (1982). *Color Science: Concepts and Methods, Quantitative Data and Formulae*, 2nd edn., Wiley: New York, pp. 130–163.

26. Jahne, B, Haußecker, H., (2000). *Computer Vision and Applications, A Guide for Students and Practitioners*, Academic Press San Diego, pp. 112–148.

27. Imai, F. H., & Berns, R. S., (1999). Spectral estimation using trichromatic digital cameras, in: *Proc. of the International Symposium on Multispectral Imaging and Color Reproduction for Digital Archives,* Chiba University, Chiba, Japan, pp. 42–49.

28. van der Meer, F. D., & De Jong, S. M., (2001). *Imaging Spectrometry, Basic Principles and Prospetive Applications,* Kluwer Academic Publisher Dordrecht, pp. 17–60.

29. Hardeberg Jon Yngve, (2001). Acquisition and reproduction of color images: Colorimetric and multispectral approaches, *Dissertation.com*, 2001, ISBN 1-58112-135-0.

30. Hunt, R. W. G., (1993). Current problems in colorimetry, *Die Farbe*, *39*(1–6), 1–12.

31. Vik, M., (2017). *Colorimetry in Textile Industry*, VUTS Liberec, pp. 111–113.

32. Eckschlager, K., (1971). *Errors in Analytical Chemistry* (in Czech), SNTL Prague, pp. 13–27.

33. Meloun, M., & Militky, J., (2002). *Matte Box of Statistical Data treatment* (in Czech); Academia, Praha, pp. 21–53.

34. Vdolecek, F., (2002). *Technical Measurement (in Czech)*, VUT, Brno, pp. 12–26.

35. Schovánek, P., & Havránek, V. (2016). *Errors and Uncertainties in Measurement* (in Czech), Modern technologies in applied physics study CZ.1.07/2.2.00/07.0018, available in: https://fyzika.upol.cz/cs/system/files/download/vujtek/texty/pext2-nejistoty.pdf.

36. JCGM 100:2008 GUM 1995 with minor corrections, Evaluation of measurement data–Guide to the expression of uncertainty in measurement, 2008.

37. Malkin, F., Larkin, J. A., Verrill, J. F., & Wardman, R. H., (1997). *Coloration Technology (J. Soc. D. Col.)*, *113*, 84–94.

CHAPTER 6

SPECTROPHOTOMETRY OF COLOR CHANGE

MARTINA VIKOVÁ

CONTENTS

Abstract..283
6.1   Specification of Device for Color Change Measurement...............284
6.2   Interactions Between Light and Media..305
6.3   Spectral Data of Chromic Materials 315
6.4   Determination of Absorption and Scattering Coefficients..............331
6.5   Measurement of Thermal Sensitivity of Photochromic
       Materials 348
6.6   Impact of Drawing Ratio on Difference in Optical Density............354
Keywords...365
References..365

ABSTRACT

In this chapter, the methods suitable and applicable for color changeable materials measurement are discussed. Systems suitable for these measurements that could measure kinetic for such kind of this color change behavior are described also in this chapter. Such device measurements have requirements such as which standard light sources we can use, setup of heating and cooling sources (thermostats), and setup of device for kinetic measurement including software and hardware. Calibration standards for this instrumentation to measure color change is also provided. It is necessary also know timing of measurement and the sequence of each individual measurement. Setup of measurement depends on requirements of colorimetric parameters or spectral parameters like absorption, reflection, transmittance, or dispersion.

6.1   SPECIFICATION OF DEVICE FOR COLOR CHANGE MEASUREMENT

Spectrophotometers are classified according to their area of application and according to whether polychromatic or monochromatic illumination is used. Analytical spectrophotometers use monochromatic illumination. Polychromatic illumination is used in certain new spectrophotometers, specifically spectrophotometers used to acquire spectral data in a short time. Unfortunately, these systems are limited in the width of the measured band pass in comparison to the systems with monochromatic illumination.

6.1.1   TRANSMITTANCE MEASUREMENT OF TRANSPARENT MEDIA

In this case, the usual measurement method involves a solution in a cuvette, which allows either lateral or axial illumination of the sample, as shown in Figure 6.1a and 6.1b, respectively. These diagrams show an experimental set-up for discontinuous irradiation with respect to time. It is evident that the same experimental set-up can be used for continuous irradiation. The light source for exposure can be, for example, a laser, discharge lamp, LED, or other sources. Analytical spectrophotometers typically measure only transparent liquids in transmission, as their 0°/0° measurement geometry makes

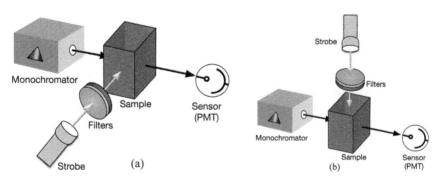

FIGURE 6.1   (a) Lateral exposure and (b) axial exposure.

them very sensitive to scattering when measuring hazy samples that scatter light. The time resolution of modern diode array spectrophotometers is around 0.5 s for a completely scanned spectrum. The measurement time for most analytical spectrophotometers is much longer, typically about a minute, due to the extended scanning range and smaller scanning wavelength interval.

Schematic diagram of the apparatus. The photomultiplier tube (PMT) monitors the intensity of the monochromatic light passing through sample (S). A flash from the strobe source is focused by lenses (L) and attenuated by the neutral density filter (F) before it falls on the sample. The sample is held in an aluminum block, through which water at a given temperature circulates. The temperature of the sample is measured directly with a thermocouple (not shown) in Williams [1].

In the case of continuous irradiation experiments, the diode array spectrophotometer measuring system allows the study of photochromic systems, when the lifetime of color change is longer than 50 ms. If systems with faster photochromic changes are to be studied, it is necessary to use the discontinuous (pulsed) irradiation measurement method. A typical example is a time-resolved method using a femtosecond laser pumped probe technique [2, 3]. If the set-up of the spectrophotometer also includes the irradiation monochromator or a system of band pass filters, it becomes possible to study the spectral sensitivity of the sample.

6.1.2 TRANSMITTANCE MEASUREMENT OF TRANSLUCENT MEDIA

The measurement of translucent media presents a specific problem that results from the combination of transmission and reflection of light. In translucent media, it is possible to measure in either transmission or reflection mode. In the case of the measurement of transmittance of light through translucent media, it is necessary to consider scattering caused by the turbidity of these media. In this case, when it is necessary to assume not only regular but also diffuse transmission, it is necessary to use the integrating sphere for measurement. Usually an adaptation of analytical spectrophotometers is used, in which the integrating sphere is placed in the cuvette space as shown in Figure 6.2a. Haze-meters for the measurement of opacity have similar construction. For the measurement of the data from the complete spectral range, it is necessary to equip the spectrophotometer with a photodiode array detector or other type of linear sensor. In this way, the spectrophotometers are equipped for the measurement of colored surfaces, and therefore, the inverted adjustment can be used as the variation, when the sample is illuminated diffusely from the integrating sphere as described in the figures.

On the basis of these reasons, for this part of the experimental work, a solution spectrophotometer ACS ChromaSensor CS-5 with spherical measuring geometry was adapted for measurement. The instrument was placed in a dark room to eliminate luminous irradiation from other light sources. Figure 6.3 shows the arrangement. Usually the concentration in solution is roughly 100 times lower than that used in this work. In the research described here, such a high concentration was used to provide the possibility to compare the behavior of photochromic pigments under conditions similar to those given by the pigments inside the polymer, as in the case of mass dyeing.

As discussed above, it is necessary to use a source of excitation to allow the measurement of photochromic materials. The main target was experimental work involving continuous illumination of the measured sample using sources operating in the UV-visible region. Suitable sources are those involving continuous discharge.

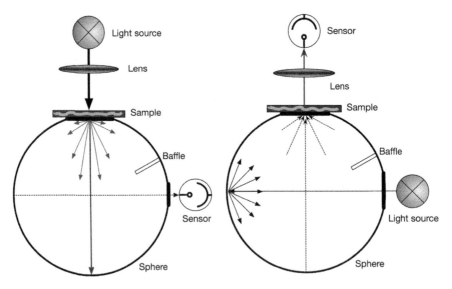

FIGURE 6.2 (a) Direct measurement of total transmission and (b) Inverted measurement of total transmission.

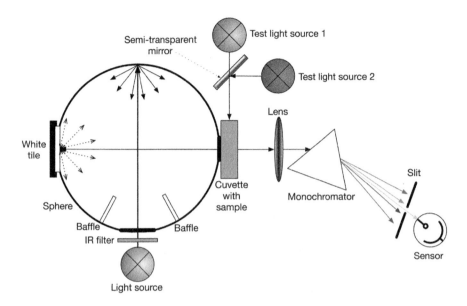

FIGURE 6.3 Optical scheme of the modified spectrophotometer CS-5 with selected light sources for the transmission measurement of photochromic solutions.

These light sources have their own individual characteristics of spectral power distribution and thus have abilities to evoke different intensities of the photochromic effect. Most discharge lamps have a bright-line spectrum, and in their spectrum, there are characteristic spectral gaps. As shown in Figure 6.3, the measuring system is equipped with two test light sources. The experimental set-ups were selected from the range of available light sources, the UV lamp, deuterium lamp, germicide lamp and xenon discharge lamp. Their spectral characteristics are illustrated in Figure 6.4. In a test of the spectral sensitivity of photochromic pigments, two independent test light sources were used. This design of this measuring system allows the use of light sources independently or in combination.

During the measurement of photochromic solutions, an interesting problem was observed, which can be described as nonuniform coloration. As shown in Figure 6.3, lateral illumination was used for the excitation. It was observed that only the upper surface was fully colored, while the rest of sample did not develop color. It appears that the colored upper surface acts as a UV filter inhibiting photochromic conversion further into the sample.

As illustrated in Figure 6.5, photochromic materials usually give rise to significant changes in spectral features in the visible and UV regions of the

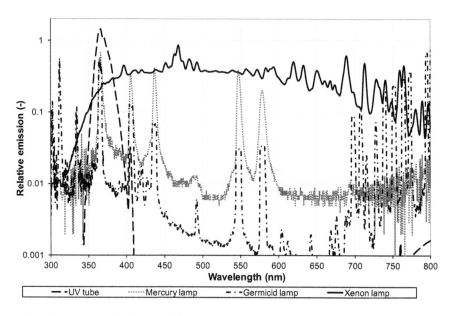

FIGURE 6.4   Spectral characteristics of the tested exciting light sources.

electromagnetic spectrum when irradiated by UV light. However, in the UV region, as the spectra reveal, only a small change in absorption intensity and shift of the absorption maximum is observed as a result of the change in the molecular conformation, which accompanies UV irradiation. From this, it follows that the higher concentration of photochromic compounds has an influence on effectively blocking UV irradiation during transmission.

As observed in this experiment, some product at the beginning of the measurement is formed in a thin upper layer of solvent, which separates from the rest of the solution in the cuvette. This results in insufficient homogenization of concentrated, saturated solutions. To eliminate this effect, a hot homogenization process was used, in which the solvent (cyclohexane, C_6H_{12}) was heated to a maximum temperature of 60°C. After heating, the photochromic solution was slowly chilled to 25°C with continuous mixing. With this technique, sedimentation was eliminated. Together with sedimentation compensation, the set-up for illumination contains excitation together with lateral and axial illumination of the sample with the help of optical mirrors.

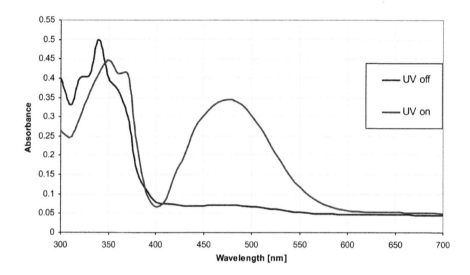

FIGURE 6.5 Spectral characteristics a of photochromic compound in the photostationary state (PS) before and after irradiation (methyl 2,2-bis(4-methoxyphenyl)-6-acetoxy-2H-naphtho-[1,2-b]pyran-5-carboxylate – pigment P3).

6.1.3 MEASUREMENT OF TRANSLUCENT MEDIA BY REFLECTANCE

In measuring colored surfaces, the basic problem with a photochromic system becomes the controlled exposure by the selected irradiation. As follows from the previously discussed description of the basic systems of illumination and observation in colorimetric devices, these systems are not usually designed for the incorporation of other light sources. For the analytical spectrophotometer, it is possible to use a simple adaptation of a Praying-Mantis accessory [4, 5], which is presented in Figure 6.6. This system allows the incorporation of a light source for irradiation in the measurement of photochromic surfaces. A disadvantage of this system is the reality that only a small area is measured and this configuration can be used only for homogenous smooth surfaces.

Based on this knowledge, the measuring system constructed contains an Avantes USB 2000 spectrometer. This spectrometer was coupled to the measuring probe by an optical cable 200 μm in diameter. The illumination

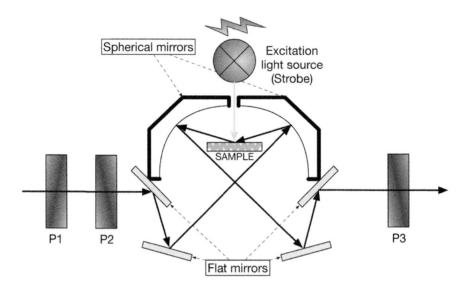

FIGURE 6.6 Modification of the Praying-Mantis optical accessory for the measurement of the photochromic property (P1, P2, and P3 are the possible positions of polarized filters).

used for excitation was the same combination of light sources as that for the measurement of concentrated solutions. There were two exciting sources.

Because the spectrophotometric measurements are based on a comparison of incident and transmitted (reflected) light at individual wavelengths in the visible region of the spectrum, it is necessary to use a special stabilized light source. This light source has added a special cut-off filter GG395, which blocks the radiation of wavelengths lower than 395 nm, to eliminate initiation of the photochromic effect by the light source of the spectrophotometer. The optical scheme of the spectrophotometer is shown in Figure 6.7a and 6.7b, which also show the detail of the measuring probe.

During the testing of this system, it was observed that for the measurement of textile samples, it is not suitable to use the angle viewing geometry. This means that in the case where illumination is realized only from one side, the reproducibility is relatively low.

Many commercially available spectrophotometers and colorimeters have cylindrical illumination, in which the illumination is realized by optical fibers or a cylindrical mirror. Nevertheless, the compact construction of the measuring head does not permit simple inclusion of an excitation light source. Optical fibers, which are used for illumination, are not suitable for excitation.

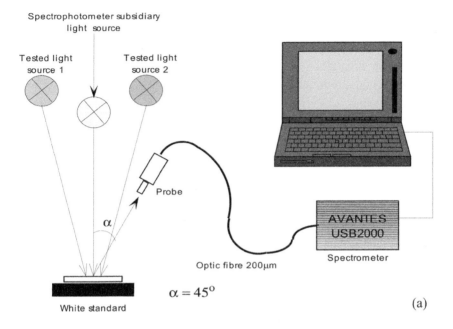

Spectrophotometer subsidiary
light source

Tested light
source 1

Tested light
source 2

α

Probe

Optic fibre 200μm

White standard

AVANTES
USB2000

Spectrometer

$$\alpha = 45^{\circ}$$

(a)

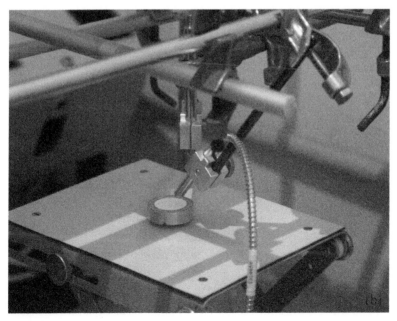

(b)

FIGURE 6.7   (a) Optical scheme of the measuring system "Photochrom I" for remission measurement of textiles with photochromic prints and mass dyed non-woven and (b) detail of the measuring probe.

A standard optical fiber causes a large reduction in intensity of UV irradiation [26–28]. In these fibers, the limitations for UV light transmission are the intrinsic attenuation (in dB/m) and additional losses (in dB) due to UV defects. The intrinsic attenuation is given by Rayleigh-scattering, electronic transitions, and a non-structured OH-absorption tail below 200 nm. Most spectroscopic applications with fiber optics have been restricted to wavelength ranges above 230 nm, because standard silica fibers with an undopped core and fluorine doped cladding are frequently damaged by exposure to deep-UV light (below 230 nm). This solarization effect is induced by the formation of "color centers" with an absorption band at 214 nm. These color centers are formed when impurities (such as Cl) exist in the core fiber material and form unbound electron pairs on the Si atom, which are affected by the deep UV radiation. Solarization resistant fibers use a modified core preform that protects the fiber from the damaging effects of deep UV. These fibers have excellent performance and long-term stability at 30% to 40% transmission (for 215 nm). Nevertheless, commercially available solarization resistant optical fibers are limited in diameter to 1000 μm [6]. The result of a small diameter of the optical fiber is a small area of measured sample, which is irradiated by the excitation irradiation. This problem may be solved by a multifiber irradiation design. Nevertheless, as mentioned above, an insufficient space in the measuring head of commercially available spectrophotometers makes this impossible. Based on this experience, the research described in this thesis employed commercial reflectance spectrophotometers as used by the textile industry. Two spectrophotometers were tested, both of which were available in our laboratory:

- Spectraflash SF300 – geometry d/8°, measuring aperture 20 mm in the SCI mode (Figure 6.8a);
- Microflash 200d – geometry d/8°, measuring aperture 5 mm in the SCI mode (Figure 6.8b).

Both spectrophotometers have the same diameter of integrating sphere (66 mm), and they allow the measurement in both modes SCI (specular component included) and SCE (specular component excluded).

These spectrophotometers are equipped with xenon lamps as the light source, and therefore, it was necessary to stop the excitation of the photochromic solution during the measurement of spectral and colorimetric characteristics. Similarly, as previously described, cut-off filters GG395 were used for this purpose. The cut-off filters were placed between the xenon

(a)

(b)

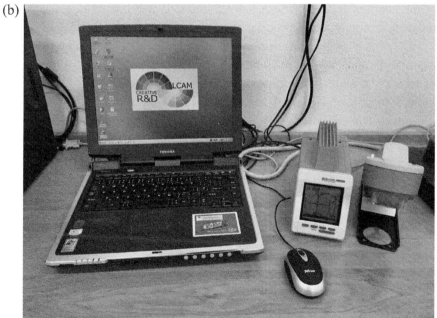

FIGURE 6.8 (a) Spectraflash SF 300 and (b) Microflash 200 d.

lamp and the measuring integrating sphere to eliminate UV irradiation and the influence on the calibration of the measuring system. The placement of cut-off filter between the sample and aperture of the device presented some problems, such as for example the uncontrolled light outlet in different directions, not only perpendicular to the sample. This type of measurement did not influence the sample during the excitation. There was also the need to solve the issue of the excitation light used to initiate the photochromism.

The first experiment with photochromic pigments applied onto a textile substrate showed that exposure and reversal is relatively long (minutes). The experimentation can be realized as an off-line system, with the exposure first controlled followed by quick measurement of the exposed sample in a stationary exposed state-in equilibrium. For initiation of a photochromic effect, a viewing box JUDGE II (Gretag Macbeth, USA) was used. To increase the effect of UV irradiation, a combination of a D65 simulator and a UV fluorescent tube was used. For illustration of the system used, the configuration of a light box is shown in Figure 6.9a, and the spectral power distribution of the experimental illumination arrangement used is given in Figure 6.9b.

The irradiance or illuminance falling on any surface varies as the cosine of the incident angle, θ. The perceived measurement area orthogonal to the incident flux is reduced at oblique angles, causing light to spread out over a wider area than it would if perpendicular to the measurement plane. For example, to measure the amount of light falling on human skin, it is necessary to mimic the skin's cosine response. Because filter rings restrict off-angle light, a cosine diffuser must be used to correct the spatial responsivity. In full immersion applications, such as the phototherapy booth, off angle light is significant, requiring accurate cosine correction optics. The cosine correctors involve spectroradiometer sampling optics, designed to collect radiation (light) over 180°, thus eliminating optical interface problems associated with the light collection sampling geometry inherent in other sampling devices Therefore, the Avantes USB2000 fiber optic spectrometer with cosine corrector was used for measurement of the spectral power distribution of the light sources used. The CC-UV/VIS cosine corrector has an active area of 3.9 mm, with a Teflon® diffusing material and is optimized for applications from 200–800 nm. These cosine correctors can be used to measure UV-A and UV-B solar radiation, environmental light, lamps and other emission sources.

Besides the cosine corrector, a solarization resistant optical fiber was used with a diameter of 200 μm. This was necessary to control the spectrometer

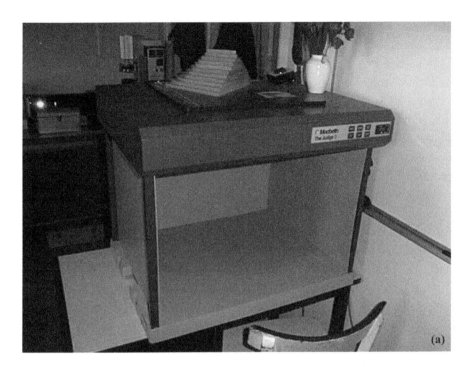

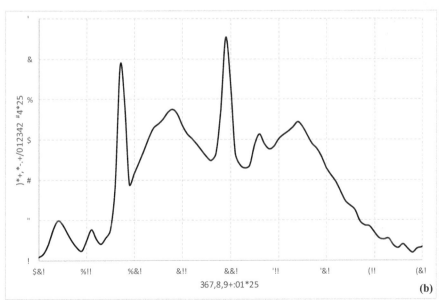

FIGURE 6.9 (a) Light box JUDGE II and (b) Spectral power distribution of the experimental illumination.

sensitivity by a calibrated light source, which was also produced by the Avantes company. The AvaLight-DH-CAL is a calibrated light source for the UV/VIS/NIR spectral range (205–1095 nm). This National Institute of Standards and Technology – traceable calibrated light source was developed for use with all AvaSpec spectrometers to be used in measuring absolute spectral intensity. The AvaLight-DH-CAL comes with a cosine corrector with a SMA 905 (SubMiniature version A) connector, which has screw, ferule diameter is 3.14 mm and usually is used for industrial lasers, military, and telecom multimode.

In addition, lighting parameters were measured in the viewing box Judge II Gretag Macbeth, as given in Table 6.1. Because the experiment has been realized in the standard viewing box, it was also necessary to establish irradiance intensity and whether the intensity is constant in all areas of the viewing box. Thus, for the measurement of illuminance, the illuminance meter Minolta IT10 was used. For the measurement of UV-A radiation intensity, an MIT Goldilux Radiometer was used. Illuminance and irradiance results are given in Figure 6.10, where the distribution of light intensity in the viewing box is documented.

Figure 6.10 Illustrates that the maximum of illuminance E was in the central position of the viewing box and the area with maximum of illuminance E was approximately a rectangle of 5 × 10 cm. In a chosen area of sample position was counted ratio of minimal and maximal illuminance equal to 0.964, which relates to central part of viewing box with highest level of illuminance as shown on graph in Figure 6.10.

The central position with maximum of illuminance E was hence used as the target for every exposed sample to obtain the same radiation level during each experiment for each sample.

Table 6.1 also documents the changes in exposure dependent on the distance of the sample from the light sources in the viewing box. To obtain constant conditions of exposure for every sample, a calibrated telescopic table was used, which was placed inside the bottom of the viewing box. Every sample was placed on the upper deck of the table, where the distance between the front layer of the sample and the light source used was controlled. This system allowed the use of different distances of sample from light source repeatable way, because the precision of the telescopic table was 10 μm.

Measurements of the photochromic substrates were carried out by separate exposure and measurement. Illumination inside the commercial reflectance

TABLE 6.1 Parameters of Light Involved in Viewing with Judge II Gretag Macbeth

Distance between light source and sample	Illumi- nance E lx	Irradi- ance Fe µWcm⁻²	Irradi- ance dose He µJcm⁻²	Correlated Color Tempera- ture CCT K	Lumi- nous flux Φv lm	Radiant flux Φe µW	Radiant energy Qe µJ
45 cm	979.3	714.6	10.00	6649	0.012	85.4	1.20
31 cm	1754.2	1149.7	16.10	6664	0.021	137.3	1.92
21 cm	2631.5	1750.4	24.51	6693	0.031	209.1	2.93
17 cm	3406.7	2077.6	29.09	6713	0.041	248.2	3.47
13.5 cm	4138.3	2436.7	34.11	6743	0.049	291.1	4.08
10 cm	5152.5	2895.0	40.53	6742	0.062	345.8	4.84
6.5 cm	6693.8	3522.5	49.32	6769	0.080	420.8	5.89
3 cm	9651.0	4621.7	64.70	6760	0.115	552.1	7.73

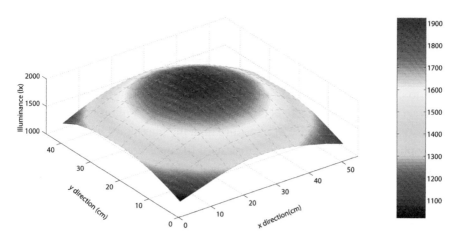

FIGURE 6.10 Distribution of light intensity in the viewing box Judge II Gretag Macbeth.

spectrophotometers was modified by cut-off filter GG395, because it was necessary to minimize the influence of the spectrophotometer light source on the photochromic effect. On the other side, it was necessary to use external exposure by the selected light source, because these spectrophotometers do not allow simple incorporation of an external light source as an accessory.

A method where the sample is exposed in one place and after exposure moved to another place for the measurement of actual color causes a time delay that is dependent on movement velocity between the place of exposure and the place of measurement. In such methodology, the results show poor reproducibility of measurement of the exposure phase of the photochromic color change.

Figure 6.11 illustrates the result for 3,3,5,6-tetramethyl-1-propylspiro [indoline-2,3'[3H] pyrido[3,2-f] [1,4] benzoxazine] (pigment P1), which was obtained as an average of 10 measurements of both phases. The measured error was less than 1.3% in the reversion phase. This explains why in the graph given in Figure 6.11 the decay curve errors bars are practically invisible in comparison with those for the exposure curve.

It is evident that the variability of measurement in the decay phase from the maximum from which the decay phase starts is lower. The lower variability during the decay phase documents that the problem of high variability during the exposure phase is caused by the variability of the time delay between exposure and color measurement, because the sample is moved from one place to another place. The variation of the measured data set for exposure is dependent also on the rate of photochromic color change as Figure 6.12 documents. There is higher variance for a sample with the pigment based on the chromene system, [34] – methyl2,2-bis(4-methoxyphenyl)-6-acetoxy-2H-naphtho-[1,2-b]pyran-5-carboxylate (pigment P3). It is possible to use the model of first order kinetics in the case of this data set, but the precision of measurement is lower. The maximum of photochromic color intensity is lower in the case of exposure than the maximum from which the decay phase starts. The reason in the case of the measurement of the sample with pigment P3 is associated with the methodology of measurement. The problem is the impossibility of exposure of the photochromic sample during measurement using a standard commercial spectrophotometer.

Due to the hemispherical illumination, the problem of the texture of the textile samples. The procedure for obtaining the data set was as follows. First, the data set in the decay phase for individual time cycles was obtained. This means that the sample was exposed for 15 min and then removed from the place of exposure directly to the measuring aperture of the spectrophotometer. Measurement of shade intensity I was conducted at 0, 30, 60, 90, 120, 180, 240, 300 and 600-s intervals. The data set for exposure was obtained via exposure of the sample followed by immediate measurement of colorimetric parameters. After exposure, the relaxation phase of the sample was continued in a black box. The time of relaxation was constant at 15 minutes. The time scale for the exposure phase was the same as for reversion phase, i.e., 0, 60, 90, 120, 180, 240, 300, and 600 s.

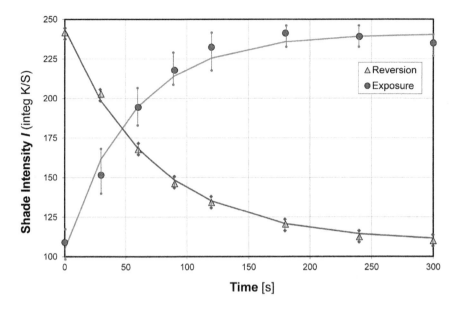

FIGURE 6.11 Off-line measurement of growth and decay processes of color change intensity for 3,3,5,6-tetramethyl-1-propylspiro [indoline-2,3'[3H] pyrido [3,2-f][1,4] benzoxazine] (pigment P1), irradiance = 714,6 μW·cm⁻².

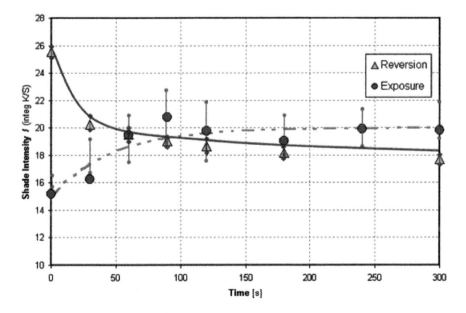

FIGURE 6.12 Off-line measurement of growth and decay processes of color change intensity for pigment P3, irradiance = 714,6 μW·cm⁻².

The method for obtaining the data set for photochromic color change on exposure is similar to the fatigue test and does not involve continuous exposure. From the above mentioned description, it is evident that instead of cyclic exposure, this method has the problem of scattering due to the time delay between the exposure and the measurement. Effort was made to minimize the delay (there was a team), but it was not possible to minimize this problem for a "quicker speed" system.

Based on this, there was a necessity to develop a special measuring system that allowed the measurement of a textured photochromic color surface. For textured photochromic surfaces, it was necessary to use the adaptation of a hemispherical illumination in the integrating sphere, so that it was possible to include another aperture for irradiation [7–9]. The optical scheme of the system, now patented, is given in Figure 6.13.

Due to the hemispherical illumination, the problem of the texture of the textile samples was eliminated. Because the reversion from the colored form to the colorless form is promoted thermally, the photochromic structure will achieve lower saturated absorbance at higher temperatures than at lower temperatures. This phenomenon is known as temperature dependency, for example, oxazine structures are more sensitive in comparison to

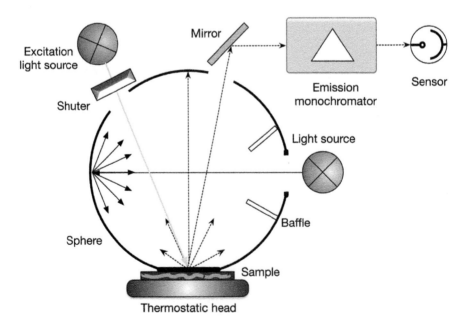

FIGURE 6.13 Optical scheme of the LCAM PHOTOCHROM2 measuring system.

naphthopyrans [10]. Figure 6.14 shows the temperature dependence of such type of photochromic pigments in which K/S values were calculated at the appropriate absorption maximum.

Recognizing that photochromic color change is affected by temperature, there was a necessity to carefully stabilize the temperature of the sample during measurement. If the temperature is not stabilized, it is possible to observe local fading of color during exposure. This phenomenon is illustrated in Figure 6.15. Light fading can be influenced by increasing the temperature of the measured sample or by photobleaching of the photochromic pigment. Nevertheless, this phenomenon was not observed for the measured samples after thermal stabilization.

A final problem that needed to be solved in the construction of the unique spectrophotometer was the question of spectral sensitivity of the selected samples of photochromic textile materials. The light sources selected in the first part of the development of a system for the measurement of colorimetric and spectrophotometric data for photochromic textiles were tested. Except for the xenon lamp, all sources had spectral lines. This is evident also for

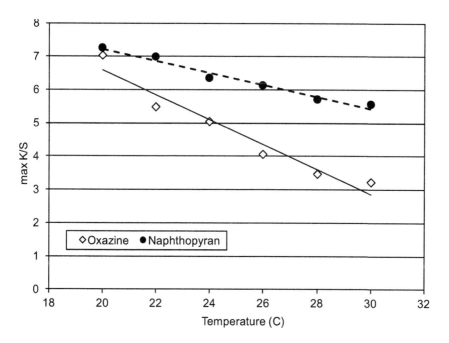

FIGURE 6.14 Temperature dependence of oxazine and naphthopyran type of photochromic pigment on textile substrate.

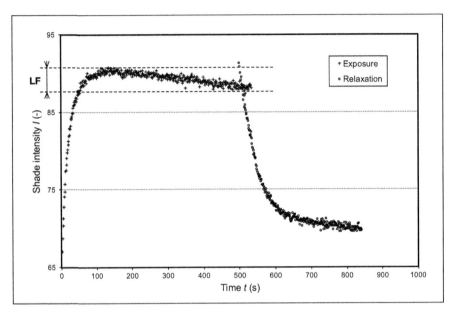

FIGURE 6.15 An example of light fading effect.

the UV lamp, which is a short band pass light source. The disadvantage of this source is the presence of many gaps in the spectrum, and these gaps can negatively influence measurement results. The absorption spectra of the photochromic systems examined are quite different, and thus, the monochromatic stimulation at specific wavelengths based on the light source affects the response obtained from the photochromic systems.

In the second part of the construction and adaptation of the measuring system, which allows the measurement of spectral sensitivity, the interference filters were used. These filters can be used as a mercury line set at a 10-nm band pass in combination with the 100-W xenon lamp. The main aim was to find out the wavelength where the pigments have highest sensitivity (i.e., the shade is the deepest) and other set up parameters are constant. After the first tests, it was evident that the intensity of transmitted light was not sufficient for excitation of the photochromic effect, and the intensity of developed shade was low. From this consideration, a 450-W Xe lamp was used as excitation light source with sufficient amount of radiance. Therefore, it was decided to use a monochromator based on a diffraction grating and a 450-W Xe lamp. This monochromator has the advantage that the intensity of excitation illumination may also be modulated by the size of slit.

The final system, LCAM PHOTOCHROM3 shown in Figure 6.17, is a system that allows the study of photochromism kinetics, the influence of exposure time, the thermal sensitivity, and in this case includes a test for the spectral sensitivity of the photochromic samples via an excitation mono-chromator [11]. The dual light source construction of the spectrophotometer with the shutter over the exciting light source makes possible continuous measurement of photochromic color change during exposure and reversion after the shutdown of the exciting light source. A xenon discharge lamp with continuous discharge was used. Pulsed discharge lamps allow using such system as a fatigue tester for photochromic systems with a half-life of the photochromic color change around 500 ms. For textile samples, which are frequently slower, it is advantageous to use an electronic shutter and a con-tinuous discharge lamp for fatigue tests (Figure 6.16).

The basic difference between the "LCAM-PHOTOCHROM3" system that is shown on Figure 6.17 and a standard spectrophotometer is in the continuous illumination of the sample against a flash source that is used as

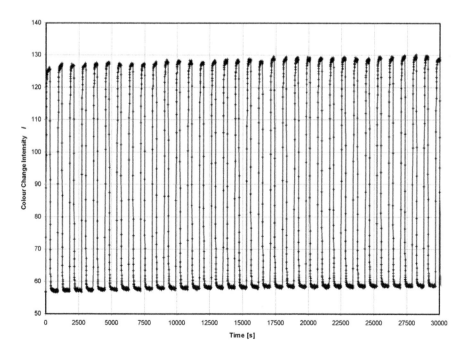

FIGURE 6.16 Selected part of a cyclic fatigue test on a photochromic textile sensor of UV-A radiation.

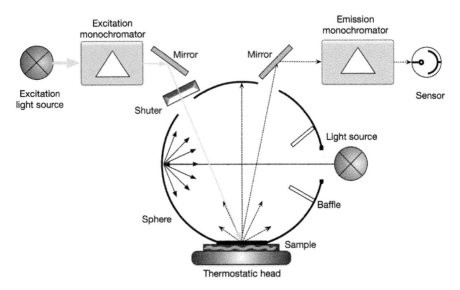

FIGURE 6.17   Simplified optical scheme of PHOTOCHROM3 device.

standard. In such a measuring system, it is possible to use a flash discharge lamp.

The reflectance values obtained are independent of whether it is a xenon, mercury, or other strobe source. This construction allows measurement of the color change of the sample in fast scans without disturbing the drift of the light source. The system developed allows 5-ms intervals between each measuring scan. Nevertheless, it was found that 5 s was a sufficient interval for the abovementioned samples (see Figure 6.18) [12].

A demonstration of the reproducibility of the PHOTOCHROM3 system is given in Table 6.2 and Figure 6.19. In this case, data sets are for 10 different individual samples. The results show the presence of experimental errors, which are due to the measuring system, are together with the influence of the method of preparation of individual samples. In the case of the repeatability of device PHOTOCHROM3, the data in Table 6.3 are presented. These data result from the short-term drift test procedure, when the white tile from the BCRA set was measured 20 times with 30s delay between each measurement. These data allow the calculation of the total color difference for the short time drift test.

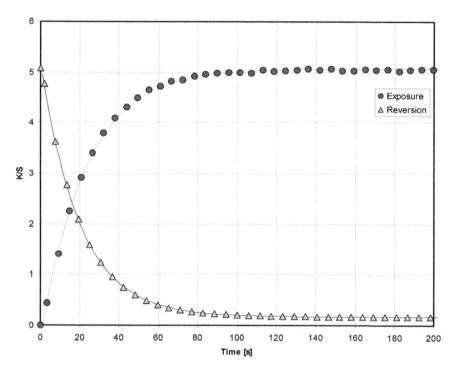

FIGURE 6.18   On/off kinetics of Photopia® AQ Ink Purple.

6.2   INTERACTIONS BETWEEN LIGHT AND MEDIA

6.2.1   *OPTICAL MODELS OF TRANSLUCENT MEDIA*

When a beam of light strikes a material medium, it is propagated through the thickness of the medium, where it experiences absorption while being transmitted according to the laws of geometrical optics. That is to say, the beam is deflected from the rectilinear path of light expected in a vacuum because of refraction, reflection, diffraction, and diffusion. This happens a number of times depending on the quality of the medium (homogeneous or not, transparent or opaque, smooth or rough). The beam is then deprived of absorbed parts at each wavelength and the resulting emitted spectrum constitutes the spectrum of reflectance. Reflectance measures the capacity of a surface to reflect the incident energy flux, given that the medium beneath this surface can absorb photons and diffuse light once or many times. Reflectance is defined as the ratio of reflected to incident radiation energy flux for a given

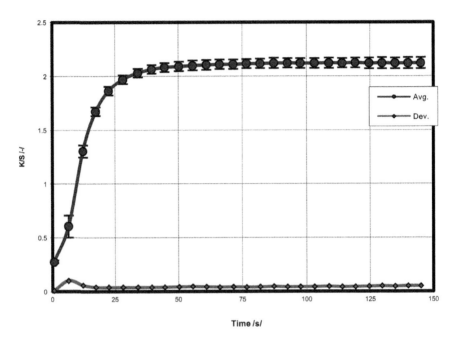

FIGURE 6.19 Reproducibility of the measurement the K/S function for Photopia Blue, t = 20°C.

wavelength and is thus a dimensionless quantity, usually expressed as a percentage [13].

The reflectance spectrum consists of a plot of reflectance values R versus wavelength λ in the visible spectrum range (from 380 to 780 nm). One curve $R(\lambda)$ uniquely characterizes one color: the reflectance spectrum objectively quantifies the color itself.

When light impinges on dispersed particles, such as pigmented fibers, in addition to being partially absorbed, it is also scattered, changing its direction arbitrarily and several times in space. Light that passes through such a turbid media loses a considerable part of its intensity, and the relationship between the colorant concentration and the observed spectral response becomes more and more complicated [14]. The analysis of light scattering in translucent, turbid media such as clouds, ice structures, human tissues, food, paper and textiles is a wide and yet increasingly active research field. Hereafter, a few of the classical approaches are reviewed. An extended survey of the reflectance theories for the reflectance of diffusing media is given in reference [15].

TABLE 6.2 Parameters of Measured Reproducibility K/S Function for Sample Photopia Blue, $t = 20°C$, No. of Samples=10

Time (s)	Avg.	Dev.	Var.	Min.	Max.
0.73	0.27764	0.009866	9.73324E-05	0.258303	0.296977
6.53	0.60607	0.102576	0.010521842	0.405021	0.807119
12.31	1.30319	0.058624	0.003436765	1.188287	1.418093
17.39	1.67229	0.040243	0.001619483	1.593414	1.751166
22.54	1.86343	0.038791	0.001504776	1.787399	1.939461
28.2	1.9707	0.039743	0.001579528	1.892803	2.048597
33.99	2.03093	0.039798	0.00158387	1.952926	2.108934
39.48	2.06344	0.040655	0.001652806	1.983757	2.143123
44.74	2.0828	0.042192	0.001780148	2.000104	2.165496
50.37	2.09062	0.044602	0.001989348	2.0032	2.17804
55.65	2.10021	0.046383	0.002151409	2.009299	2.191121
61.05	2.10499	0.0445	0.001980291	2.017769	2.192211
66.23	2.10937	0.042407	0.001798344	2.026253	2.192487
71.55	2.10994	0.043612	0.001902002	2.024461	2.195419
76.68	2.11581	0.042303	0.001789523	2.032897	2.198723
82.07	2.11694	0.042731	0.001825916	2.033188	2.200692
87.41	2.12045	0.047027	0.002211554	2.028277	2.212623
92.72	2.1215	0.0433	0.00187493	2.036631	2.206369
97.73	2.12006	0.044299	0.00196244	2.033233	2.206887
103.13	2.11952	0.045088	0.002032948	2.031147	2.207893
108.82	2.1228	0.047632	0.002268794	2.029442	2.216158
114.05	2.12144	0.044026	0.001938322	2.035148	2.207732
119.17	2.12385	0.0456	0.002079338	2.034474	2.213226
124.64	2.11964	0.047705	0.002275802	2.026137	2.213143
129.9	2.12185	0.051068	0.002607929	2.021757	2.221943
135	2.1197	0.048493	0.002351562	2.024654	2.214746
140.08	2.12063	0.052327	0.002738164	2.018068	2.223192
145.13	2.12228	0.051864	0.002689864	2.020627	2.223933

Rayleigh scattering is a fundamental theory that provides a tool to analyze the phenomenon of light scattered by air molecules [16]. According to this theory, the scattering power is strongly wavelength-dependent, which explains the blue appearance of the sky away from the sun. The theory can

TABLE 6.3 Results of Short-Term Drift Test of PHOTOCHROM3 at Selected Wavelengths

Wavelength/nm/	Max.	Dev.	Var.	Avg.	Min.
400	82.924	0.019676	0.000387	82.889	82.853
410	85.292	0.013793	0.00019	85.269	85.248
420	86.399	0.012472	0.000156	86.379	86.358
430	87.258	0.011302	0.000128	87.238	87.208
440	87.639	0.013204	0.000174	87.61	87.585
450	87.997	0.012006	0.000144	87.968	87.944
460	88.518	0.008822	7.78E-05	88.503	88.483
470	88.933	0.011291	0.000127	88.912	88.893
480	89.277	0.011487	0.000132	89.248	89.228
490	89.408	0.010947	0.00012	89.388	89.372
500	89.648	0.010547	0.000111	89.633	89.613
510	89.776	0.010234	0.000105	89.759	89.737
520	89.844	0.008499	7.22E-05	89.826	89.81
530	89.96	0.009867	9.73E-05	89.939	89.918
540	90.026	0.008814	7.77E-05	90.009	89.995
550	90.097	0.009828	9.66E-05	90.079	90.06
560	90.109	0.010585	0.000112	90.093	90.074
570	90.043	0.009835	9.67E-05	90.023	90.007
580	90.001	0.01206	0.000145	89.973	89.955
590	90.084	0.01413	0.0002	90.056	90.032
600	90.208	0.012253	0.00015	90.186	90.165
610	90.271	0.010178	0.000104	90.254	90.235
620	90.299	0.009605	9.22E-05	90.278	90.263
630	90.3	0.011661	0.000136	90.277	90.254
640	90.317	0.010453	0.000109	90.298	90.279
650	90.404	0.008649	7.48E-05	90.379	90.365
660	90.483	0.00923	8.52E-05	90.461	90.446
670	90.55	0.00773	5.97E-05	90.536	90.518
680	90.69	0.010254	0.000105	90.671	90.642
690	90.761	0.009345	8.73E-05	90.742	90.719
700	90.812	0.009195	8.45E-05	90.793	90.773

be extended to light scattered from particles with a diameter of up to a maximum of a tenth of the radiant wavelength. This restriction makes the theory impractical for analyzing the scattering of light in technical opaque or translucent media.

The Mie theory [16], on the other hand, describes the scattered light from a single spherical particle with a diameter that may be even larger than the scattered wavelength. Mie scattering is less wavelength-dependent than Rayleigh scattering and explains the almost white glare around the sun and the neutral colors of fog and clouds. It is a single scattering theory that ignores any rescattered light from neighboring particles. Therefore, without being further extended, this approach does not strictly apply to light scattered from an assembly of particles.

To circumvent this disadvantage, the multiple scattering approaches are proposed for particle crowding with an average separation between particles greater than three particle diameters. However, when the particle crowding gets tighter, the problem of dependent scattering begins to increase as Mie scattering starts to fail, due to wavelength-dependent interference between the neighboring scatterers. Nevertheless, most colorimetric problems do not require such elaborate handling and, in technology, more emphasis is placed on simpler calculation methods.

One of the most important simplified scattering approaches is based on a theory known in astrophysics as radiative transfer [17]. In its original form, it studies the transmission of light through absorbing and scattering media such as stellar and planetary atmospheres. The radiative transfer equations are rather complex, and the multichannel technique is a suitable approach that overcomes the imposed complexity. This technique subdivides the analyzed medium into as many channels as needed, each of them covering a different range of angles from the perpendicular to the horizontal [18]. Each channel is supplied with specific absorption and scattering coefficients, which determine how much light is being absorbed and scattered into other channels. The interesting aspect of this concept is its ability to connect the series of coefficients directly to Mie's theory. This helps to incorporate fundamental properties of the scatterer into the model, such as the particle size and the refractive index.

Reducing the number of channels considered even further, one arrives at the case of two-flux models proposed by Schuster [19] and other authors. However, the Kubelka and Munk model, which is presented in the following section, is probably the most recognized approach and best suits the purpose of analyzing light scattered within textile fabrics and by colorant particles.

6.2.2 KUBELKA–MUNK'S THEORY

The original theory of Kubelka and Munk [20] was developed for uniform colorant layers. Due to its simple use and its acceptable prediction accuracy, the model has become widely used in industrial applications (Figure 6.20) [21–23]. It is suitable for turbid layered media. The expressive dimensionality of the scattering problem is reduced by assuming that light is being absorbed and scattered only in two directions, up and down. Thus, no special account needs to be taken of fluxes proceeding parallel to the boundaries, and only two vertical fluxes need to be considered. The illumination of the top face of the turbid medium is expected to be homogeneous and diffuse. Besides being infinitely extended, the concept assumes that the medium forms a plane layer of constant, generally finite thickness. The material is presumed to be homogeneous, i.e., the optical inhomogeneity is incomparably smaller than the thickness of the specimen and uniformly distributed in the material. Finally, the material is assumed to emit no fluorescent radiation. In addition, it is assumed to have the same refractive index as the medium from which the light comes [23, 24].

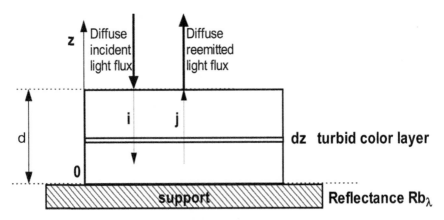

FIGURE 6.20   Scheme of Kubelka and Munk's theoretical model.

As depicted in Figure 6.19, the colorant layer has a total thickness D and is in direct optical contact with a backing of reflectance $R_{b\lambda}$. The model analyzes two fluxes through an elementary layer of thickness dz, the flux $i_\lambda(z)$ proceeding downward, and $j_\lambda(z)$ simultaneously upward. The depth parameter z is considered to be zero at the background of the layer and D at the illuminated interface of the layer.

During its passage through dz, the upward flux $j_\lambda(z)$ is attenuated by absorption $\alpha_\lambda dz$ and by scattering downwards $\sigma_\lambda dz$. At the same time, $j_\lambda(z)$ gains by scattering from the opposed flux amount to $\sigma_\lambda dz$ yielding the flux variation $dj_\lambda(z)$ per thickness element dz:

$$\frac{dj_\lambda(z)}{dz} = \sigma_\lambda \cdot i_\lambda(z) - (\alpha_\lambda + \sigma_\lambda) \cdot j_\lambda(z) \tag{6.1}$$

The corresponding variation of the downward flux $i_\lambda(z)$ is derived in the same manner with the exception that dz is negative, as the process is occurring in the reverse direction

$$-\frac{di_\lambda(z)}{dz} = \sigma_\lambda \cdot j_\lambda(z) - (\alpha_\lambda + \sigma_\lambda) \cdot i_\lambda(z) \tag{6.2}$$

The differential equations (6.1) and (6.2) obtained describe the net changes of both light fluxes, $i_\lambda(z)$ and $j_\lambda(z)$. The equations are classified as the two-constant theory, referring to the introduced colorant optical coefficients, absorption, and scattering.

Kubelka and Munk solved this system of equations and obtained the reflectance value associated with a given wavelength in the case of an infinite $R_{\infty\lambda}$ or finite R_λ painted layer, in terms of the absorption and scattering coefficients and the reflectance of the support $R_{b\lambda}$, as expressed in Eq. (6.3):

$$R_{\infty\lambda} = \frac{\dfrac{R_{b\lambda} - R_{\infty\lambda}}{R_{\infty\lambda}} - R_{\infty\lambda}\left(R_{b\lambda} - \dfrac{1}{R_{\infty\lambda}}\right)\exp\left[\sigma_\lambda d\left(\dfrac{1}{R_{\infty\lambda}} - R_{\infty\lambda}\right)\right]}{R_{b\lambda} - R_{\infty\lambda}\left(R_{b\lambda} - \dfrac{1}{R_{\infty\lambda}}\right)\exp\left[\sigma_\lambda d\left(\dfrac{1}{R_{\infty\lambda}} - R_{\infty\lambda}\right)\right]} \tag{6.3}$$

In practice, $R_{\infty\lambda}$ is the reflectance of a layer that is thick enough to completely hide the support [24], that is, the limiting reflectance that is not modified by any additional thickness of the same material [24]. Moreover, the calculation of $R_{\infty\lambda}$ allows the ratio $\alpha_\lambda/\sigma_\lambda$ to be obtained by inversion of equations (6.4) and (6.5):

$$R_{\infty\lambda} = 1 + \frac{\alpha_\lambda}{\sigma_\lambda} - \sqrt{\frac{a_\lambda^2}{\sigma_\lambda^2} + 2\frac{\alpha_\lambda}{\sigma_\lambda}} \tag{6.4}$$

$$\frac{\alpha_\lambda}{\sigma_\lambda} = \frac{\left(1 - R_{\infty\lambda}\right)^2}{2R_{\infty\lambda}} = K/S \tag{6.5}$$

where K/S function is the Kubelka–Munk function.

6.2.3 SHADE INTENSITY

Considering that significant current and future research is directed toward the development of smart textile sensors with photochromic pigments, which can react under UV irradiation, it is necessary for the purposes of their calibration to provide an objective description of visual color change. The Kubelka–Munk function forms the basis of one possible solution for spectrophotometric description of color appearance. For photochromic pigments, the changes of spectral characteristic before and after illumination, expressed by the integral equation 6.6, can be considered [12, 25]:

$$I = \int_{380}^{780} (K/S)_\lambda \, d\lambda \tag{6.6}$$

The value I represents shade intensity. In practice, it is obtained by integration using Eq. (6.6) expressed by the sum in Eq. (6.7):

$$I = \sum_{380}^{780} (K/S)_\lambda \, \Delta d\lambda \tag{6.7}$$

where $\Delta\lambda$ depends on the band pass of spectrophotometer (usually 10 nm).

For photochromic substances, I depend on both time and intensity of illumination. For the purpose of the work, described in this thesis, the term is the color intensity developed by irradiation for time t.

6.2.3.1   Kinetic Model of Photochromic Response During Exposure

The first order kinetic model of photochromic response during exposure is proposed, which describes simply the process of changes within a sample of an applied photochromic pigment from I_0 (color intensity at time t_0, or sample without exposure) to I_∞ (color intensity after infinite time t_∞) as illustrated in Figure 6.21.

The proposed model is based on the view that the process follows first order kinetics with the rate described by Eq. (6.8). The rate of color intensity development dI/dt is directly proportional to the difference of color intensity at time t and in equilibrium:

$$\frac{dI_t}{dt} = -k\left(I - I_\infty\right) \tag{6.8}$$

For $t = 0, I = I_0$.

Integration of this equation between the limits $(0,t)$ (I_0, I) leads to the exponential equation (6.9):

$$I_t = I_\infty + \left(I_0 - I_\infty\right)e^{-kt} \tag{6.9}$$

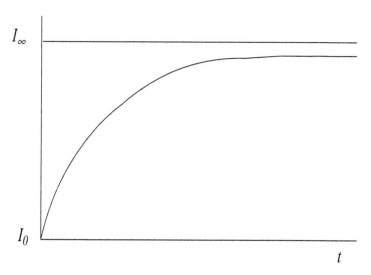

FIGURE 6.21   Curve of color intensity change for a photochromic dye during exposure with time.

6.2.3.2 Kinetic Model of the Photochromic Response During Reversion

The kinetic model for relaxation and change of color intensity I for the photochromic pigment (photochromic response during reversion) is based on fact that during reversion, the change of color intensity I_t reduces from I_∞ to I_0 (see Figure 6.22).

The basic presumption is again of first order kinetics as for exposure. The rate of color intensity changes dI/dt is also directly proportional to the difference in color intensity after time t and at equilibrium:

$$\frac{dI_t}{dt} = -k\left(I_\infty - I\right) \tag{6.10}$$

For $t = 0\, I = I_\infty$.

Again, through the integration, the solution of the differential equation within the limits $(0,t)$ (I_∞, I) gives the exponential Eq. (6.11):

$$I_t = I_0 + \left(I_\infty - I_0\right)e^{-kt} \tag{6.11}$$

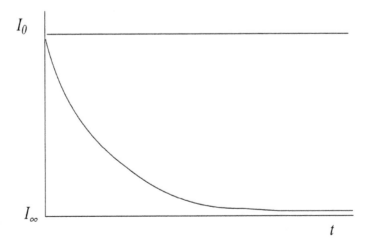

FIGURE 6.22 Curve of color intensity change for a photochromic dye during reversion with time.

6.3 SPECTRAL DATA OF CHROMIC MATERIALS

Photochromic pigments and other functional dyestuffs that are used presently as component of smart materials (textiles), and they are able to indicate environmental variations by color changes. For photochromic sensors, the color change indicates the amount of UV irradiation [27, 28]. Many of them are based on the pair comparison. There are mainly two types, namely color stable and color changeable. Color changes are evaluated using various standard spaces. These color spaces are based on principle of concrete perception of color differences, the description of color perception, and color illusions or light conditions used for visualizing color sensation. Some of these spaces are the CIE XYZ, CIE L*a*b*, CIE L*c*h* and CIE CAM02. For photochromic sensors such as smart materials, there is a need to design a stable material, which exhibits high fatigue resistance against UV irradiation.

In the creation of the sensorial system, it is necessary to use not only one comparative segment defining a UV level, but it is necessary to at least use three comparative segments with a constant level of color perception. If at least three standards are used causing no detectable color change during variations of UV intensity, then color evaluation and visual evaluation of change in intensity of UV irradiation is correct and sufficiently accurate. In this case, color matching can be used as demonstrated in Figure 6.23.

From scheme in Figure 6.23, it is evident that the left part of the sensorial system has a light tint or left part is white. For low level of UV irradiation, the color match is found between the changeable part of sensor and reference standard A. For middle level of UV irradiation, the color match is observed after comparing the changeable part of the sensor and reference standard B, and finally for high level of UV irradiation, the color match between the changeable part of sensor and referent standard C is found.

The stable parts are graded according to the intensity of pigment coloration. The first level (low) corresponds to UV index 2 meaning that the person can stay under direct sun without limitation. The second level (middle) corresponds to UV index 4 indicating that when the sensorial (changeable) part is colored in this shade intensity, the person can remain under direct sunlight for limited time. The third level is designed to give an UV index 6 and higher. Under these conditions, the exposure to sunlight should be limited to as short time as possible.

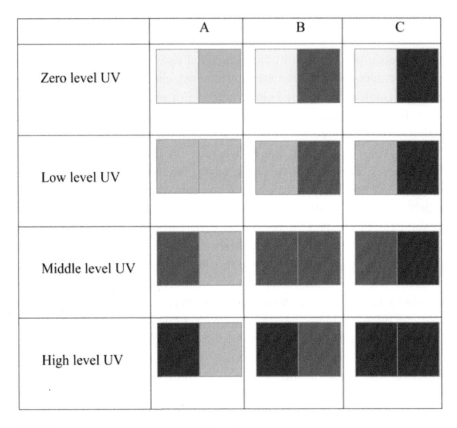

FIGURE 6.23 An example of simple UV indicator.

The stable section is used as a reference scale to which the changeable photochromic part can be compared. The sensorial part is divided in four segments; three segments are stable comparative parts and the fourth one is photochromic changeable segment. However, not all photochromic pigments are colored in the same shade or hue and for different reason, the shade or the hue is shifted in tint. This is visible in Figure 6.24 and in the picture where an evident shift from greenish to intensive blue tint is observed. In Figure 6.30, curves of photochromic color change are visible during the decay phase. Curves for individual pigments appear similarly as set of concentrations of standard dyestuffs, where it is possible to recognize change of hue angle with increasing concentration. In the data presented, pigment P2 with purple shade is more reddish in the beginning than in mid time of decay. Consequent effect is connected to chroma,

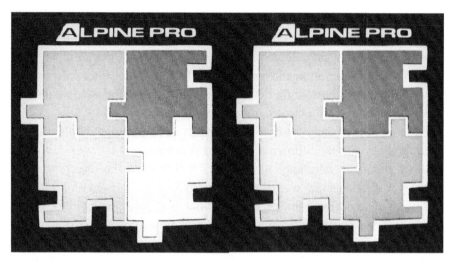

FIGURE 6.24 Smart sensor (a) before irradiation and (b) after irradiation.

where the increase and subsequent decrease of chroma is visible. In the Figures 6.28 and 6.29, curves of photochromic color change are visible during decay phase. Curves for individual pigments appear similarly as set of concentrations of standard dyestuffs, where it is possible to recognize change of hue angle with increasing concentration. In the data presented, pigment P2 with purple shade is more reddish in the beginning than in mid time of decay. Consequent effect is connected to chromaticity of pigment P1, where the increase and subsequent decrease of excitation purity is visible in Figure 6.31. Such specific character of photochromic color change during discoloration is caused by shift of distribution of individual readings out of direction to achromatic point as shown in the Figure 6.27 (pigment P1 is present as "Blue High"). That means color changeable part of sensor has own shade in inactive form of pigment P1 in comparison to solution of pigment P1, which is without specific shade. Light scattering on the particles of pigment P1 probably causes this effect in state of weak opalescence, which brings green yellowish tint.

Due to the problems of measuring kinetic photochromic color change during the exposure phase, a special device named PHOTOCHROM was developed at LCAM (Laboratory Color and Appearance Measurement, Technical University of Liberec, Czech Republic). The PHOTOCHROM spectrophotometer allows measuring the colorimetric and the spectral characteristics of photochromic textiles as photochromic sensors and also the fatigue test for the control of color change stability [3, 4]. This concept of colorimetric and spectral parameters also allows finding the dependence of color change on intensity of UV irradiation and temperature. This part of chapter presents the advantages of this measuring device when obtaining colorimetric data and their kinetic behavior in basic color spaces.

To create such kinds of sensor we can use three commercial photochromic pigments with the following chemical structures.

Such pigments were applied on a PET substrate by the screen printing method. A PET substrate was used according to standard ISO 105-F04:2001 in form of plain weave. Pigment concentration was 100 g per kg of the printing paste. After printing, all samples were dried for 10 min at 105°C and then cured for 2 min at temperature 150°C.

Measurement of colorimetric and spectral parameters of photochromic textiles was made by the original measuring device LCAM Photochrom due to two reasons. Without hardware adaptation of standard reflectance spectrophotometers, it is impossible to measure growth phase of photochromic color change. Based on this, a second problem was found. Measured data are influenced by delay between the activation of photochromic materials by the external light source and initial series of measurements with constant time interval. This problem affects the validity of the measured data. Standard reflectance spectrophotometers in "time measurement" mode is limited by time delay of 5 s. This means that if decay part of photochromic color change is faster than this interval, it is impossible to measure kinetic data correctly. The original measuring system LCAM Photochrom solves these problems via complex construction [14]. This device allows studying color photochromic kinetics, the influence of exposure time, and the thermal sensitivity if the spectral sensitivity of the excitation monochromator is included as shown in Figures 6.25 and 6.26 [25]. The dual light source construction of the spectrophotometer with a shutter over exciting light source makes the continuous measurement of photochromic color change possible during reversion after switching off the exciting light source. A xenon discharge lamp with continuous discharge is used. The pulsed discharge lamp allows using such system as a fatigue tester for photochromic systems with a halftime of photochromic color change around 500 ms. For textile samples, which are frequently slower, it is recommended to use an electric shutter and a continuous discharge lamp for the fatigue tests.

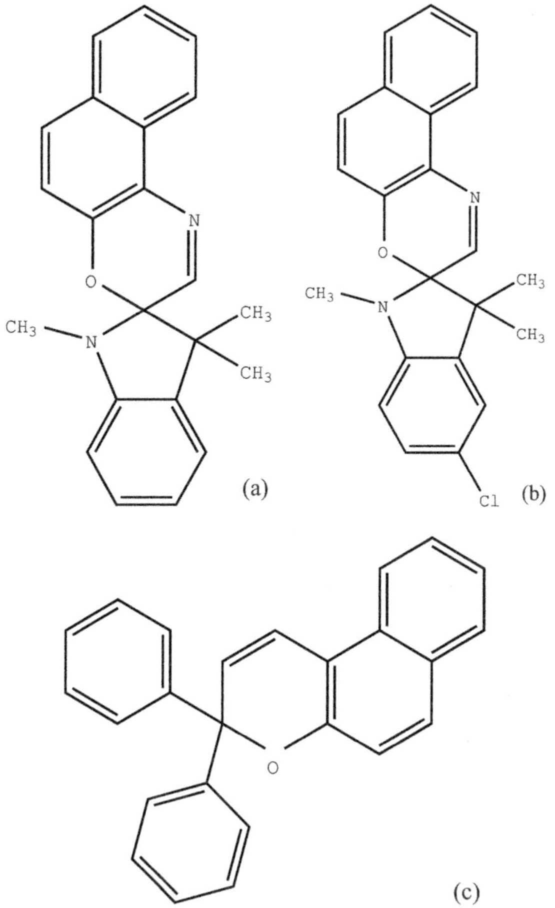

FIGURE 6.25 (a) Pigment P1:1,3,3-trimethylspiro[indolino-2,3′-(3H)naphtho (2,1-b) (1,4)-oxazine] CAS: 27333-47-7, (b) Pigment P2: 5-chloro-1,3,3-trimethylspiro[indoline-2,3′-(3H)naphtho(2,1-b) (1,4)-oxazine], CAS: 27333-50-2, and (c) Pigment P3: 3,3-diphenyl-3H-naphtho[2,1-b]pyran, CAS: 4222-20-2.

6.3.1 PHOTOCHROMISM IN THE CIE CHROMACITY DIAGRAM

The color in CIE chromaticity diagram is defined by coordinates x and y, but the Helmholtz numbers can also be used. These values inform us about the percentage of clear spectral color corresponding to the relevant wavelength. This description is near to coloristic practice where expressions such as the color is lighter – darker, purer – cloudier or the hue is reddish, yellowish are

FIGURE 6.26 Prototype of the measuring system PHOTOCHROM, which was developed at the Laboratory of Color and Appearance Measurement, Faculty of Textile Engineering, Technical University of Liberec.

used. The dominant wavelength λ_D (in case of purple shade λ_{DK} – complementary dominant wavelength) and excitation purity p_E for study of behavior in CIE chromaticity diagram was used [28–31].

It was found that for pigment P1, the dominant wavelength λ_D and excitation purity p_E were 479 nm and 0.3798, respectively. These values correspond to 38% of pure spectral color shade at a wavelength 486 nm. Figure 6.27 shows the change in shade for the pigment P1 in CIE chromaticity diagram and the line of blue shade generation does not point to W as in the case of pigments P2 and P3 (in this situation for light source D65). However, the shade is shifted in its line to shade with green yellowish tint.

The reason is that the inactivated form of the photochromic pigment P1 shows distinct shade coloration (to green yellowish) than the other pigments P2 and P3. This situation is shown in Figure 6.28, where minimum reflectance is documented at 400 nm, which is independent of activation of pigment P1.

After comparing the spectral curves for the discoloration of pigments P1 and P2 (Figures 6.28 and 6.29), the time of pigment P2 is two times slower than that for pigment P1. There is also a hypsochromic shift of the reflectance minima for the pigment P2 relative to its blue counterpart P1. The behavior

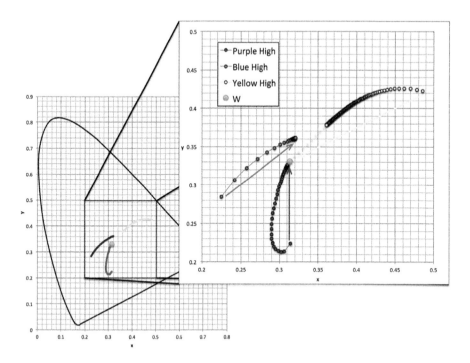

FIGURE 6.27 Color change of the decay phase of photochromic pigments in CIE xy chromaticity diagram.

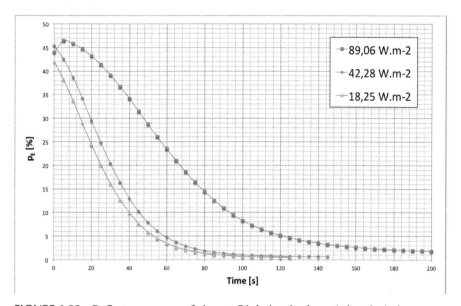

FIGURE 6.28 Reflectance curves of pigment P1 during the decay (relaxation) phase.

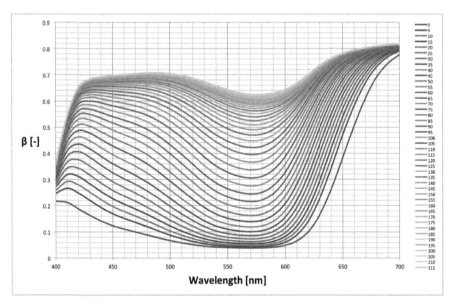

FIGURE 6.29 Reflectance curves of pigment P2 during the decay (relaxation) phase.

is the effect of the chlorine atom at the fifth position in indoline molecule. In the case of pigment P2, chlorine acts as auxochrom and results in deepening of shade as shown in Figure 6.29. Since the chlorine is at the para-position relative to the indoline nitrogen, the chlorine shifts the maximum of absorbance in a hypsochromic by about 37 nm. The first reflectance minimum is shifted toward the UV area and exerts influence on the total color perception only to 420 nm. When studying the discoloration time, different decreasing curves of the exciting purity are observed (Figures 6.30–6.32).

Figures 6.30–6.32 show that all investigated pigments have a longer time of discoloration for higher intensity of irradiation. This is well described in the case of pigment P1, where the exciting purity is coming nearest to the achromatic point W in its minimum. Also, the fixation constant value in different time can be observed, and it can be predicted that a higher exciting purity during exposition means longer reaction time. Pigment P3 has a time of discoloration in hours. The trend of discoloration for pigment P3 to the achromatic point is shown as yellow broken line (Figure 6.27). It is also evident that the inactivated forms of pigment P3 and P2 end in achromatic point W. Other significant differences include the curve of excitation purity for the pigment P1 for the highest value of irradiation, where differences are visible

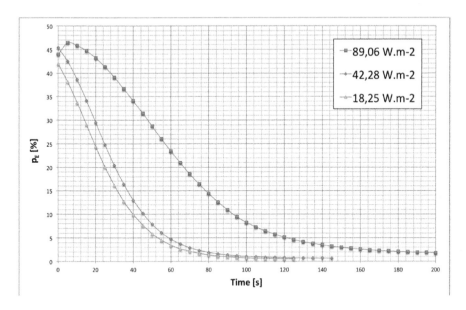

FIGURE 6.30   Excitation purity during the decay (relaxation) phase of pigment P2.

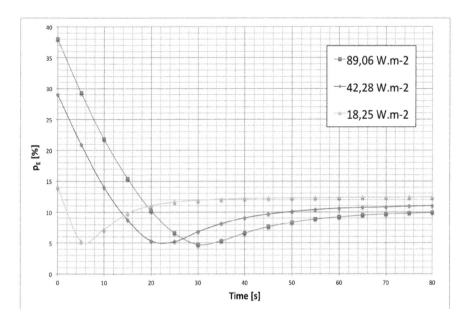

FIGURE 6.31   Excitation purity during the decay (relaxation) phase of pigment P1.

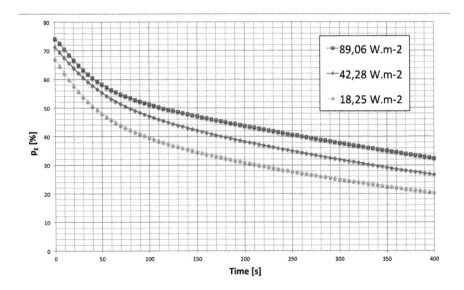

FIGURE 6.32   Excitation purity during the decay (relaxation) phase of pigment P3.

(Figure 6.31). This is caused by the effect of dependence of shade intensity on concentration from classical dyestuff. The characteristic curvature is visible showing a butterfly effect for the pigment P2 (Figure 6.27).

6.3.2   PHOTOCHROMISM IN THE CIELAB COLOR SPACE

The CIELAB space can be used to describe photochromism. This system uses an orthogonal 3-axis system for describing colors. The shift in hue is described using the chromatic plane $a*b*$, while the shift in lightness can be recorded using the $L*$ axis. In Figure 6.33, the color shift of photochromic pigments on the chromatic plane $a*b*$ is presented. This shift does not inform about change in lightness. However, there is a visible achromatic white shade in central position "0" corresponding to the area of white color in Figure 6.33. The level of lightness decreases, while the intensity of shade increases with exposure time.

The change in lightness records for the three pigments under investigation is shown in Figure 6.34. The difference in kinetics of discoloration for every tested pigment to colorless shade is observed when the lightness

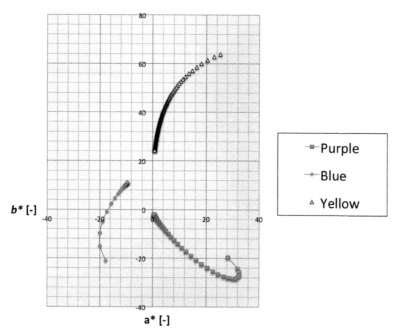

FIGURE 6.33 Projection of discoloration on the photochromic plane a*b*: time of activation 120 s, irradiation 89.06 W·m⁻².

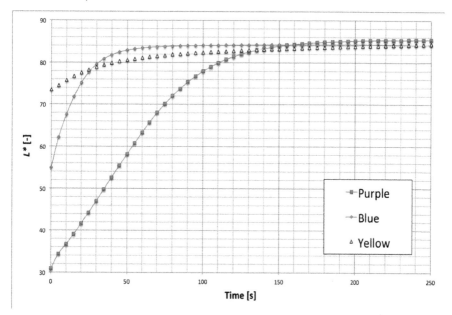

FIGURE 6.34 Dependence of lightness on the discoloration time: time of activation 120 s, irradiation 89.06 W·m⁻².

reaches a maximum value. The speed of coloration and discoloration is different, and the half-life of discoloration depends on the structure of the photochromic molecule and the additives used in the final stages of photochromic dye production. This is also the reason why different commercial photochromic compositions exhibit different light fastness properties.

Another important parameter for the evaluation of color constancy in colorimetry is the shift in hue expressed by the color difference dE^*. The color difference dE^* is also suitable for defining the hue shift of a photochromic reaction in the CIELAB space. As a complex parameter, the color difference dE^* includes information about the change in shade intensity and the change in lightness. The color difference dE^* can explain the visibility of color change dependent on physiology of human vision. This parameter is suitable for the evaluation of the stability of photochromic effect by individual subjective evaluation and assessment of photochromic reaction by the human eyes [32]. Photochromic pigments are also used as a sensorial system for the evaluation of UV radiation and its harmful effects. Therefore, an important requirement is the ability to recognize the concrete color differences during visual evaluation. In other words, how the shade intensity changes during the photochromic reaction on a textile substrate that undergoes a transition from colorless to colored shade. The color difference for individual assessment could be higher than 0.4 dE^*. This is the limit for color differences discernable by the human eyes. In Figure 6.35, it can be observed that the color difference for the tested pigment was higher than 40 dE^* units. Also, the speed of discoloration is different for the tested pigment as can be seen from the dependency on the lightness parameter (Figure 6.34).

The CIELAB color space is very well established for the evaluation of color changes by the use of the colorimetric parameters $L^*a^*b^*$. However, if the photochromic effect shows a color or tint shift, expressions such as more yellowish, reddish, or greenish are used. Thus, for this type of evaluation, the cylindrical system CIELCH could be used. The cylindrical system CIELCH uses the parameter h (hue angle) to describe a color shift.

The main disadvantage of mathematical description employed by the CIELAB space is the definition of tint and coloration shift. In colorimetry, the cylindrical system CIELCH is employed as it is analogous to the Munsell system. The color shift can be described as hue angle. The description of hue angle is used when working in the design of smart photochromic sensors as shown in Figure 6.22.

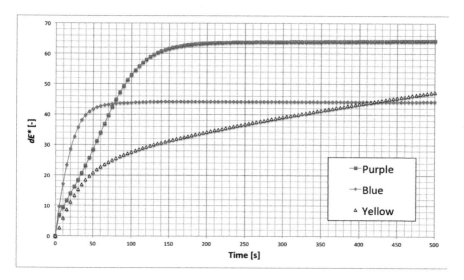

FIGURE 6.35 Dependence of color difference on the discoloration time: time of activation 120 s, irradiation 89.06 W·m⁻².

In the case of a photochromism study and a photochromism description of differential chromatic planes $dC*dH*$ (Figure 6.36), the differences between the various pigments tested can be evaluated by comparing hue and chroma differences. Pigment P1 exhibits large hue versus chroma differences relative to other pigments. This is in agreement with the assumption that photochromic pigments mainly experience changes associated to their chroma values.

The reason behind these differences could be the opalescence caused by the particles fixed on the textile substrate due to the presence of pigment P1. A slight coloration can be observed in the inactivated form, which is usually colorless.

When designing sensorial systems on textile substrates it is necessary to consider the change in hue as a complex factor describing change in hue angle as well. This situation is described in Figure 6.37, where it is evident that all tested pigments exhibit different curves of $dH*$ in time comparable to the $dE*$s shown in Figure 6.35. It can also be seen in Figure 6.35 that pigment P3 was not fully stabilized to constant value of $dE*$, while pigments P1 and P2 are fully changed to the leuco form of the photochromic dyestuff.

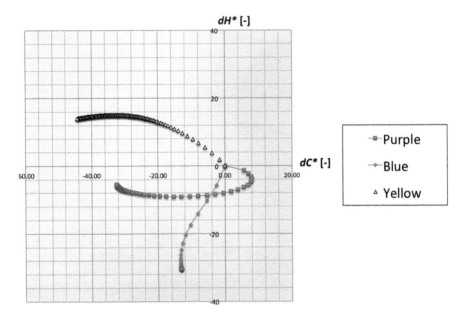

FIGURE 6.36 Change in chroma and hue of photochromic pigments in decay: time of activation 120 s, irradiation 89.06 W·m⁻².

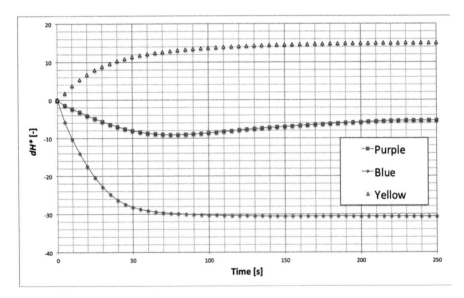

FIGURE 6.37 Change in chroma and hue of photochromic pigments in decay: time of activation 120 s, irradiation 89.06 W·m⁻².

6.3.3 PHOTOCHROMISM IN THE CIE CAM 02

The reactions of photochromic substances were studied under different light sources showing various spectral power distributions. The experiments were aimed at studying the chromatic adaptation of visual differences and the shift in colorimetric parameters caused by the use of different illuminants such as D50, D65, D75, and D90. These sources describe and simulate the spectral power distribution of natural daylight. This simulation corresponds with the effect of using photochromic sensors during an entire day and the shift in colorimetric parameters together with the chromatic adaptation can be evaluated. These differences are caused by changes in spectral power distribution that have occurred over the day.

The construction of photochromic sensors must include a reference hue and a color changeable part. This will vary depending on the intensity of the exciting radiation.

Because colorimetric parameters depend on the light source employed, it is necessary to consider these influences together with the chromatic adaptation meaning the change of perception and adaptation of human vision to changes in types of illumination.

Complex models for evaluating color appearance such as CIE CAM 02 [33–37] allow predicting colorimetric changes under different light sources. Figure 6.38 shows the shift of colorimetric parameters for pigment P1 under D50, D65, D75, and D90 illumination. Figure 6.38 shows the shift observed when comparing two data sets corresponding to sunlight and cloudy skies. The shift is approximately 3 dE units, which is approximately 10 times higher than the resolution of the human eye to color differences [38]. The advantage is the constant shift of colorimetric data for the stable part of the sensorial system and will not cause significant measurement error. Table 6.1 documents the influence of the light source on color differences of samples corresponding to agreement between the stable and changeable parts of the sensorial system.

The sensitivity of human eyes to color differences is approximately 0.3 units DE in the CIEDE2000 color difference formula; due to nonuniformity of the CIELAB color space the first noticeable threshold varies from 0.2–0.5 [62] in CIECAM02. Therefore, light source changes may induce color difference, which are higher than the limit of human vision (Table 6.4).

It can be discussed that color differences for D65 were not ideally 0. However, color differences were near the sensitivity limit of human vision. If this value is used for normalizing the measurements conducted using

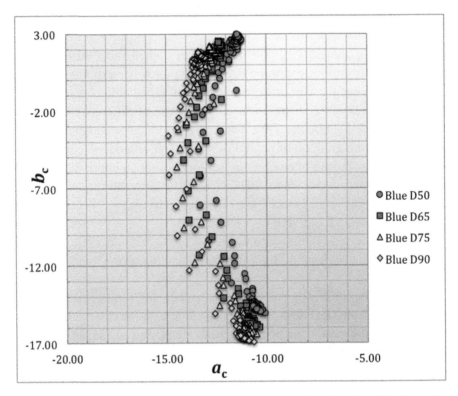

FIGURE 6.38 Shift of colorimetric parameters on plane ab of CIECAM02 after using different illuminants: pigment P1, time of activation 300 s, irradiation 89.06 W·m⁻².

TABLE 6.4 Calculated Color Difference for Different Photochromic Pigments

dE dE*	D50	D65	D75	D90
Blue	0,32	0,17	0,26	0,34
Purple	0,39	0,24	0,31	0,42
Yellow	0,35	0,29	0,22	0,28

different light sources, then the color differences also remain under the threshold value. This result indicates that in case of photochromic sensors, the change of illuminant in the temperature range from 5000 to 9000 K has no influence on the evaluation of UV irradiation by the method of color match between two sensor parts (stable and changeable).

The speed of color change is a complex parameter that can be described by the change from achromatic to chromatic hue, when all three colorimetric coordinates are evaluated. In this article, a complex description of photochromic colorants in selected color spaces is provided. Parameters such as excitation purity, color, and hue difference and also the shift in lightness provide the main description of photochromic coloration of samples from a visual point of view. The observation and evaluation of photochromic effects together with the stable part of a sensorial photochromic system is carried out. The precise measurement of the colorimetric parameters allows computer-match prediction of the stable part of smart UV sensors based on the relevant intensity of UV irradiation. It is also possible to develop specific sensors especially for different skin types using MED scale factor. The specific sensors for different skin types will have the photochromic part of sensor containing different amounts of sensitivity modulator (UV absorbers, etc.). This means that the color changeable part of the photochromic sensor should be more sensitive for skin type 1 than to skin type 4. The results show different kinetics of photochromic color change for the tested pigments and the possible modulation speed of UV sensor reaction. The evaluated standardized color spaces show that pigment P1 is affected by colored residuals, which affect its colorimetric and photochromic properties.

6.4 DETERMINATION OF ABSORPTION AND SCATTERING COEFFICIENTS

The main recent research on textile coloration has focused on improving dye uptake by textiles, producing new colors, improving process economics, enhancing environmental performance, and improving the quality of colorants in terms of color fastness, such as to light, washing, and rubbing. The concept involved in the development of color changeable textiles is based on applying special colorants to textile materials. These particular colorants possess the ability to undergo reversible color change under the influence of external stimuli such as light, heat, or chemical processes. Considering that the current and future development of new color changeable textiles with photochromic pigments, which can react under UV irradiation, it is necessary to provide an objective description of visual color change for improvement of its production. Fresnel's postulates and equations of energy transfer, including approximate Kubelka–Munk theory, also allow the determination of other optical parameters for measured materials, for example absorbance coefficient k and scattering coefficient s [39–43].

For colorants, it is possible to determine molar coefficient ε_0 by absorption spectrophotometry and the Lambert–Beer Law from a calibration set with a different concentration of specific substance. For photochromic pigments, it is necessary to use suitable solvents (toluene, cyclohexanone, trichlorobenzene) with limited solubility. At present, the commercial photochromic pigments are not available as a pure substance, but in most cases, the photochromic pigment contains a number of auxiliaries. Photopia™ by Matsui is an example of photochromic pigments; the method of pigment encapsulation was used for higher light stability by HALS (hindered amine light stabilizer) [44, 45], as the sebac acid`s derivates are (for example, bis(2,2,6,6-tetramethyl-4-piperidinyl)sebacate), or polycondensation`s products of piperidine (for example, poly[{6-(1,1,3,3-tetramethylbutyl) amino-1,3,5-triazine-2,4-diyl} {(2,2,6,6-tetramethyl-4-piperidinyl)imino} hexamethylene 2,2,6,6-tetramethyl-4-piperidinyl)imino]), and other auxiliary substances. Due to presence of above mentioned components in photochromic capsules of commercial products is complicated to prepare clear solution of photochromic pigment, which is used for common spectrophotometric method of measurement growth and decay phase in form of liquids. Resulting liquid form of photochromic pigment is turbid and common use of Lambert – Beer Law for measurement of resulting extinction is problematic [43].

The solubility of photochromic pigments is the maximum of 5% at 20°C; the rate of turbidity is relatively high, and the result of the shade is pure (A≈0.2). An advantage of final composition photochromic pigments applied on textile or another surface can be determined by scattering and absorption coefficients with the method of black and white background.

As mentioned in the previous section, we also use three commercial photochromic pigments P1–P3 in this study.

These pigments were applied on a cotton (CO) substrate by the screen printing method. A cotton substrate was used according to standard ISO 105-F04:2001 in the form of plain weave. Pigment concentration was 100 g per kg of the printing paste. After printing, all samples were dried for 10 min at 105°C and then cured for 2 min at a temperature of 150°C.

Measurement of the colorimetric and spectral parameters of photochromic textiles was conducted on the original measuring device LCAM

PHOTOCHROM by the contrast ratio method, which is frequently called the black/white background method [44].

If the black and white background method uses the white standard (WS) and black trap (BT), we can measure the reflectance and absorbance data on the spectrophotometer. It is important to use white standard with CIE brightness Y higher than 90; such standards are made from barium sulfate (BaSO4), white ceramic tile, Labsphere white standard, etc. The graphs described on Figure 6.39a–6.39c are spectral reflectance data for photochromic inactivated pigments (OF) and activated pigments (ON).

The graphs in Figure 6.39a–6.39c also show that constant reflectance minimum is in blue area of visible spectrum. It is an important fact that the reflectance minimum is practically independent on used background, and this behavior complicates the calculation of absorbance coefficients k, because the black and white background method is based on measured spectral differences when extreme difference in the reflectance allows the determination of total light transmittance.

If we take consider the Kubelka–Munk theory [45, 46], after substitution $k=2\alpha/(\alpha+\sigma)$, $s=\sigma/(\alpha+\sigma)$, it is possible to rewrite Eqs. (6.1) and (6.2) into:

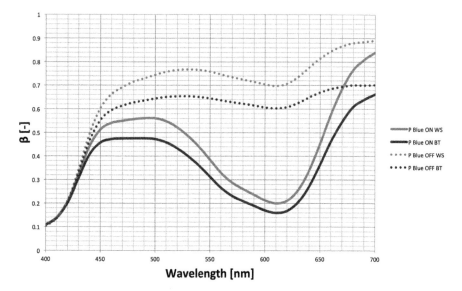

FIGURE 6.39A Spectral data of Photopia Blue on white and black background in the activated and inactivated state.

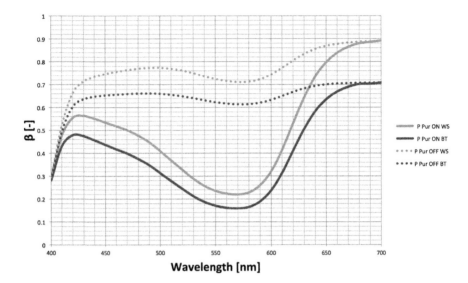

FIGURE 6.39B Spectral data of Photopia Purple on white and black background in activated and inactivated state.

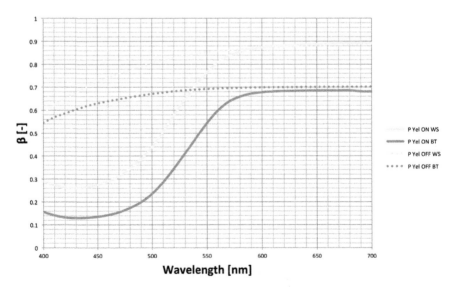

FIGURE 6.39C Spectral data of Photopia Yellow on white and black background in activated and inactivated state.

$$+dj = -2kjdx - 2sjdx + 2sidx \qquad (6.12)$$

$$-di = -2kidx - 2sidx + 2sjdx \qquad (6.13)$$

we can, together with the substitution $K = 2k$, $S = 2s$ and $a = 1+K/S$, also define parameter b via equation:

$$b = \sqrt{a^2 - 1} \qquad (6.14)$$

Thus, for the edge condition, when flow $J = 1$ and flow $I = 0$, we can calculate the reflectance and light transmittance from the following Eqs. (6.15) and (6.16):

$$R = \frac{1}{a + b \ \coth(kd)} \qquad (6.15)$$

$$T = \frac{b}{a \ \sinh(kd) + b \ \cosh(kd)} \qquad (6.16)$$

If we apply the black and white background method and together with Eqs. (6.15) and (6.16), we obtain the following relations for the reflectance one layer–textile substrate measured on white R_w and black R_b background:

$$R^w = \frac{1 - R_{bg,w}\left[a - b \ \coth(kd)\right]}{a - R_{bg,w} + \coth(kd)} \qquad (6.17)$$

$$R^b = \frac{1 - R_{bg,b}\left[a - b \ \coth(kd)\right]}{a - R_{bg,b} + \coth(kd)} \qquad (6.18)$$

From Eqs. (6.17) and (6.18), we can calculate parameter a:

$$a = \frac{1}{2} \frac{\left(R^b - R_{bg,b}\right)\left(1 + R_{bg,w}R^w\right) - \left(R^w - R_{bg,w}\right)\left(1 + R_{bg,b}R^b\right)}{R_{bg,w}R^b - R_{bg,b}R^w} \qquad (6.19)$$

We can continue with the calculation of the reflectance degree for the infinitely thin layer of materials β_∞ from Eq. (6.19). Parameter b is calculated from Eq. (6.14) so that we can determine the apparent scattering coefficient by following modification of Eq. (6.17)

$$S_w = \frac{1}{bd} \coth^{-1} \left[\frac{1 - aR_{bg,w} + \left(R_{bg,w} - a\right)R^w}{b\left(R^w - R_{bg,w}\right)} \right]$$

(6.20)

And analogically, we can use also this modification for Eq. (6.18):

$$S_b = \frac{1}{bd} \coth^{-1} \left[\frac{1 - aR_{bg,b} + \left(R_{bg,b} - a\right)R^b}{b\left(R^b - R_{bg,b}\right)} \right]$$

(6.21)

The agreement of apparent scattering coefficient S_w, S_b, calculated from Eqs. (6.21) and (6.22) show accuracy of assignment. The apparent absorbance coefficient is given by substitution of the abovementioned inverse calculation:

$$K = \left(a - 1\right)s$$

(6.22)

The graph in Figure 6.40 documents the dependency of apparent absorbance coefficient on wavelength.

From the calculated absorbance coefficients k in Figure 6.40, we can see also the influence of textile substrate in the blue region of the spectrum corresponding with the yellowish tint of used printed cotton textile by photochromic pigments. The yellowish tint is also visible for textile substrate with blind printing paste corresponding with the resulting film as the covering layer. Except for pigment P1 (Photopia Blue), the remaining pigments have relatively low absorbance in their inactivated state. For pigment P2 (Photopia Purple), we can find two maximums, and the main maximum is in the UV region in 376 nm; this maximum is not visible in the figure. But we can see the second maximum hypsochromic shift in the visible region of spectrum $\Delta\lambda = -7.5$ nm together with hyperchromic shift of the absorbance. Pigment Photopia Blue has marked tint in the inactivated state, and on the absorbance coefficients curves, we can see two maximums lying in the visible region of the spectrum. The maximum at 420 nm is artifact, because this maximum is not visible in absorbance spectrum, and moreover, if we

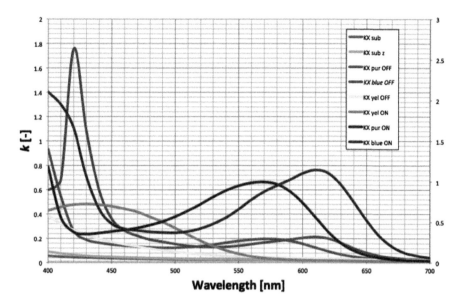

FIGURE 6.40   The dependency of calculated absorbance coefficients k on wavelength.

consider the influence of reflectance on interface for ambience with different optical density, this maximum disappears. The changes in spectral behavior for the scattering coefficient s are shown in graph in Figure 6.41.

The graphs in Figure 6.41 show the increase in the scattering coefficient for printed textile materials with inactivated photochromic pigments and the decrease in scattering coefficient for the textiles with activated photochromic pigments. Photochromic pigments in the capsule shape, as shown in Figure 6.42, cause this situation.

The microscopic picture confirms the producer's information that the mean size of capsules is around 5 μm, and it is similar for tested photochromic pigments. Also, we can see the relatively high concentration of used pigments for the total covering of textile substrate. For scattering coefficient s, there is no minimum of scattering shift, and we can only record the marked deepening of curves corresponding with the decrease in scattering coefficient s on wavelength for absorbance maximums for the measured pigments. An interesting phenomenon is the decrease in scattering coefficient s for blind printed substrate (the substrate is printed by blind paste–paste containing all chemical components except photochromic pigments). This

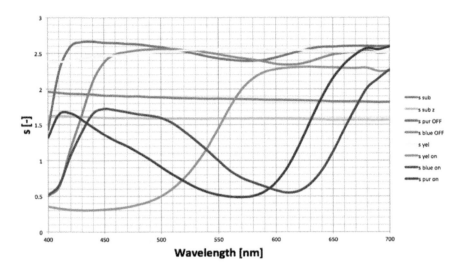

FIGURE 6.41   The dependency of calculated scattering coefficients s on wavelength.

effect corresponds with a decrease in hairiness and elementary fibers glued to the body of yarn [47].

For the confirmation of this hypothesis, we have also measured the optical porosity of the tested textile sample by Hewlett–Packard HP ScanJet 5530 flat scanner and by HP ScanJet TMA transmittance adapter. Evaluation of captured images was made at 2400 DPI resolution, used data format was tiff without compression. Image analysis was performed in software ImageJ Fiji version 1.48b.

First, we transferred the data to grey scale followed by threshold to the black background, and we evaluated the ratio (in %) of translucency trough for the tested textile. Following the constant condition, the fixed threshold to the black background was used in range 150–255 unit. For the comparison of the optical porosity OP (%) with individual coefficients obtained from K/M function, the "INTEG" method was used as the standard in colorimetry. This method uses the sum of K/S, weighted at each wavelength by the illuminant/observer data [48].

The graphs in Figures 6.43 and 6.44 show that the dependence between optical porosity OP and absorbance coefficient k is linear. We also confirmed in our experiment that the substrate printed with blind paste has lower filling and higher porosity. This means the printing paste bonded elementary fibers

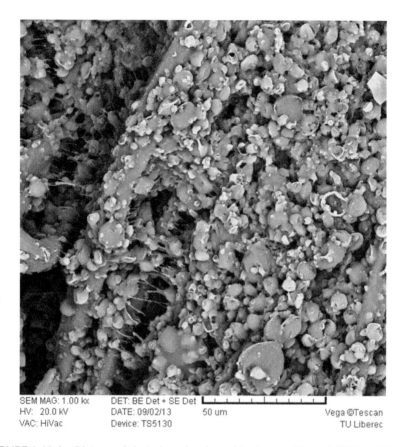

SEM MAG: 1.00 kx   DET: BE Det + SE Det
HV: 20.0 kV   DATE: 09/02/13   50 um   Vega ©Tescan
VAC: HiVac   Device: TS5130   TU Liberec

FIGURE 6.42 A   Pictures of photochromic prints with pigments Photopia™ Matsui Purple.

together with body of yarn and the number of protruding fibers is decreased with the blocking of inter-bonding pores.

The spectral curve of apparent absorbance coefficient k for pigment P1 (Photopia Blue) in Figure 6.45 shows as artifact the maximum at 420 nm, and its maximum was not measured by the absorbance spectrophotometry as it is documented by spectral curves in Figure 6.45. Also, the turbidity influence of decanted additives used in commercial photochromic pigments for their high stability is visible in the level of absorbance $A = 0.12$.

For the measurement of colorimetric parameters of photochromic textiles, the diffused illumination and integrating sphere was used, and we can use Fresnel's postulates for calculation and prediction: Eqs. (6.23)–(6.26) describe light behavior on the interface of two ambiences [45–47]:

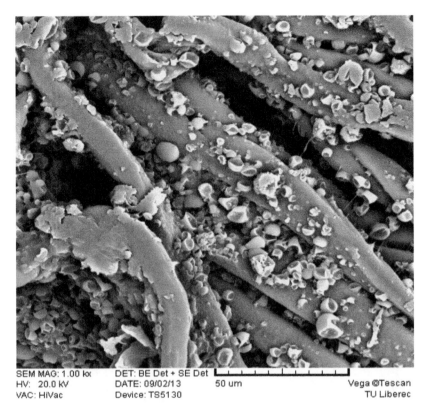

SEM MAG: 1.00 kx   DET: BE Det + SE Det
HV: 20.0 kV   DATE: 09/02/13   50 um   Vega ©Tescan
VAC: HiVac   Device: TS5130   TU Liberec

FIGURE 6.42 B   Pictures of photochromic prints with pigments Photopia™ Matsui Blue.

$$r_p = \frac{n_t \cdot \cos\alpha - n_i \cdot \cos\beta}{n_t \cdot \cos\alpha + n_i \cdot \cos\beta} = \frac{\tan(a-\beta)}{\tan(\alpha+\beta)} \tag{6.23}$$

$$t_p = \frac{2n_i \cdot \cos\alpha}{n_t \cdot \cos\alpha + n_i \cos\beta} = \frac{2\cos\alpha \cdot \sin\beta}{\sin(\alpha+\beta) \cdot \cos(\alpha-\beta)} \tag{6.24}$$

$$r_s \frac{n_i \cdot \cos\alpha - n_t \cos\beta}{n_i \cdot \cos\alpha + n_t \cdot \cos\beta} = -\frac{\sin(\alpha-\beta)}{\sin(\alpha+\beta)} \tag{6.25}$$

$$t_s = \frac{2n_i \cdot \cos\alpha}{n_i \cdot \cos\alpha + n_t \cos\beta} = \frac{2\cos\beta \cdot \sin\alpha}{\sin(\alpha+\beta)} \tag{6.26}$$

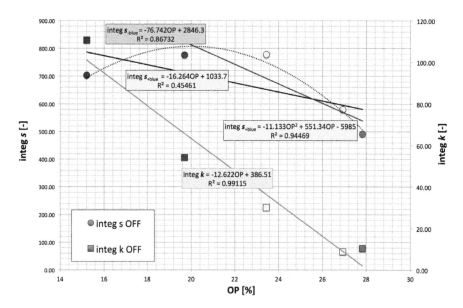

FIGURE 6.43 The dependency between optical porosity OP and integral value of scattering and absorbance coefficients for textiles printed by inactivated photochromic pigments.

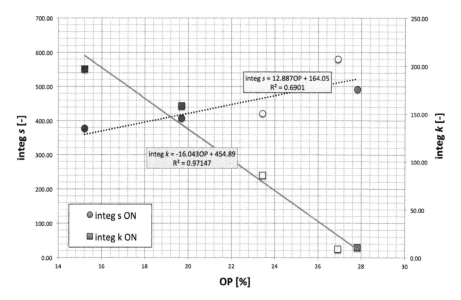

FIGURE 6.44 The dependency between optical porosity OP and integral value of scattering and absorbance coefficients for textiles printed by activated photochromic pigments.

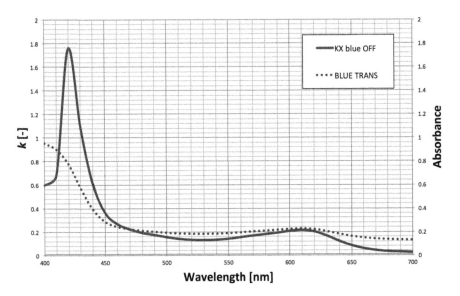

FIGURE 6.45 The spectral curves of absorbance coefficient and absorbance measured from 0.5% solution of inactivated pigment Photopia Blue in toluene and 25°C by transmittance spectrophotometry.

where r is Fresnel amplitude reflectance coefficient and t is the Fresnel's amplitude transmittance coefficient. The indices p and s indicate polarization in the incident plane and polarization perpendicular to the incident plane, respectively. Angle α is the angle of incident and β is the angle of refraction, and n_i and n_t are the corresponding indices of the refraction, respectively [48].

If we extend the Fresnel's conception from one beam illumination to diffusion illumination (Figure 6.46) and we assume two diffusion light flux (internal and external) and we also assume the layer with the isotropic refractive index n_2 and the layer is limited by media with isotropic refractive indices n_1 and n_3, we will obtain corresponding Fresnel's coefficients for diffusion illumination r_d, r^*_d and o_d, as follows from Eq. (6.27):

$$\frac{1-r_b}{n_1^2} = \frac{1-r_d^*}{n_2^2} = \frac{1-o_d}{n_3^2} \tag{6.27}$$

If we calculate Fresnel's index for external reflection r^*_d, we have to use the index of the specific textile fiber in the formula. For cotton, the transversal

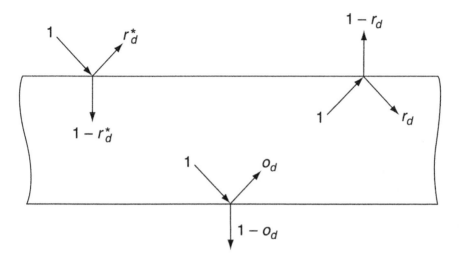

FIGURE 6.46 The illustration of reflectance coefficients for diffusion illumination.

refractive index is usually at a level of 1.530 with variability of 0.003 according to the cultivation area [49, 50]. If the refractive index of ambient air environment surrounding the cotton fiber is n_1 and $n_3 = 1$, we can illustrate the transmittance and reflectance of light by cotton fiber as presented in Figure 6.47.

On the basis of the abovementioned model conception, we can implicate Eq. (6.28) for the external reflectance and transmittance via the modification of equation for the combined reflectance and diffuse illumination [51]:

$$R_E = r_d^* + \left(1 - r_d^*\right) \cdot \left(1 - r_d\right) \cdot \frac{\left(1 - o_d \cdot R\right) + o_d \cdot T^2}{\left(1 - o_d \cdot R\right) \cdot \left(1 - r_d \cdot R\right) - o_d \cdot r_d \cdot T^2} \quad (6.28)$$

$$T_E = \frac{\left(1 - r_d^*\right) \cdot \left(1 - o_d\right) \cdot T}{\left(1 - o_d \cdot R\right) \cdot \left(1 - r_d \cdot R\right) - o_d \cdot r_d \cdot T_2} \quad (6.29)$$

Eq. (6.28) for the borderline case of non-translucent, opaque materials, when $T = 0$ is coming to well-known Sanderson's correction [51, 52]:

$$R_E = r_d^* + \frac{\left(1 - r_d^*\right) \cdot \left(1 - r_d\right) \cdot R}{\left(1 - r_d \cdot R\right)} \quad (6.30)$$

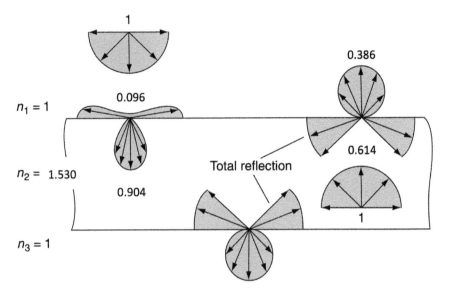

FIGURE 6.47 The illustration of the transmittance and reflectance on model cotton fiber in diffused illumination.

The cotton fibers are the birefringence system, and the transversal refractive index of cotton fiber is 1.530, while the axial refractive index is higher at approximately 1.577 [53, 54]. For the determination of refractive index, we need to consider that the refractive index depends not only on temperature but also on wavelength of the illumination used. The refractive indices are usually related to the Fraunhofer's D line of sodium at 589.3 nm and at a temperature of 25°C. The reason is that the sodium lamp is very often used as the light source and the sodium lamp is presently replaced by LED with an interference filter at 589 nm.

For the abovementioned application, we need to obtain a data set of the refractive indices for different wavelengths, and we have used the experimental method of Beck's lines for the investigation [55, 56], which were measured using the research microscope NIKON Eclipse ME600P. The set of 31 samples of α-monochlornaphtalene and kerosene solution according to Fox was investigated in diascopic illumination with a set of 16 interference filters in 400–700 nm with a spectral resolution 20 nm. If we apply the method of Beck's lines, when the immerse liquid has the same refractive index as measured substances (in our case – textile fiber), the usually visible optical interference becomes invisible under the microscope. Thus, we use the method when the direction movement of Beck's line is investigated, and this direction of movement indicates the interface of refractive indices in which the wavelength is also determined, when the fiber "is loosened" from the microscope field. Table 6.5 documents measurement results for the refractive indices of cotton fiber as the used textile substrate for printed photochromic textiles.

If the abovementioned method is used, it is necessary to point out that refractive indices of cotton fibers can be obtained only at a specific wavelength. Saunderson [52] presented a conventional adjustment of the estimate value to the experimental results taking account of the ignored reflectance r_d and transmittance o_d factors at the inner top boundary surface. Because refractive index and other optical parameters are obviously measured at specific wavelength and are in principle monochromatic. It is necessary to predict refractive index of wavelength that absent in Table 6.5 by use of suitable fitting function. Due to smooth and monotone shape of refractive index dependence in range of wavelengths of visual spectrum it is possible to use a centered 3rd order polynomial function (see Eq. 6.31). On the basis of this model, we can calculate the remaining the refractive indices with the same spectral band pass as in the previous measurement of reflectance (10 nm). From the point of view of the data set, we use the linear regression centered by third order polynomial function:

$$n_j = B_o + B_1\left(\lambda_j - \lambda_c\right) + B_2\left(\lambda_j - \lambda_c\right)^2 + B_3\left(\lambda_j - \lambda_c\right)^3 \qquad (6.31)$$

where λ_c is the centered wavelength of polynomial function

TABLE 6.5 The Measured Value for Refractive Index of Cotton Fiber by Beck's Lines Method at Temperature $25 \pm 0.5°C$

Wavelength (nm)	n
400	1.546±0.005
420	1.543±0.004
460	1.539±0.003
480	1.537±0.005
540	1.533±0.004
590	1.530±0.003
680	1.527±0.004

This polynomial function was used for a relatively small number of points and for eigenvalues of covariance matrix. The values of elementary parameters are shown in Table 6.6 together with basic statistical characteristics.

On the basis of the calculated refractive index values for elementary wavelengths and the tabular refractive index values with the same band pass for acrylic films, we have revised the influence of the external reflectance in the system: fiber–photochromic pigment and acrylic film. The result is the representation of revised spectral graph for the absorbance coefficients k in Figure 6.48.

TABLE 6.6 Centered Third Order Polynomial (Cubic) Parameters Characterizing the Dependency of Cotton Fiber Refractive Index on Wavelength

Centered third order polynomial (cubic)	Best-fit values	Std. Error	Goodness of Fit	-
B0	1.535	0.0001602	Degrees of Freedom	3
B1	−0.00007187	0.000002877	R square	0.9993
B2	2.211E−07	2.113E−08	Absolute Sum of Squares	2.128E−07
B3	−3.912E−10	1.888E−10	Sy.x	0.0002663
λc	510	–	AICc	−51.16

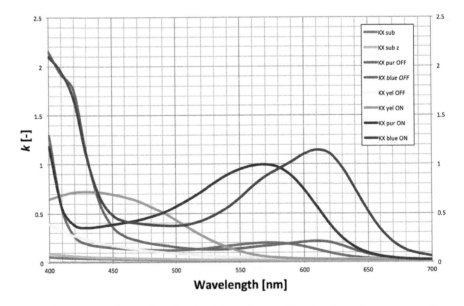

FIGURE 6.48 The illustration of revised dependency of the absorbance k on wavelength.

The absorbance maximum at 420 nm, which is visible in Figures 6.40 and 6.46 was indicated as artifact, is missing. It is evident that for the activated and also for the inactivated form of pigment, Photopia Blue (P1) was obtained a similar shape of function in the blue area of the visible spectrum continued to the maximum of absorbance in the UV part of the spectrum. A similar situation was also observed for the pigment Photopia Purple (P2).

The additivity of Kubelka–Munk function and the absorbance and reflectance coefficients allow the elimination of the influence of used textile substrate and acrylic film, and we can determine the change of absorbance (also the change of optical density $\Delta OD_{\lambda max}$) in absorbance maximum for the individual pigments [56–61]. As shown in Table 6.7, the values of the absorbance maximums and the change in the optical density determined by the modified method of black and white background are in good agreement with the values found from the literature sources measured by classical transmittance spectrophotometry for pure photochromic pigments without HALS as common for commercial photochromic pigments [62].

The test method of black and white background with the usage of unique spectrophotometer PHOTOCHROM has a good correlation with the data set for pure photochromic substance by the method of absorbance spectrophotometry. By using an optical integrator in the spectrophotometer PHOTOCHROM and calibration data set, we can measure and calculate the absorbance and scattering coefficients for the photochromic pigments and for the substrate and disperse medium (textile substrate, thin film layer, other translucent materials, etc.). The Kubelka–Munk theory, though it remains the most used in practice, has some disadvantages. If Sanderson`s correction is used, it is possible to determine the scattering characteristics not only for nontranslucent materials such as car varnish, but also for special cases of translucent materials (textile, paper, etc.). Based on the presented results,

TABLE 6.7 Comparison of Absorbance Maximums Values and the Change in Optical Density ΔOD Measured by Absorbance Spectrophotometry (LITER) and by the Black and White Method (MEASURED)

Tested pigment	λ_{max} liter	λ_{max} measured	ΔOD liter	ΔOD measured
Photopia Blue	605	610	0.9	0.91
Photopia Purple	568	570	0.8	0.77
Photopia Yellow	430	430	0.36	0.43

the method of black and white background can be used for quality control of color changeable materials.

6.5 MEASUREMENT OF THERMAL SENSITIVITY OF PHOTOCHROMIC MATERIALS

Our previous sections described a unique device PHOTOCHROM for photochromic measurement in a reflectance mode together with a methodology. This device allows photochromic sensors to test the colorimetric and spectral characteristics of photochromic textiles and also conduct fatigue tests for control of color change stability. This concept of colorimetric and spectral parameters is also related to dependence of color change on the intensity of UV irradiation and temperature. In this paper, we describe the dependence of color change on temperature for the photochromic Photopia AQ Ink system (Blue, Purple, and Yellow) produced by Matsui Shikiso Chemical Co. Ltd. It is known that the reversion of photochromic compounds from colored to colorless is promoted thermally. The photochromic structure can achieve a lower level of saturated absorbance at higher temperatures when thermo-reversible photochromic systems, such as spirooxazines and chromenes, increase the rate of thermal bleaching reaction and thus decrease light stimulated coloration.

Some examples of active intelligent textiles can react by changing their own color dependent on external stimuli [63–66] and are therefore called chameleonic textiles or heat-containing textiles, which are able to gain or lose energy according to the external temperature. Textile-based sensors and active protective textiles are easily customizable by sewing, thermal bonding, or gluing. They are also easy to maintain (e.g., washing and chemical drying) and light with good strength, tenacity, and elasticity. It is possible to integrate these types of sensors into the system of protective clothes at a reasonable price. This paper is dedicated to research of textile-based sensors with photochromic behavior for studying the dynamic behavior and modulation of sensitivity photo chromic sensors.

In this part of chapter, the thermal dependency of used photochromic substances is described by a kinetic model, which defines the speed of color change initiated by external stimuli (i.e., UV light). A newly proposed kinetic model of thermal sensitivity photochromic substances is discussed.

6.5.1 KINETIC DESCRIPTION

The chemical and physical kinetics are described by the first-order equations, and the conversion rate is proportional to the concentration of unreacted species. Thus, the momentary concentration of the reactant ($n_R(t)$) should follow Eq. (6.32) [67]:

$$n_R(t) = n_R(0) \cdot e^{-kt} \tag{6.32}$$

where ($n_R(0)$) is the initial concentration and k stands for the constant rate. In principle, the kinetics of a photo induced to a colored form can be deduced from monitoring one of the characteristics of absorption bands. The absorbance $A = -\log(T)$ is determined from the measurement of transmission $T = I(d)/I(0)$ of the sample thickness d [68]. Here, $I(0)$ is the light intensity measured at input face ($z = 0$) of the sample and $I(d)$ is the transmitted light intensity measured behind its output face ($z = d$). For first-order kinetics, Eq. (6.33) is fulfilled as follows:

$$\frac{t}{\tau} = -\ln\left(\frac{n_R(t)}{n_R(0)}\right) = -\ln\left(\frac{A(\infty) - A(t)}{A(\infty) - A(0)}\right) \tag{6.33}$$

Here, the time-constant (τ) is defined as $\tau = k_R^{-1}$, and $A(0)$, $A(t)$, and $A(\infty)$ are the initial, momentary, and final values of absorbance, respectively. Sometimes, it is difficult to determine $A(\infty)$, especially in long-time relaxation processes where instability in measurement can play an essential role in UV degradation of material. Equation (6.33) is frequently written in the following form:

$$A(t) = \left(A(0) - A(\infty)\right) \cdot e^{-kt} + A(\infty) \tag{6.34}$$

In terms of translucent media, it is possible to replace Lambertian absorption A by the Kubelka–Munk function K/S. Viková and Vik [9] have showed that Eqs. (6.33) and (6.34) are suitable for the calculation of kinetic data, which means that shade intensity ɩ can be used as absorption A or a Kubelka–Munk function K/S (see Eqs. (6.9) and (6.11), respectively). From these equations, it is possible to calculate the half-life time of color change $t_{1/2}$:

$$t_{1/2} = \frac{\ln 2}{k} \qquad\qquad (6.35)$$

From the previous experiments, it is evident that the experimental data incorporate a first-order kinetic model according to Eq. (6.9) for exposure and reversion. It is evident that the proposed model functions fit the experimental data well. The main task of our research is to develop a simple textile sensor, which is sensitive to UV light and to study kinetic behavior.

If the temperature during experiment is not stabilized, it is possible to observe local color fading during exposure [62, 64]. Local fading can be influenced by the increased temperature of the measured sample or by photobleaching of the photochromic pigment. This fading effect was shown in Figure 6.14.

The repeatability test for real measurement produces a graph as shown in Figure 6.18, in which the thermal stabilization for 20°C is well illustrated. In this figure, the error bars show the variability of experimental data that are about 10-time repetition from 10 different samples. This experimental data set documents the precision and accuracy of the presented measurement. It is evident that the deviation of measurement forming a mean value on the graph is practically constant despite a minor exception in the second measured value. The highest deviation at this point is caused by the used time sequence of measurement, in which there are 5 intervals between individual measurements. Slight improvement is obtained by shortening the reading intervals to 1 s.

The most considered and studied factor is temperature. Because the reversion from the colored to the colorless form is promoted thermally, the photochromic structure achieves a weak shade at higher temperatures than at lower temperatures. This phenomenon is known as temperature dependency. For example, oxazine structures are more sensitive than naphthopyrans [69]. Temperature can influence the kinetic data on photochromic color change as described in Figure 6.49, in which the decreasing of the equilibrium level color shade is visibly evident.

This aspect of thermal sensitivity is well described in the graphs shown in Figure 6.50. The Kubelka–Munk values in equilibrium K/S_{eq} were measured for the temperature intervals from 20°C to 30°C. The K/S values were calculated at the appropriate absorption maximum for each pigment.

The linear dependence of the measured data shows a relationship with back isomerization and activation energy as described by the Arrhenius idea.

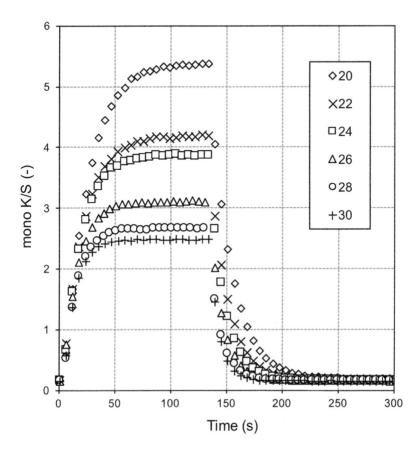

FIGURE 6.49 K/S temperature dependence of kinetic of photochromic color change, Photopia® AQ Ink Purple.

The Arrhenius relation (Eq. (6.36)) developed theoretically for chemical and biochemical reaction has been experimentally shown to hold empirically for a number of more complex chemical and physical phenomena (e.g., viscosity, diffusion, and sorption).

$$k_{(T)} = k_0.e^{-\frac{Ea}{k_B T}} \tag{6.36}$$

with k_B as the Boltzmann factor, k_0 (i.e., the frequency factor), and Ea (i.e., activation energy) are required to overcome the barrier. Linear Arrhenius plots are obtained (Figure 6.51) when Eq. (6.36) fits the data at different temperatures.

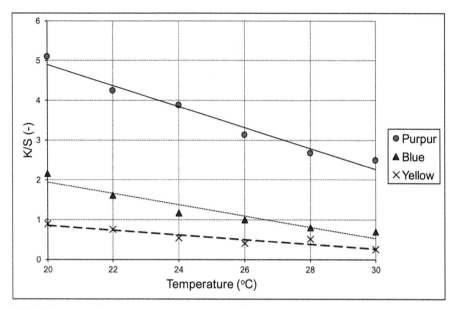

FIGURE 6.50 K/S_{eq} temperature dependence of three photochromic inks – Photopia® AQ Ink on textile substrate.

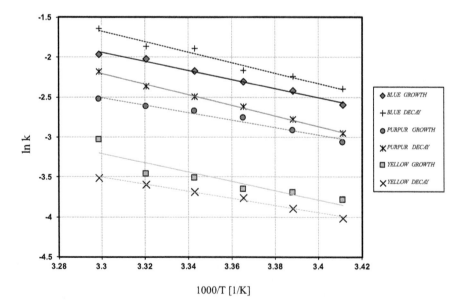

FIGURE 6.51 Arrhenius plot of thermal ring opening reaction for Photopia® AQ Ink photochromic systems.

As shown in Tables 6.8 and 6.9, the half-life values of t$_{\frac{1}{2}}$ decrease very rapidly when the temperature increases (e.g., yellow – from 31 s at 20°C to 14.43 s at 30°C), indicating relatively high activation energy (*Ea*) for thermal back isomerization. In order to eliminate time dependence, all measurements must be performed with the same annealing time *t*.

This chapter attempts to continue and introduce the possibility of computing activation energy directly from reflectance measurement. The

TABLE 6.8A Half-Lives and Activation Energies of Tested Photopia Pigments – Growth Phase—Part 1

Pigment	20°C [s]	22°C [s]	24°C [s]	26°C [s]
Blue	9.25	7.76	6.92	6.08
Purple	14.90	12.81	10.94	10.08
Yellow	30.43	27.97	26.66	23.34

TABLE 6.8B Half-Lives and Activation Energies of Tested Photopia Pigments – Growth Phase—Part 2

Pigment	28°C [s]	30°C [s]	Ea [kJMol^{-1}]
Blue	5.25	4.95	46.235
Purple	9.50	8.70	39.787
Yellow	22.04	14.43	55.094

TABLE 6.9A Half-Lives and Activation Energies of Tested Photopia Pigments – Decay Phase

Pigment	20°C [s]	22°C [s]	24°C [s]	26°C [s]
Blue	7.61	6.50	6.02	4.60
Purple	13.25	11.09	9.53	8.36
Yellow	38.56	33.91	29.75	27.62

TABLE 6.9B Half-Lives and Activation Energies of Tested Photopia Pigments – Decay Phase

Pigment	28°C [s]	30°C [s]	Ea [kJMol^{-1}]
Blue	4.46	3.57	55.848
Purple	7.38	6.12	57.029
Yellow	25.29	23.17	37.619

reaction rate of photoisomerization of each tested photochromic pigment was strongly temperature-dependent. The above results clearly demonstrate that the difference among the tested pigments evokes significant changes in the potential energy surface and particularly in the ground-state energy barrier. Concerning this phenomenon, the thermal sensitivity can be used for calculation during development and construction of textiles based on photochromic sensors, and thermal sensitivity can play an important role in the development of UV textile sensors according to the changing temperature in different climatic conditions [71–75].

6.6 IMPACT OF DRAWING RATIO ON DIFFERENCE IN OPTICAL DENSITY

Photochromic textiles have been used in the sensible materials (sensor) in the smart textiles and esthetic materials in the fashion industry due to the color changing properties, which gives immense inspiration to prepare photochromic textile materials. In this regard, we focused more in preparing the mass-dyed polypropylene filaments with photochromic pigments. Metallocene polypropylene (mPP) filaments were doped by different concentration of photochromic pigment. In our experimental work, the optical and mechanical properties of these multifilaments depending on their drawing ratio were investigated. 5-Chloro-1,3-dihydro-1,3,3-trimethylspiro [2H-indole-2,3'-(3H)naphtha[2,1b](1,4)oxazine] was used as photochromic pigment. With regard to the optical properties, our investigation confirms the known effect of the primary reflectance on the depth of shade of colored multi-filaments as shown by linear increase of difference in optical density with increase in the ratio of fineness.

One of the most promising and flourishing field of the photochemistry is photochromism, and it has been studied more about its photochromic properties in both organic and inorganic materials [82]. Photochromism is defined by IUPAC as a "Light-induced reversible change of color"; these reversible color changing can be caused by the exposure of electromagnetic radiation (mainly under ultraviolet irradiation) [77, 78]. Functional textiles have renewed interest on the field of smart and fashionable textiles, and photochromic textile is one among them [80, 98]. Recent research manuscripts rapidly increase the interest of the application of photochromic compounds on textiles, though there are comparatively few reports in this specified area [77, 83, 86–88, 91, 95–98].

Polypropylene (PP) is a polymer that is a member of the "polyolefin" family; it has good impact strength, surface hardness, excellent abrasion resistance, and resistant to a wide variety of acids, alkalis, and solvent. Because of the many advantages, it can be widely used for the production of medical products, carpets, apparel and household textiles, filtering media, agro and geotextiles, automotive interior, and many other technical textiles [81, 84, 85, 89]. Coloration of polypropylene is least possible to carried out in classical methods (in a dye bath); the reason underlying this is fiber having the non-polar, aliphatic structure as well as high crystallinity and high stereo-regularity, which limit the accessibility of dye molecules. Commonly, this issue can be solved by mass coloration techniques, which give very good dispersion and homogenization properties of dyed polypropylene goods [85].

Based on these advantages and the fact behind the dyeability of polypropylene, we decided to produce photochromic polypropylene filaments through mass coloration techniques. The incorporation of coarse pigment particles on the polypropylene not only impairs tensile and strength, but also jeopardizes the subsequent processing as a whole, such as optical and physical properties. The additives used during mass coloration can also influence the optical and mechanical properties of produced multifilament. These additives may hinder (HALS) for light fastness enhancement used in photochromic capsules. However, the commercial spectrophotometer cannot measure the whole color changing propertied of photochromic materials during the ultraviolet exposure without interruption of illumination of the samples [92, 98]. However, the previous experiments [76, 90] can be done in Datacolor Spectraflash SF600 spectrophotometer with specular component excluded and ultraviolet component included. In their research, the samples first undergo ultraviolet irradiation, and the fabric is then immediately transferred to the spectrophotometer to evaluate the fading behavior; however, it will not give the accurate kinetic data, because the photochemical reaction can occur between attoseconds to nanoseconds. To solve this issue, it is necessary to use spectrophotometer with modified design of integrating sphere, which has the additional aperture for the UV light source.

Application of photochromism in textile materials is being investigated for developing and designing with the functional effects on textile-based

UV-sensor, camouflage, and smart application in textiles. In this scenario, we published many research manuscripts about applications of photochromic material on textiles. For this work, we have produced a photochromic pigment-incorporated polypropylene multifilament with different drawing ratios. The goal of this study is to find out the relationship between the concentration of photochromic pigments and drawing ratio with respect to the mechanical properties and optical properties of mass-colored polypropylene filaments.

In this study, metallocene polypropylene PP HM 562 R abbreviated as mPP, chips and ground, Melt Flow Index (MFI) = 26.6 g/10 min, which was purchased from Lyondell Basell, Italy, was used. The photochromic pigment 5-chloro-1,3-dihydro-1,3,3-trimethylspiro[2H-indole-2,3'-(3H)naphth[2,1-b](1,4)oxazine], CAS: 27333-50-2 abbreviated as pBl, blue photochromic pigment in the form of ink with 50 wt. % concentration of pure pigment, was purchased from Matsui Shikiso Chemical Co., Ltd, Japan. Before the preparation of photochromic pigment-colored mPP filament, the colorless mPP (without photochromic pigments) filaments were produced. For the preparation of both colored and colorless mPP, the dried mPP chips was used as raw material with standard conditions, as specified by the manufacturer. In the former, the 100% colored mPP was produced like a tape or ribbon form and converted into chips form. Finally, these colored chips were mixed with colorless chips at four different concentrations (0.25, 0.50, 1.50, and 2.50 on wt % of the filament) to produce the colored filament, which has been used for this study. For the mass colored tape production the relevant components were mechanically mixed and melted in a single-screw ribbon extruder-GÖTTFERT (Germany) and having the three temperature zones. The temperature of each zone was maintained the same (i.e., T1=T2=T3=220°C) for this study. After the production, mass colored tape or ribbon was cooled immediately in cold water and cut into chips. The air-dried colored chips were finally vacuum-dried for 2 h at T4= 105°C and used for filament preparation.

Spinning of the dried colored chips was performed in the laboratory melt spinning equipment with an extruder of D=16 mm at T5 = 220°C, feeding weight = 15.6 g/10 min, take up speed = 50 m.min^{-1}, and a spinneret nozzle with 13 fibrils (holes). After the spinning, the filaments were drawn at T6 = 120°C by seven different drawing ratios: DR1= 1.0; DR2 = 2; DR3 = 2.5; DR4 = 3; DR5 = 3.2; DR6 = 3.5, and DR6 = 4.0.

6.6.1 MECHANICAL PROPERTIES

After the production, the mechanical properties of the filaments were investigated. The result is reported in Table 6.9. According to the ISO 2062:1993, the Instron instrument was used to determine the tenacity, elongation at break, and Young's modulus. For each specimen, the above test was repeated 20 times, and the mean value was reported in Table 6.10. During the test, we maintained the clamping length of 12.5 cm and speed of clamp shift is 350 m.min^{-1}.

6.6.2 COLORIMETRIC PROPERTIES OF THE FIBERS

After the drawing process, each sample was wound with uniform tension on gray cardboard. We maintain the sufficient thickness during the winding process to prevent show-through. In this case, we used six layers of filaments as shown in Figure 6.52. The color of the card was gray. In the test for presence of fluorescing agents, the result was negative, which allowed good correlation with visual assessment [93].

The produced filaments were analyzed for spectral and colorimetric properties under UV exposure using LCAM-PHOTOCHROM3, which allows to determine the decay phase of photochromic color change. PHOTOCHROM3 has been made with the activation light source of Edixon UV LED EDEV-3LA1 with radiometric power $\Phi V = 350$ mW, minimum peak wavelength is 395 nm, and the maximum peak wavelength is 410 nm. Radiometric power was measured with an accuracy of $\pm 10\%$. For spectral and colorimetric analysis, each sample was measured 10 cycles with respect to decay and exposure. The condition of the decay cycle was 10 min and 5 min. The exposure was at temperature of $22 \pm 1°C$ and the relative humidity of $50\% \pm 3\%$. The reflection wavelengths were measured with an accuracy of ± 0.5 nm. The reflection values were used to determine the color strength (K/S) by using of Kubelka–Munk functions. The optical density and the difference in optical density respectively, were computed using the following Eq. (6.37).

$$\Delta OD = K/S_{\lambda \max G \max} - K/S_{\lambda \min D \min} \tag{6.37}$$

where $K/S_{\lambda maxGmax}$ is the maximal value of Kubelka–Munk function [96] at the wavelength of absorption maxima during the exposure phase and $K/S_{\lambda maxDmin}$ is the minimal value of Kubelka–Munk function at the wavelength of absorption maxima during the decay phase.

TABLE 6.10 Mechanical and Optical Properties of Polypropylene Photochromic Multifilaments

DR λ	Concentration of pBI	T [tex]	σ [cN/tex]	CV$_\sigma$ [%]	ε [%]	CV$_\varepsilon$ [%]	E [N/tex]	CV$_E$ [%]	ΔOD [-]	u$_{\Delta OD}$ [-]
1.0	0.00 wt.%	45.3	10.51	22.1	260.3	16.6	0.56	23.2	0.0003	0.0008
	0.25 wt.%	39.1	9.47	19.8	232.2	15.8	0.49	22.8	0.2201	0.0041
	0.50 wt.%	35.5	9.64	21.3	210.1	17.2	0.57	24.1	0.3584	0.0044
	1.50 wt.%	35.1	9.83	23.5	190.0	15.6	0.53	21.0	0.7418	0.0287
	2.50 wt.%	34.2	8.55	24.0	173.5	15.1	0.48	23.2	1.1017	0.0352
2.0	0.00 wt.%	24.6	22.40	14.8	185.9	8.9	0.99	16.5	0.0084	0.0036
	0.25 wt.%	21.3	11.81	11.6	159.3	10.1	0.87	13.2	0.1443	0.0047
	0.50 wt.%	18.1	12.03	12.8	142.5	12.1	1.02	14.9	0.2705	0.0044
	1.50 wt.%	18.2	12.26	16.9	145.0	13.9	0.94	11.2	0.5413	0.0126
	2.50 wt.%	18.4	10.67	18.1	132.4	9.6	0.84	15.6	0.8078	0.0129
2.5	0.00 wt.%	20.0	22.41	15.2	140.5	9.9	1.50	17.9	0.0061	0.0021
	0.25 wt.%	16.6	17.97	13.7	124.1	12.4	1.42	14.2	0.1331	0.0069
	0.50 wt.%	13.9	15.50	19.2	119.5	10.2	1.50	16.7	0.2574	0.0046
	1.50 wt.%	14.2	16.57	14.3	112.7	11.6	1.56	14.5	0.5467	0.0085
	2.50 wt.%	16.1	13.17	11.2	86.8	12.3	1.39	13.9	0.8322	0.0232
3.0	0.00 wt.%	15.4	28.39	11.8	97.9	11.9	1.54	12.1	0.0034	0.0037
	0.25 wt.%	11.9	21.13	16.9	85.4	17.8	2.06	16.4	0.1301	0.0018
	0.50 wt.%	10.4	19.44	13.5	80.4	13.6	1.93	13.2	0.2340	0.0027
	1.50 wt.%	11.9	19.42	14.1	79.3	14.1	1.80	13.9	0.5143	0.0192
	2.50 wt.%	14.8	19.34	15.6	71.7	15.2	1.78	14.9	0.7993	0.0271
3.2	1.50 wt.%	13.2	19.03	14.9	56.8	15.1	1.84	15.0	0.4962	0.0309
	2.50 wt.%	10.7	21.54	17.2	55.3	17.6	2.62	16.8	0.8209	0.0014
3.5	0.00wt.%	14.8	29.34	17.4	74.3	17.2	2.27	17.6	0.0031	0.0034
	0.25 wt.%	12.9	23.15	18.1	67.1	17.6	2.40	17.7	0.1251	0.0089
	0.50 wt.%	10.2	27.59	16.7	54.9	16.5	2.61	17.1	0.2309	0.0028
4.0	0.00 wt.%	14.0	31.94	18.6	39.9	18.9	2.60	17.8	0.0023	0.0023

Note: Fineness (T), tenacity (σ), elongation at break (ε), Young's modulus (E), change in optical density (ΔOD) and its variations, Coefficient of Variations (CV).

6.6.3 EFFECT OF MECHANICAL PROPERTIES

The impact of photochromic pigment concentration and drawing ratios of the filaments with respect to their mechanical parameters are shown in

FIGURE 6.52 Samples winded on a card wound (left UV OFF), (right UV ON).

Table 6.10. Graph in Figure 6.53 explains the dependence for the mechanical properties of photochromic colored filament with respect to the drawing ratio. From all curves, it was noticeable that the used m_{pp} filaments were in the range of low-oriented filaments. The maximum attainable mechanical properties P_{max} (in our case tenacity (σ)) could be predicted by using the following equation. It can be utilized to fit the data for all experiment, which against 1/DR, and Eq. (6.38) [77].

$$P = P_{max} - k_p \left(1/DR \right) \qquad (6.38)$$

where k_p is a constant taking into account the sensitivity property for the drawing process. Based on this equation, the maximum attainable tenacity of uncolored filament was calculated, and the tenacity is 37 cN/tex.

The interesting fact behind this was the addition of photochromic pigment, which could decrease the elongation at break, which is linearly dependent on the drawing ratio (Figure 6.54). Increasing the drawing ratio can reduce the change of this property. A possible explanation of these phenomena can be a pre-orientation of the crystalline segments of the polymeric chain around the photochromic pigment capsules, which also causes with slight decreasing of its melt viscosity. However, the drawing process can increase the parallelization of the polymeric molecules with respect to the filament axis. This molecular arrangement

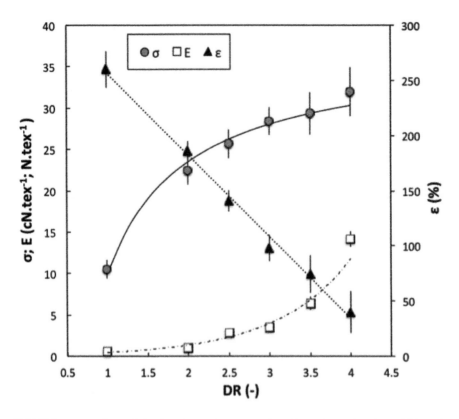

FIGURE 6.53 Effect of the drawing ratio on tenacity, elongation at break, and Young's modulus of colorless filament (without photochromic pigment) (DR = drawing ratio).

toward to the parallelization can be increased with increase in the draw ratio; finally, the highest draw ratio (i.e., some extents) filament gets more strength than its lowest ratio. These structural rearrangements can also be another reason for decreasing the elongation at break of filaments. In this case, Young's modulus increased with increase in the draw ratio and the growth was exponential. This is due to the parallelization of the molecular arrangements, which increased the crystalline region in the fiber, which in turn increased many of the physical properties but reduced the property of elasticity. The results showed no impact on the concentration of photochromic pigment with respect to the young's modulus. The CV for the tenacity decreased with an increase in pigment concentration. But, in some cases, the results were not affected as such, for example, at the concentration of 0.5% and the drawing ratio of 2.5. In case of Contrary

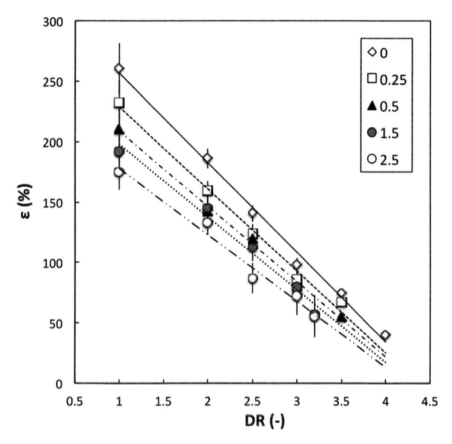

FIGURE 6.54 Effect of the drawing ratio and photochromic pigment concentration on elongation at break (DR = drawing ratio).

to CV of tenacity is CV for the elongation and Young's modulus variable, without specific trend.

6.6.4 EFFECT OF PHOTOCHROMIC RESPONSE AND OPTICAL DENSITY

Generally, the photochromic response is related to the Kubelka–Munk's functions. In this case, the results explain the relationship between Kubelka–Munk's function with respect to the drawing ratio along with the concentration of photochromic pigment. It was clearly shown that increase in the drawing ratio caused decrease in the Kubelka–Munk value due to the increase in the mass per volume of the filaments. Based on the Kubelka–Munk's functions, the optical density is shown in Table 6.11.

Based on Eq. (6.37), the optical density ΔOD was reported in Table 6.11. The optical density can be measured on undrawn filaments and with respect to drawn one, and the results are shown in Figure 6.55. The results show a

TABLE 6.11 Maximum Attainable Values of Difference in Optical Density (ΔOD) of Colored Multifilaments

Pigment concentration [wt.%]	ΔOD_{max} [-]	k_p [-]
0.25	0.082±0.004	−0.136±0.009
0.50	0.180±0.005	−0.179±0.008
1.50	0.415±0.028	−0.298±0.048
2.50	0.651±0.042	−0.436±0.073

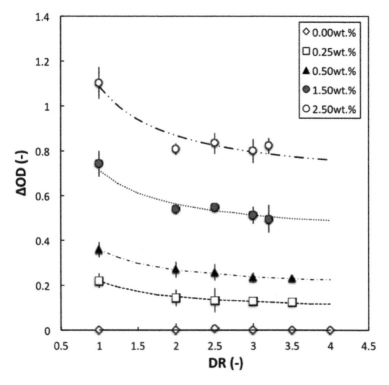

FIGURE 6.55 Effect of drawing ratio and photochromic pigment concentration on difference in optical density ΔOD.

decrease in optical density with increase in the draw ratio. Equation (6.38) was used to fit the data points of ΔOD; the tenacity sensitivity constant k_p was negative and linearly dependent on the concentration of the photochromic filament, as given in Table 6.11.

The geometrical and physical properties of filament can influence optical properties, and it can be seen through the Kubelka–Munk (K/S) analysis (i.e., light reflectance) during the spectral measurement of light rays in the presence of colorant particle (photochromic pigment). The results show reduction of their intensity during light absorption, which they reflected, back to the observer. It is known that absorption coefficient K expresses the selective absorption of light by the colorant particle and it is only slightly influenced by the interaction with the filament. However, the scattering coefficient S, the opposite of absorption coefficient K, depends on the optical properties of media (filament). It is influenced by the following physical specifications:

- Refraction coefficient of the boundary air/polymer and polymer/particle (colored pigment or delustering particle) together with its concentration.
- Filament fineness, shape cross section, smoothness of surface (effect of shaping).
- Distribution of colorant particle fiber structure and its size and geometry.

Generally, increasing the scattering coefficient S, which is nothing but increase the colorless reflection, may lead to decrease in the shade intensity at the same average of dye concentration present in the filament. The results showed that the main influence of our experiment related to the fiber fineness influenced by the drawing ratio. Theoretically, the fineness of filaments does not impact the trajectory of light rays in fiber structure. But, the color of textile is determined by the length of the light rays' trajectory in fibers due to light absorption [73, 90, 94]. Therefore, increasing the primary reflectance on the boundary of air/polymer due to the high specific fiber surface, and it also increased the reflected white light. By using Kubelka–Munk theory, it is possible to predict the difference between the shade intensity of thicker and thinner filament with circular cross-section, which is equal to the ratio of diameter d or the square root of fineness explained in Eq. (6.39),

$$\frac{(K/S)_1}{(K/S)_2} = \sqrt{\frac{T_1}{T_2}}\frac{d_1}{d_2} = \frac{A_{s1}}{A_{s2}} \tag{6.39}$$

where A_S [m².g⁻¹] is the specific fiber surface and T is the thickness of the filament in tex. Eq. (6.39) is valid for the same dye and concentration in fibers. But the linear dependence for the Kubelka–Munk function on colorant concentration can normalize the difference in optical density (ΔOD) into one unit concentration of photochromic pigment for all tested concentration by using Eq. (6.40).

$$c_2 = \sqrt{\frac{T_1}{T_2} \cdot c_1}$$ (6.40)

From Figure 6.56, the relationship to square root of the fineness ratio is linear. Therefore, it is possible to measure these properties of photochromic colorants with respect to the similar trends of standard colorants. For very fine filament (microfiber), the measurement of shade depth will probably coincide with Eq. (6.40) [79].

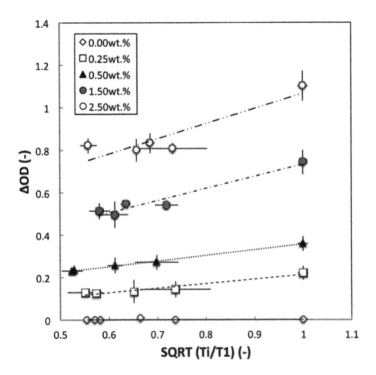

FIGURE 6.56 Effect of square root of the fineness ratio for normalized concentration of the photochromic pigment on difference in optical density ΔOD.

In this study, the photochromic filaments were produced with different concentrations and different drawing ratios. We analyzed the mechanical and optical properties by using Instron tensile testing machine and dynamic spectrophotometer FOTOCHROM, respectively. We observed that elongation at break of photochromic filament linearly decreased with increase in the drawing ratio. Effect of the drawing ratio on difference in optical density was slightly nonlinear and decreasing. Regarding the optical properties, our investigation confirmed that the known effects of the primary reflectance on the depth of shade of colored multifilament is shown by linearly increasing difference in optical density with an increased ratio of fineness (drawing ratio). Less light comes to the photochromic dye molecules in finer fibers, and greater part of light is reflected out of the textile, thus forming colorless reflection. On comparing the two fibers containing the same quantities of dyes, the finer fibers seemed to be lighter in point of view of observed color change.

KEYWORDS

- colorimetry
- Kubelka-Munk function
- photochromism
- spectrophotometry
- thermal sensitivity
- UV sensors

REFERENCES

1. Williams, T. P., (1964). Photoreversal of rhodopsin bleaching, *J. Gen. Physiol. 47*(4), pp. 679–689.
2. Rulliere, C., Amand, T., & Marie, X., (1998). *Spectroscopic Methods for Analyze of Sample Dynamics* (Femtosecond Laser Pulses), ed. Springer-Verlag, Berlin, 203–259.
3. Ern, J., Petermann, M., Mrozek, T., Daub, J., Kuldov, K., & Kryschi, C., (2000). Dihydroazulene/vinylheptafulvene photochromism: dynamics of the photochemical ring-opening reaction, *Chemical Physics, 259*, 331–337.
4. Hogue, R., (1991). 'Praying Mantis' diffuse reflectance accessory for UV-Vis-NIR spectroscopy, *Fresenius J. Anal. Chem., 339*, 68–69.

5. Boroumand, F., Moser, J. E., & Van den B., (1992). Quantitative diffuse reflectance and transmittance infrared spectroscopy of non-diluted powders, *Appl. Spectroscopy, 46*(12), 1874–1886.

6. Polymicro Technologies, available from: http://www.polymicro.com/products/.optical-fibers/products_opticalfibers_sr_uvmi_uvm_uvi.htm.

7. Vik, M., & Viková, M. Equipment for monitoring of dynamism of irradiation and decay phase photochromic substances (in Czech), *Czech Patent no. 304865.*

8. Viková, M., (2005). *Selected Problems of Measurement of Photochromic Colorants,* Book of papers AIC Color 05–10th Congress of the International Color Association, Granada, pp. 1135–1138.

9. Viková, M., & Vik, M., (2007). *Accurate Measurement Photochromic Materials,* Book of proceedings 6th International Conference - TEXSCI 2007, June 5–7; TUL:Liberec, pp. 278–282.

10. Van Gemert, B., & Kish, D. G., (1999). The intricacies of color matching organic photochromic dyes, *PPG Informations, 5*(14), 53–61.

11. Viková, M., & Vik, M., (2007). Measurement of photochromic textiles, *ISOP07,* Vancouver, Canada, 7–10.

12. Viková, M., (2004). Visual assessment UV radiation by color changeable textile sensors, *AIC, Color and Paints,* the Interim Meeting of the International Color Association, Porto Alegre, Brasil, 1–5.

13. Vik, M., (2017). *Colorimetry in Textile Industry,* VÚTS, a.s. Liberec, pp. 36–42.

14. Viková, M., (2011). Photochromic textiles, PhD Dissertation, Heriot-Watt University, Edinburgh, pp. 53–147.

15. Philips-Invernizzi, B., Dupont, D., & Caze, C., (2001). Bibliographical review for reflectance of diffusing media, *Optical Engineering, 40*(6), 1082–1092.

16. Grum, F. C., & Bartleson, C. J., (1980). *Optical Radiation Measurements: Color Measurement,* Academic Press.

17. Chandrasekhar, S., (1960). *Radiative Transfer,* Dover, New York, pp. 8–51.

18. Mudgett, P. S., & Richards, L. W., (1971). Multiple scattering calculations for technology, *Journal of Applied Optics, 10*(7), 1485–1502.

19. Schuster, A., (1905). Radiation through a foggy atmosphere, *Astrophysics Journal, 21*(1), pp. 1–22.

20. Judd, D. B., & Wyszecki, G., (1975). *Color in Business, Science, and Industry,* Third edition, John Wiley and Sons, New York, pp. 397–437.

21. Green, P., & MacDonald, L., (2002). *Color Engineering,* John Wiley and Sons: Chichester, pp. 9–12.

22. Schultze, W., (1966). *Education in Color and Colorimetry (in German),* Springer Verlag, Berlin, pp. 61–71.

23. Kubelka, B., (1954). New contributions to the optics of intensely light-scattering materials. Part II, *J. Opt. Soc. Am., 44,* 330–335.

24. Kubelka, B., (1948). New contributions to the optics of intensely light-scattering materials. Part I, *J. Opt. Soc. Am., 38,* 448–457.

25. Viková, M., Christie, R. M., & Vik, M., (2014). A unique device for measurement of photochromic textiles, *Research Journal textile and Apparel, 18*(1), pp. 6–14.

26. Vik, M., & Viková, M., (2016). US patent: 20160299054 A1 A method and device for fatigue testing of photochromic, fluorescent or phosphorescent dye/dyes, or of a mixture of at least two of them, and a device for carrying out this method, Published.

27. Hunter, R. S., & Harold, R. W., (1987). *The Measurement of Appearance*, John Willey & Sons, New York, pp. 13–54.
28. Vik, M., (1995). *Fundamentals of Color Measurement (in Czech)*, vol. *1*, TUL: Liberec, pp. 37–48.
29. Schanda, J. ed., (2007). *Colorimetry, Understanding the CIE System*, John Wiley & Sons, Hoboken, pp. 29–37.
30. Ohta, N., & Robertson, A. R., (2005). *Colorimetry, Fundamentals and Applications*, John Wiley & Sons: Chichester, pp. 63–113.
31. Kuehni R. G., (2013). *Color, An Introduction to Practice and Principles*, 3rd Edition; Wiley, Hoboken, pp. 121–126.
32. CIE 15–2004 Colorimetry, 3rd Edition.
33. Hunt, R. W. G., & Pointer, R. M., (2011). *Measuring Color*, 4th edition, Wiley, Chichester, (The Wiley-IS&T Series in Imaging Science and Technology), pp. 293–324.
34. Luo, M. R., & Hunt, R. W. G., (1998). The structure of the CIE 1997 color appearance model (CIECAM97s), *Color Res. Appl., 23*, 138–146.
35. Moroney, N., (2002). The CIECAM02 Color appearance model, *Proceeding of the Tenth Color Imaging Conference: Color Science, System and Applications*, pp. 23–27.
36. CIECAM02 model: the standardized CIECAM02 model as described in Fairchild MD. *Color Appearance Models*, 2nd edition. Chichester, UK, Wiley, 2005.
37. Wen, S., (2012). A color difference metric based on the chromaticity discrimination ellipses, *Optics Express, 20*(24), pp. 26441–26447.
38. Johnston-Feller, R., (2001). Color science in the examination of museum objects: *Nondestructive Procedures*, The Getty conservation Institute, Los Angeles, pp. 33–39.
39. Liu, L., Gong, R., Huang, D., Nie, Y., & Liu, C., (2005). Calculation of emittance of a coating layer with the Kubelka-Munk theory and the Mie-scattering model, *J. Opt. Soc. Am. A., 22*(11), pp. 2424–2429.
40. Richmond, J. C., (1963). Relation of emittance to other optical properties, *J. Res. National Bureau Standards-C, 67C*(3), pp. 217–226.
41. Kamata, M., Suno, H., Maeda, T., & Hosikawa, R., (1994). *Reversibly Variable Color Patterning Composition for Synthetic Resin Articles*, Patent US5431697 A, US 08/255,999.
42. Kamada, M., & Suefuku, S., (1990). *Photochromic Materials*, Patent US5208132, US 07/828,951.
43. Viková, M., & Vik, M., (2014). Colorimetric properties of photochromic textiles, *Applied Mechanics and Materials,* Switzerland, vol. 440, pp. 260–265.
44. Viková, M., Periyasamy, A. P., Vik, M., & Ujhelyiova, A., (2017). Effect of drawing ratio on difference in optical density and mechanical properties of mass colored photochromic polypropylene filaments, *J. Text. Inst., 108*(8), 1365
45. Koleske, J. V., (1995). *Paint and Coating Testing Manual*, 14th Edition, ASTM International.
46. Bass, M., (1995). *Handbook of Optics*, sec. ed., McGraw-Hill, Inc., New York, pp. 7.1–9.20.
47. Kortuem, G., (1969). *Reflectance Spectroscopy* (in German); Springer-Verlag, Berlin, pp. 106–168.
48. Khajeh, M. M., Mortazavi, S. M., Mallakpour, S., Bidoki, S. M., Vik, M., & Viková, M., (2012). Effect of carbon black nanoparticles on reflective behavior of printed cot-

ton/nylon fabrics in visible/near infrared regions, *Fibers and Polymers*, vol. *13*(4), pp. 501–506.

49. Baumann, W., Groebel, B., Krayer, M., Oesch, H., Brossman, R., Kleinemefer, N., & Leaver, A., (1987). Determination of relative color strength and residual color difference by means of reflectance measurements, *J. Soc. D. Col. (Coloration Technology)*, *103*, pp.100–105.

50. Gordon, S., & Hsieh, Y. L., (2007). *Cotton: Science and Technology*, Woodhead Publishing, Ltd., Cambridge, pp. 59–60.

51. Klein, G. A., (2010). *Industrial Color Physics*, Springer, New York, pp. 335.

52. Saunderson, J. L., (1942). Calculation of the color of pigmented plastics, *J. Opt. Soc. Am.*, *32*(12), pp. 727–736.

53. McDonald, R., (1997). *Color Physics for Industry*, sec. ed.; SDC, Bradford, pp. 306.

54. Heyn, A. N. J., (1952). Observations of the birefringence and refractive index of synthetic fibers with special reference to their identification, *Textile Research Journal*, *22*, pp. 513–522.

55. Norwick, B., (1953). The rapid identification of fibers by the Christiansen effect, *Textile Research Journal*, *23*, pp. 259–261.

56. Fox, K. R., (1939). Refractive indices of textile fibers: Double variation method for the determination of, *Textile Research Journal*, *10*(2), 79–93.

57. Crano, J. C., & Gugliemetti, R. J., (1999). *Organic Photochromic and Thermochromic Compounds*, main photochromic families, vol. *1*, Kluwer Academic Publishers, New York, pp. 101–117.

58. Crano, J. C., & Guglielmetti, R. J., (1999) editors. Organic photochromic and thermochromic compounds, *Physicochemical Studies, Biological Application and Thermochromism*, Plenum Press, New York, vol. *2*.

59. Viková, M., & Vik, M., (2011). Alternative UV sensors based on color-changeable pigments, *Advances in Chemical Engineering and Science*, *1*(4), pp. 224–230. doi: 10.4236/aces.2011.14032.

60. Bouas-Laurent, H., & Durr, H., (2001). Organic photochromism, *Pure Appl. Chem.*, *73*(4), pp. 639–665.

61. Crano, J. C., & Gugliemetti, R. J., (1999). Organic photochromic and thermochromic compounds, physicochemical studies, *Biological Application and Thermochromism*, vol. *2*, Kluwer Academic Publishers, New York, pp. 77–81.

62. Viková, M., & Vik, M., (2014). Photochromic textiles and measurement of their temperature sensitivity, *Research Journal of Textile and Apparel*, *18*(3), 15–21.

63. Viková, M., & Vik, M., (2014). Description of photochromic textile properties in selected color spaces, *Textile Research Journal*, Published online before print September 29.

64. Jocic, D., (2008). 'Smart textile materials by surface modification with biopolymeric systems', *Research Journal of Textile and Apparel*, *12*(2), pp. 58–65.

65. Klukowska, A., Posset, U., Schottner, G., Jankowska-Frydel, A., & Malatesta, V., (2004). 'Photochromic Sol-Gel derived hybrid polymer coatings: the influence of matrix properties on kinetics and photodegradation', *Materials Science-Poland*, *22*(3), 187–199.

66. Liu, X., Cheng, T., Parhizkar, M., Wang, X., & Lin, T., (2010). 'Photochromic textiles from hybrid silica coating with improved photostability', *Research Journal of Textile and Apparel*, *14*(2), pp. 1–8.

67. Lopes, F., Neves, J., Campos, A., & Hrdina, R., (2009). 'Weathering of microencapsulated thermo-chromic pigments', *Research Journal of Textile and Apparel, 13*(1), pp. 78–89.

68. Maafi, M., (2008). 'Useful Spectrokinetic methods for the investigation of photochromic and thermo-photochromic spiropyrans', *Molecules*, vol. *13*, pp. 2260–2302.

69. McLaren, K., (1983). *The Colour Science of Dyes and Pigments;* Adam Hilger Ltd.: Bristol, pp. 10–198.

70. Van Gemert, B., & Kish, D. G., (1999). 'The intricacies of color matching organic photochromic dyes', *PPG Informations, 5*(14), pp. 53–61.

71. Viková, M., & Vik, M., (2005). 'Color shift photochromic pigments in color space CIE L*a*b*', *Molecular Crystals and Liquid Crystals, 431*, pp. 103–116 [403–417].

72. Viková, M., (2010). In: Somani, P. R., editor, Chromic materials, phenomena and their technological applications, chapter 15: 'Methodology of measurement of photochromic materials', *Applied Science Innovation*, pp. 137–163.

73. Yang, H., Zhu, S., & Pan, N., (2010). 'On the Kubelka-Munk Single – Constant/Two-Constant Theories', *Textile Research Journal, 80*(3), pp. 263–270.

74. Bamfield, P., & Hutchings, M. G., (2010). *Chromic Phenomena Technological Applications of Color Chemistry*, Sec. ed., Royal Society of Chemistry, Cambridge.

75. Bilal, M. B., Viallier-Raynard, P., Haidar, B., Colombe, G., & Lallam, A., (2011). A study of the structural changes during the dyeing process of IngeoTM fibers of poly (lactic acid). *Textile Research Journal, 81*(8), 838–846.

76. Billah, S. M. R., Christie, R. M., & Morgan, K. M., (2008). Direct coloration of textiles with photochromic dyes. Part 2: The effect of solvents on the color change of photochromic textiles, *Coloration Technology, 124*(4), 229–233.

77. Billah, S. M. R., Christie, R. M., & Shamey, R., (2008). Direct coloration of textiles with photochromic dyes. Part 1: Application of spiroindolinonaphthoxazines as disperse dyes to polyester, nylon and acrylic fabrics, *Coloration Technology, 124*(4), 223–228.

78. Dabrowska, I., Fambri, L., Pegoretti, A., Slouf, M., Vackova, T., & Kolarik, J., (2015). Spinning, drawing and physical properties of polypropylene nanocomposite fibers with fumed nanosilica, *Express Polymer Letters, 9*(3), 277–290.

79. Dorion, G. H., Canaan, N., Loeffler, K. O., & Conn, N., (1967). Photochromic cellulose paper, synthetic paper and regenerated cellulose. *US Patent 3314795.*

80. Dorsch, P., Wilsing, H., & Peters K. H., (1981). Parameters influencing the depth of dyed shades, particular on Acrylic fibers. *Melliand Textilberichte, 62*, 188–192.

81. Ferrara, M., & Bengisu, M., (2014) *Materials that Change Color*, Smart Materials, Intelligent Design, Springer, Cham, pp. 3–8.

82. Hirte, R., (1983). Polypropylene fibers – Science and technology, *Acta Polymerica, 34*(9), 1–594.

83. Khudyakov, I. V., Turro, N. J., & Yakushenko, I. K., (1992). Kinetics and mechanism of the photochromic transformations of N-salicylidene-4-hydroxy-3,5-dimethylaniline and its complex with uranium(VI) dioxide, *Journal of Photochemistry and Photobiology, A: Chemistry, 63*(1), 25–31.

84. Lee, S. J., Son, Y. A., Suh, H. J., Lee, D. N., & Kim, S. H., (2006). Preliminary exhaustion studies of spirooxazine dyes on polyamide fibers and their photochromic properties, *Dyes and Pigments, 69*(1–2), 18–21.

85. Marcincin, A., & Jambrich, M., (1999). Textile polypropylene fibers: fundamentals. In: Karger-Kocsis, J., (ed.), *Polypropylene: An A-Z Reference*, Dordrecht: Springer Netherlands, Dordrecht, 172–177.

86. Marcincin, A., & Jambrich, M., (1999). Textile polypropylene fibers: fundamentals. In: Karger-Kocsis, J., (ed.), *Polypropylene: An A-Z Reference*, Dordrecht: Springer Netherlands, Dordrecht, 813–820.

87. Parhizkar, M., Zhao, Y., & Lin, T., (2014). Photochromic fibers and fabrics. In: Xiaoming Tao (ed.), *Handbook of Smart Textiles*, Springer International, 155–182.

88. Periyasamy, A. P., Viková, M., & Vik, M., (2017). A review of photochromism in textiles and its measurement *Text. Prog.*, *49*(2), 53.

89. Periyasamy, A. P., Viková, M., & Vik, M., (2016). Problems in kinetic measurement of mass dyed photochromic polypropylene filaments with respect to different color space systems, *Proceedings of the 4th CIE Expert Symposium on Color and Visual Appearance*, 325–333.

90. Sen, K., (1997). Polypropylene fibers. in V. B. Gupta, V. K. Kothari (Eds.), Manufactured FibreTechnology. Springer, Netherlands: Dordrecht, pp. 457–479.

91. Shah, C. D., & Jain, D. K., (1983). Dyeing of modified polypropylene: Cationic dyes on chlorinated polypropylene, *Textile Research Journal*, *53*(5), 274–281.

92. Son, Y. A., Park, Y. M., Park, S. Y., Shin, C. J., & Kim, S. H., (2007). Exhaustion studies of spirooxazine dye having reactive anchor on polyamide fibers and its photochromic properties, *Dyes and Pigments*, *73*(1), 76–80.

93. Springsteen, A., (1999). Introduction to measurement of color of fluorescent materials. *Analytica Chimica Acta*, *380*(2–3), 183–192.

94. Vik, M., (2003). Color difference formula evaluation on LCAM textile data, *Vlakna a Textil, 10*(3), 126–129.

95. Viková, M., & Vik, M., (2014). Photochromic textiles and measurement of their temperature sensitivity, *Research Journal of Textile and Apparel*, *18*(3), 15–21.

96. Viková, M., & Vik, M., (2015). Description of photochromic textile properties in selected color spaces, *Textile Research Journal*, *85*(6), 609–620.

97. Viková, M., & Vik, M., (2015). The determination of absorbance and scattering coefficients for photochromic composition with the application of the black and white background method. *Textile Research Journal*, *85*(18), 1961–1971.

98. Viková, M., Vik, M., & Christie, R. M., (2014). Unique device for measurement of photochromic textiles, *Research Journal of Textile and Apparel*, *18*(1), 6–14.

CHAPTER 7

TESTING OF CHROMIC MATERIALS

ARAVIN PRINCE PERIYASAMY and MARTINA VIKOVÁ

CONTENTS

Abstract...371
7.1   Introduction 372
7.2   Definition of Color Fastness..372
7.3   Definition of Fatigue Resistance..375
7.4   Light Fastness 385
7.5   Wash Fastness 398
7.6   Rubbing Fastness..400
7.7   Sources of Further Information..403
Keywords..404
References...404

ABSTRACT

This chapter discusses the key concepts that influence fatigue resistance and color fastness of chromic textile materials, which have been developed and applied for the laboratory test, to predict the behavior of color changeable textile materials. Generally, *fatigue* describes the decrease in the efficiency of *chromic* compounds. For photochromic materials, fatigue depends on photodegradation, photobleaching, photooxidation, and other side reactions. Therefore, *fatigue resistance* is an essential property for photochromic materials. This chapter reviews the principle of fatigue resistance of color changeable materials and various color fastness and its evaluation. The color fastness evaluation is related to color resistance and durability of the dyes. The chapter then discusses the standardized methods for the assessment of color fastness and comments on the future trends and advances of these methods.

7.1   INTRODUCTION

Innumerable factors can influence the performance of a color in a textile fabric. The performance of color may be assessed in several ways including the levels of fading, change of hue, staining on white objects, and change of saturation. It is the property of a dye or print that enables it to retain its depth and shade throughout the wear life of a product. Functional dyes are considered fast when they resist the deteriorating influences such as laundering or dry cleaning, exposure to sunlight, perspiration, etc. It has been observed that the fabric loses color resulting from detergent solution and abrasive action during hand or machine washing. A garment without color fastness stains other garments when washed together. Quality is of prime importance for every industry or business to get increased sales and better brand image amongst consumers and competing companies. Generally, quality control standards for export are very stringent as this business also holds the prestige of the country of the exporting company. In this competitive scenario, it is mandatory to perform the correct tests and use accurate measurement systems that can help to determine the real performance of colored textile materials, which also helps to survive amongst competition in the markets. This chapter describes the measurement of color fastness on color changeable materials, fatigue resistance, factors affecting the photostability, and wash fastness of color changeable materials.

7.2   DEFINITION OF COLOR FASTNESS

Color fastness is the property of a pigment or dye, or the leather, cloth, paper, ink, etc., containing the coloring matter, to retain its original hue, especially without fading, running, or changing when wetted, washed, cleaned, or stored under normal conditions when exposed to light, heat, or other influences. Essentially, this means that different dyes have different fastness on different materials. Color fastness has been defined in the ISO 105 part A01 section as "Color fastness means the resistance of the color of textiles to the different agents to which these materials may be exposed during manufacture [e.g. mercerizing] and their subsequent use [e.g. domestic laundering]." Generally, the ISO definition is based on two factors [1–5]:

- Variation of color;
- Agents involved in the variation of color.

7.2.1   VARIATION OF COLOR

Generally, the variation of color is defined by the significant visible fading when it is exposed to the influence of an agent. There are two key forms of color variation, and it is required to restrict change in color and staining of adjacent fabrics. However, the change in color defined the variation in lightness and darkness (light/dark), saturation or chroma (deeper/ paler), change in hue (visible into another color), or combined variation of these factors. Therefore, these changes are due to the decomposition of dye molecule in the fiber (i.e., damage of dye-fiber bods). The decomposition of dye molecules may occur during sunlight, washing treatment, abrasion, perspiration, etc. [1–5].

7.2.2   AGENTS INVOLVED IN THE VARIATION OF COLOR

The visible modification of colored textile (either functional dyes or classical dyes) may be caused by different agents, namely sunlight, weathering, washing, dry cleaning, and other aqueous agents. These structures have been published in ISO 105 and listed in Table 7.1. The current ISO nomenclature and classification were adopted at the 1978 ISO/TC38/SC1 meeting in Ottawa, Canada. Presently, 13 sections have been published with each section designated by letters A, B, C, D, E, F, G, J, N, P, S, X, and Z. Each letter is designated to a common agent that influence change in color. Each section contains two-digit partitions, and thus, each part can be expanded to include up to 99 methods or tests if necessary. It is important to point out that standardized ISO 105 tests are designed to determine the color fastness under an effect of a single agent and do not usually allow for combination of two or more agents, even though textiles may be subjected to the influence of simultaneous agents during processing or end-use [1–8].

TABLE 7.1 Structure and Sections of ISO 105 Standard (Adapted from ISO 105-A01 (2010))

Section	Title
A	General Principles
B	Color Fastness to Light and Weathering
C	Color Fastness to Washing and Laundering
D	Color Fastness to Dry Cleaning
E	Color Fastness to Aqueous Agents
F	Specification of Standard Adjacent Fabrics
G	Color Fastness to Atmospheric Contaminants
J	Measurement of Color and Color Differences
N	Color Fastness to Bleaching Agencies
P	Color Fastness to Heat Treatments
S	Color Fastness to Vulcanization
X	Miscellaneous tests
Z	Colorant characteristics

The property of color fastness on colored textile materials are based on the dye-fiber system. However, there are multiple factors that directly influence the final fastness properties. The influencing factors may vary according to different fastness. There are some known factors that directly influence the fastness properties:

- The molecular structure of the color changeable dyes/pigment.
- The type and structure of the fiber.
- The coloration process (dyeing/printing/mass coloration).
- The depth of color.
- The molecular structure of a dye is crucial for its fastness properties.

In general, the choice of dyes is an important parameter, which decides the color fastness properties. However, choice of good chemicals and auxiliaries along with the perfect dyeing process may not help for high quality fastness properties, until the right dyes suitable to the respective textile fibers are chosen. In the current market, series of dyes are available with different color fastness properties. Therefore, the choice of dye is based on the product requirements (i.e., color fastness required to the respective product). For example, the color fastness of the respective dye itself can only reach 2 to 3 or even 1 to 2. Regardless of how good the additives and dyeing process, it cannot bring the color fastness of 4 to 5. Therefore, color fastness is directly proportional to the binding force between photochromic dye and the fiber, which means the structure of the dyes. If the bond between dyes and fibers is not securely strong, it may cause poor color fastness. The few examples of bonding between photochromic dyes and different fibers are well explained in Chapter 3 (see Scheme 3.1 on page 132 and Figure 3.24 on page 134). Depth of shade is defined as the amount of dyestuff present in the fiber. Generally, heavy shades show less color fastness than the light shades. This is due to the higher quantity of dyestuff per unit area. Sometimes, unfixed dyes may show poor fastness properties. Unfixed dyes and its impact on color fastness are discussed in the soaping section. Therefore, heavy shades require proper rinsing and washing off process to increase their fastness properties. However, due to entrapped dye particles within the fiber structure, some unbound dye molecules can remain and contribute to color loss and dye transfer.

From the economic point of view, dyers generally use different dyes from different manufacturers. Therefore, to achieve the expected target, people mix different dyes from different manufactures. But most of them are not aware of the incompatibility of dyes in the above mixture, which leads to serious issues with respect to leveling, exhaustion, and fastness properties and spoils the reproducibility %. Therefore, choosing the right dyes with suitable compatibility is important.

7.3   DEFINITION OF FATIGUE RESISTANCE

One of the major problems associated with organic photochromic systems is the irreversible formation of non-photochromic products by photochemical or thermal reactions of the colored form. In photochromic materials, fatigue refers to the loss of reversibility by processes such as photodegradation, photobleaching, photooxidation, and other side reactions. All photochromic suffer fatigue to some extent, and its rate is strongly dependent on the activating light and the conditions of the sample. The more important requirements are thermal stability of both isomers and fatigue-resistant characteristics. Recently, a new class of photochromic compounds, which fulfill these requirements, has been developed. The compounds, named diarylethenes, undergo the following reversible photochromic reactions [9, 10]. The main disadvantage of organic materials for various applications is their lack of durability. Although the durability problem has already been overcome for liquid crystals and organic photo-conductors, it remains for photochromic materials. Photochromic reactions are always attended by rearrangement of chemical bonds. During the bond rearrangement, undesirable side reactions occur to some extent. This limits the durability of photochromic materials. Extensive examination of the fatigue resistance of various diarylethenes partly solved the durability problem. More than 10 diarylethenes that have benzothiophene aryl groups undergo fatigue-resistant photochromic reactions [9, 10]. The photostimulated reversible coloration/decoloration cycles can be repeated more than 10^4 times [9–11].

7.3.1   TESTING OF FATIGUE RESISTANCE

The lifetime of photochromic molecules is mostly affected by many factors, including concentration, environment, stabilizers, and strength and spectral characteristic of the illuminant. However, it is very difficult to predict the expected life time of the respective materials. As explained in the previous chapters, the photochromic reactions are always accompanied by rearrangement of chemical bonds. During the rearrangement, undesirable side reactions occur to some extent. This limits the number of cycles in the photochromic reactions. The difficulty in obtaining fatigue-resistant photochromic compounds can be understood by considering the following reaction sequence, in which a side reaction to produce B' is involved in the forward process.

$$B' \xleftarrow{\Phi_s} A \underset{\lambda_2}{\overset{\lambda_1}{\rightleftharpoons}} B \qquad\qquad (7.1)$$

Even if the side-reaction quantum yield, Φs, is as low as 0.001 and B perfectly converts to A, 63% of the initial concentration of A will have decomposed after 1000 coloration/decoloration cycles. Thus, the quantum yield for conversion to by-products should be less than 0.0001 to enable the dyed sample to repeat the cycles more than 10,000 times. Photostability is defined as the stability of dyes or pigments during the influence of light. As discussed early, photochromic materials should have photostability properties. Generally, four important factors could influence the photostability of normal and functional colors:

- Physical and chemical nature of substrate (polymer).
- Physical and chemical nature of coloring materials.
- Environment and type of system used.
- Presence of antioxidants and light stabilizers.

Generally, the first factor may change the physical properties of colored textiles; this is due to the UV light irritation and followed by the photochemical reaction. Therefore, it is important to analyze the stability of photochromic dyes. The second factor is related to the structure of photochromic dyes, particle distribution, and chemical bonding between fiber and dyes; therefore, all these may influence the stabilizing or destabilizing effects on the photochromic dye-incorporated fibers. The chemical bonding between photochromic dyes and different fiber is explained in Chapter 3. The third factor states the environment conditions and the type of system used, namely temperature, humidity, oxygen, type of light source (spectral power distribution), and the intensity of UV light.

Generally, UV irradiation is required to produce the photochromic effects, and if the photochromic materials are required to be exposed to UV irradiation for a prolonged period, there is a chance of degrading the photochromic materials. Therefore, fatigue resistance (photostability) is an important property to study for the photochromic materials [12–14]. Generally, for light fastness testing in the textile industry and other industries, accelerated light fastness tests based on the artificial light sources are used, typical examples of which are the ATLAS Xenotest® Fadeometer or ATLAS Weather-O-meter® (Atlas Material Testing Technology, Illinois, USA) in which xenon discharge lamps are used to provide the irradiation. Generally, for photostability tests, photochromic pigments are exposed to xenon arc lamp for several hours with an optical filter system that cuts off the wavelength below 310 nm to follow the standard BS EN ISO 105-B02:1999 [15].

Parhizkar et al. [13] studied the photostability and durability of a photochromic organosilica-coated fabric; in their previous studies, they found the continuous UV irradiation drastically reduced the photochromic properties; so, they attempted to overcome this issue by incorporating the UV photostabilizers and antioxidants to modify the surface wettability of coated fabric with the help of low surface energy silane and blockading the dye-encapsulating pores with an additional silica coating layer. The photochromic dye (5-chloro-1,3-dyhydro-1,3,3-trimethylspiro(2H-indole-2,3'-(3H)-nahth (1-b) (1,4) oxazine)) was applied to the wool fabric by the sol-gel method. To measure photostability, the sol-gel-coated samples were exposed to intense UV irradiation from a mercury metal halide lamp. The result of absorbance was recorded for each exposure period of 3.2 h, showing that the treatments (which were intended to provide some protection) did indeed bring about increased photostability without adversely affecting the coloration and decoloration process of the photochromic compound. The treatments applied to the photochromic organosilica-coated fabric, indicated in Figure 7.1, were as follows:

- UVQ-2,2'-thiobis(4-tert-octylphenolato)-N-butylaminenickel (II);
- HALS-poly (4-hydroxy-2,2,6,6-tetramethyl-1-piperidine ethanol-alt-1,4-butanedioic acid);

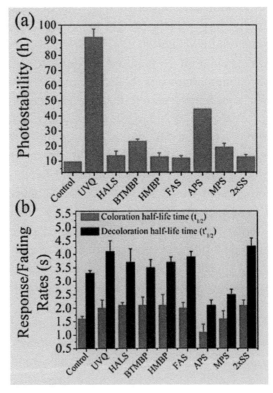

FIGURE 7.1 (a) Photostability and (b) photochromic response and fading half-life times of the photo-chromic organo silica-coated fabric subjected to various treatments. (Reprinted from Marzieh Parhizkar et al. [13]. Reproduced courtesy of Journal of Engineered Fibers and Fabrics, P.O. Box 1288, Cary, North Carolina 27512-1288, USA. Tel: (919) 459-3700 Fax: (919) 459-3701 Internet: www.jeffjournal.org.)

- BTMBP-2-(2H-benzotriazol-2-yl)-4-(1,1,3,3-tetramethylbutyl) phenol;
- HMBP-2-hydroxy-4-methoxy-benzophenone;
- FAS-Tridecafluorooctyltriethoxysilane;
- APS-3-Aminopropyl) triethoxysilane;
- MPS-(3-Trimethoxysilyl) propylmethacrylate and,
- 2xSS- 2-layer coating with octyltriethoxysilane:phenyltriethoxysil ane).

Usually, photodegradation is very high in the first few cycles of UV exposure due to a superficial layer of photochromic dye that degrades faster than that which has penetrated to the inner parts of the fibers and fabric. Billa et al. [16] produced photochromic effects on wool by direct coloration of photochromic acid dyes. The original yellow color of the dyed fabric was changed to blue by UV irradiation, and maximum absorbance was obtained at 620 nm. The durability of the photochromic acid-dyed fabric was analyzed with respect to its wash fastness and photostability properties. Generally, the photostability of dyed textile goods is measured against that shown by standard blue wool samples prepared specifically for the purpose of light-fastness testing, but it is not possible to utilize the same method for photochromic materials, due to their dynamic color-changing nature. Instead, photostability was determined from the ratio of ΔE before and after UV irradiation (expressed as a percentage).

Aldib [17] investigated the photostability process for photochromic textiles. Six commercial photochromic dyes were applied to polyester fabric, but the trade name of dyes was not disclosed for proprietary reasons. Dyed fabrics were exposed in the Xenotest light-fastness tester for a controlled period and then left in the dark for 2 hours to ensure that whatever colors developed on Xenotest light exposure had faded to yield only background colors, and the results were noted as (ΔE_1). The photostability of six commercial photochromic dyed polyester fabric was assessed by comparing the percentage decrease in the degree of photocoloration of the tested samples with the percentage decrease in color of blue wool reference standards for light fastness exposed to Xenotest light for the same period. Viková [15] exposed photochromic printed and nonwoven fabrics to artificial sunlight (Xenon arc lamp with an optical filter system that cut off the below 310 nm wavelength).

The test was conducted according to BS EN ISO 105-B02:1999 standards. The results (Figures 7.2 and 7.3) showed a decrease in the shade intensity

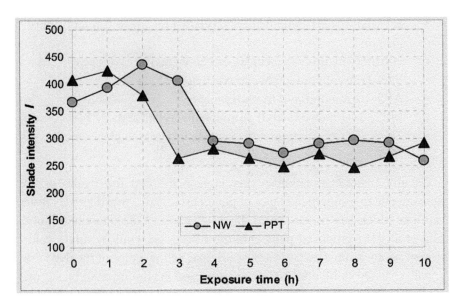

FIGURE 7.2 Dependence of fatigue resistance on exposure time in the XENOTEST concentration 3.0 g (Blue) PPT (printed) and NW (nonwoven) samples [15].

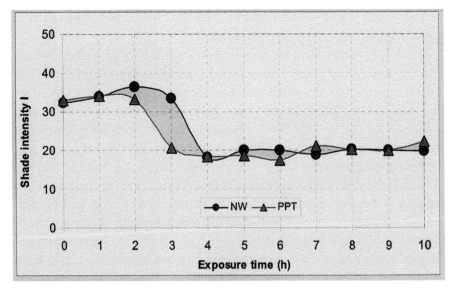

FIGURE 7.3 Dependence of fatigue resistance on exposure time in the XENOTEST concentration 3.0 g (Purple) for PPT (printed) and NW (nonwoven) samples [15].

(I) after two to three exposure cycles (one cycle/one hour), i.e., a significant loss of photochromic behavior with respect to increasing doses of radiation. The author also conducted the photostability test for the photochromic pigment (see Figure 7.4) and showed 5% loss after testing for 40 h. Viková and Vik [18] and Little and Christie [14, 19, 20] concluded that the incorporation of UV absorbers could increase the photostability of photochromic dyes in some cases and that improvement depends on the UV absorber/dye combinations. Generally, there is a competition for the UV light between the dye and the UV absorber, and it may reduce the degree of photocoloration of the dye. Tong Cheng [21] studied the photostability properties of sol-gel coated photochromic wool fabric. It was found that incorporation of the UV absorber into the photochromic dye containing the sol-gel matrix may decrease the absorption intensity. However, it depended on the type of UV stabilizer. Cheng reported that the addition of UV stabilizers slows down the optical responsiveness of the photochromic effect; they found that samples treated with HALS showed degradation of 50% of the photochromic effect after 36 hours. Chowdhury et al. [22] studied the thermochromic pigments for improving their photostability. In this study, they used HALS, which acts as a UV absorber; certainly, it improves the photostability with photochromic and thermochromic pigments. There are few literatures available for the HALS with respect to the photochromic pigment [23–26].

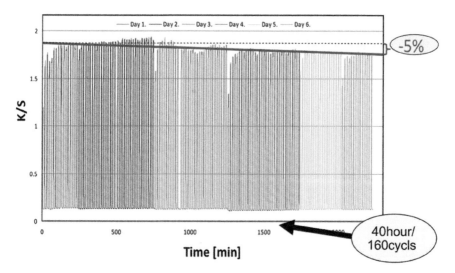

FIGURE 7.4 Photostability test of the photochromic pigment.

In this research, the authors carried out the dyeing of cotton fabric by using the pad-dry-cure method. The padding bath contained the thermochromic pigments, cationic agent, non-ionic dispersing agent, leveling agent, acrylic soft binder, and water. The fabric was padded with the solution (wet pick-up 80%), dried at 80°C for 3 min, and cured at 140°C for 3 min. The UV absorber and HALS was already mixed with padding solutions. The color strength values of thermochromic dyed cotton fabric are shown in Figure 7.5; the fabric dyed without incorporation of UV absorbers and HALS. The color strength values rapidly decreased after exposure to Xenotest, but red color provide better stability than other colors; thus may due to long absorbance band of red colors in the electromagnetic spectrum. Perhaps, this result is very poor than the UV absorber and HALS-incorporated samples (see Figures 7.6 and 7.7). Incorporation of the UV absorber and HALS

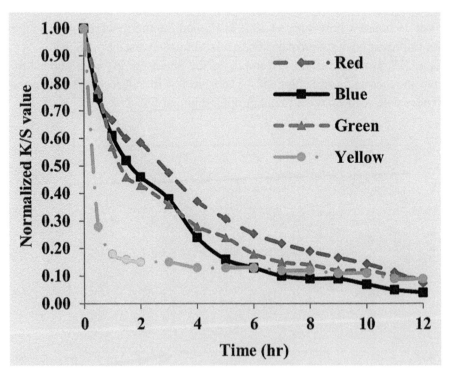

FIGURE 7.5 Normalized photostability curves for thermochromic dyed-cotton fabric without incorporation of UV absorbers. (reprinted from Chowdhury et al. [22] with the permission of Journal of Textile and Apparel Technology Management).

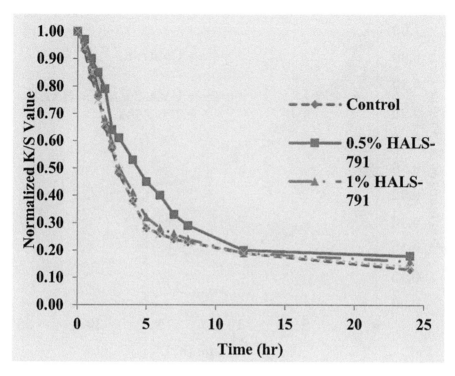

FIGURE 7.6 Impact of HALS on the photostability properties of a red colorant (reprinted from Chowdhury et al. [22] with the permission of Journal of Textile and Apparel Technology Management).

increases the photostability of cotton dyed goods, but the combination of the UV absorber and HALS gives better result.

Aldib and Christie [27, 28] dyed polyester fabric by the exhaustion method with six photochromic dyes based on spironaphthooxazine and naphthopyran. The results showed that the photocoloration was dependent on the photochromic dye and its concentration. There was little difference in the aqueous- and solvent-based processes in terms of ΔE values, and the half-life of fading was shown to be significantly higher in spironaphthoxazine dyes than for naphthopyrans. It should be noted that naphthopyrans fading much more slowly than the spironaphthoxazine dyes has been a consistent feature of this group's investigations. The fatigue resistance of different photochromic dyes was determined after 20 cycles of UV light exposure, and it has been compared with their first exposure (Figure 7.8). Results show

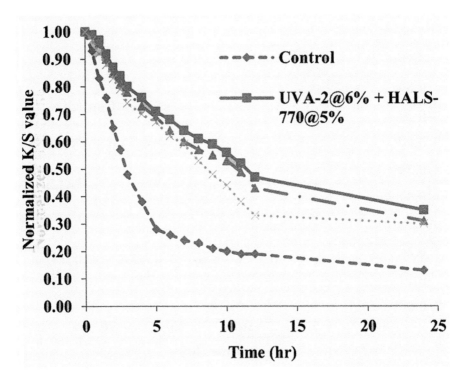

FIGURE 7.7 Combination of UV absorbers and HALS on the photostability properties of a red colorant (reprinted from Chowdhury et al. [22] with the permission of Journal of Textile and Apparel Technology Management).

the difference in the fatigue resistance with respect to the dyes. The aqueous form of dyeing form a rigid environment for the dyes, which provides some protection to the dyes during photodegradation with respect to the UV irradiation.

Thermochromic colors are more sensitive to the heat; therefore, it is necessary to determine the thermal stability of thermochromic dyed/printed textiles. Figure 3.9 shows the results of thermal stability with respect to the leuco dye-based thermochromic pigments. Aldib and Christie [28] tested the stability of the photochromic dyed fabrics after 170 days. The retention of photocoloration after storage in the dark was in the range of 64–96%, with Dyes 1 and 3 retaining more than 90% of their original level of photocoloration. Dye 6 possessed the lowest storage stability, retaining only 64% of its degree of photocoloration (Figure 7.9).

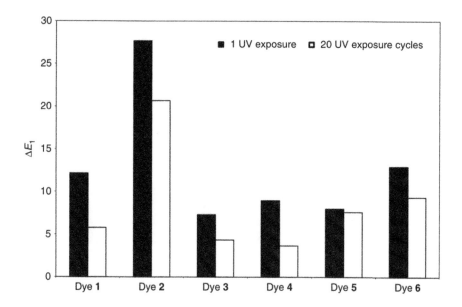

FIGURE 7.8 Fatigue resistance of different photochromic dyes on polyester (Dye 1= Naphth[1,2-b]oxazine; Dye 2= Naphth[1,2-b]oxazine; Dye 3= Purple Naphth[2,1-b]oxazine; Dye 4= Red Naphth[2,1-b]oxazine; Dye 5= Ruby Naphtho[2,1-b]pyran; Dye 6 =Yellow Naphtho[2,1-b]pyran). (reprinted from Aldib et al. [28] with permission of Coloration Technology).

7.4 LIGHT FASTNESS

The UV radiation that reaches the Earth's surface accounts for only about 6% of the total solar radiation at maximum exposure, and it has wavelengths from 290 to 400 nm [29]. Sun electromagnetic spectrum consists of gamma to radio waves. Of these, the radiation ranging from UV to IR reaches the earth surface. UV rays have high energy that accelerate the fading of dye. UV rays can be divided into UV-A, UV-B, and UV-C. UV-A have long wave length of 320–400 nm, which is not absorbed by the atmospheric ozone; UV-B has medium wavelength of 280–320 nm, which is partly absorbed; and UV-C have short wavelength of 100–280 nm, which is completely absorbed. Because UV-A and UV-B are not completely absorbed by ozone in the atmosphere, they reach the Earth's surface with photon energy of 315–400 (kJ/mol), which exceeds the carbon-to-carbon single bond energy of 335 (kJ/mol). This results in the fading of dyes and pigments [15, 30–32].

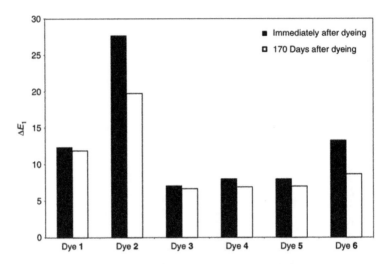

FIGURE 7.9 Storage stability for different photochromic dyed PET fabric by the solvent-based pad dry-heat dyeing method (Dye 1= Naphth[1,2-b]oxazine; Dye 2= Naphth[1,2-b] oxazine; Dye 3= Purple Naphth[2,1-b]oxazine; Dye 4= Red Naphth[2,1-b]oxazine; Dye 5= Ruby Naphtho[2,1-b]pyran; Dye 6 =Yellow Naphtho[2,1-b]pyran). (reprinted from Aldib et al. [28] with permission of Coloration Technology).

Light fastness is the resistance of dyes and pigments to the effect of electromagnetic energy.

7.4.1 MECHANISM OF FADING BY LIGHT

Textile dyes when applied to fabrics are subject to the action of a range of external influences like light, gases such as nitrogen oxides, and moisture that may contain dissolved atmospheric chemicals. In general, dyes and pigments were absorbing the certain wavelength of light and reflecting or transmitting the rest of the wavelength. Therefore, while absorbing the photon (with gains the energy), the electrons can be excited to a higher energy state. Similarly, the absorption of light by photochromic dyes, as explained in previous chapters, may initiate many physicochemical interactions. However, absorbed photons may induce the dyes to the excited state, which is more reactive than the ground state. Therefore, the promotion of electron from the HOMO (Highest Occupied Molecular Orbital) to the LUMO (Lowest Unoccupied Molecular

Orbital) level. Later, the excitation energy is released, while returning the promoted electrons to their original state. There are three possibilities while returning to their original state, excitation energy may have dissipated in to heat due to the radiation-less transitions between the energy states, possibilities to converted by fluorescence or phosphorescence. Apart from these reaction, excitation energy may have converted in to chemical energy. The chemical energy may be developed due to the intramolecular arrangement, redox process, and other photochemical reactions. However, because the excited state is a high-energy state, it has the potential to undergo a chemical reaction, breaking a covalent bond or otherwise irreversibly reacting with another molecule that may act like photosensitizer, which results in the photofading of dyes (i.e., reduce its durability). Therefore, the photofading of photochromic dyes by the light of different wavelengths may expected to some extent. In some cases, it is more sensitive when the UV light was absorbed by the textile fibers (particularly synthetic fibers). Therefore, these changes the electronic structure of the molecule, which changes its absorption properties: e.g., many dyes that absorb visible light have large systems of conjugated double bonds and if these are broken, the absorbance can shift to much shorter wavelengths. How likely this kind of destructive chemistry depends on the nature of the dye. Organic dyes tend to be more susceptible to photobleaching than things like quantum dots and inorganic pigments. When UV light falls on dyes and pigments, the UV photons with high energy excites the electrons of dye from the ground state to the exited state. The dye at the exited state is highly reactive and unstable, and so, it comes back to the original ground state. During the quenching of dyes from the exited state to the ground state, atmospheric triplet oxygen reacts with dyes to form singlet oxygen. Under the presence of visible light, the excited dye molecules react with the atmospheric oxygen to form superoxide radical. How quickly a color fades in sunlight depends on many factors. The most important factor is the chemical makeup of the dye or pigment itself. Certain colors are made up of relatively stable molecules, making it difficult for ultraviolet radiation to break the bonds; these molecules provide long-lasting defenses against fading. Others, especially those created to be environmentally friendly, may be more unstable and easier to breakdown when exposed to light. Over time, virtually all pigments fade in the presence of light. Therefore, many important artifacts such as paintings or documents are kept in an environment with controlled lighting and ultraviolet protection. The Declaration of Independence and the US Constitution are protected under glass panes designed to block as much ultraviolet radiation as possible. Even something as simple as a photo album often contains UV-blocking capabilities that protect personal memorabilia from the effects of sunlight. The singlet oxygen and superoxide radical formed is highly reactive and capable of destroying the dyes. The exited dye molecules undergo various processes that results in fading. Standard light fastness tests have been devised for traditional dyes and pigments. For photochromic materials, because of their dynamic color change properties, fatigue tests are relevant.

7.4.2 PHOTODECOMPOSITION

Photodecomposition is defined as the decomposition of dyes with the influence of light. However, the absorption of light is based on the concentration of colored particles. Figure 7.10 explained the relationship between the concentration of colored particles, absorption of photons, and its impact on color loss (fading).

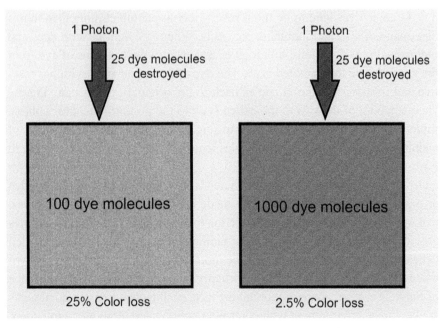

FIGURE 7.10   Concentration of color with respect to absorption of photos and its color loss.

$$Photodegradation \rightleftharpoons CCI \leftrightarrows PMC \rightarrow PMC$$

SCHEME 7.1   Photodegradation of spiropyran (PMC = primary merocyanine; CCI = cis-cisoid colored isomer).

The photodegradation of spiropyran (see Scheme 7.1) may be due to the *cis-cisoid* colored isomer.

7.4.3   PHOTOOXIDATION OF PHOTOCHROMIC DYES

In the last few decades, photochromic and other chromic materials have been widely used in the coloration of textiles and other materials for many commercial applications [17–27]. Photochromic dyes require UV light to produce the color; on the other side, continuous irradiation of UV light may weak the polymer system (particularly synthetic fibers). Spirooxazines has good resistance to the photochemical degradation (compared to spiropyran) by UV irradiation. Photooxidation is the major factor that degrades the spirooxazines. Some study of the photooxidation of some spiropyrans and spironaphthoxazines indicates that the spiro and open forms of these dyes are singlet oxygen quenchers and that the colored form does not act as a sensitizer. A mechanism is proposed that involves the formation of a superoxide radical anion by photoinduced electron transfer to oxygen from a merocyanine form of the dye, followed by nucleophilic attack of the radical anion on the radical cation of the dye. Therefore, the degradation of spirooxazines purely depends on the excited state of oxygen (Schemes 7.2 and 7.3).

Oxidation is the major factor responsible for the photodecomposition of photochromic compounds. It mainly affects the double bond in the open form of isomer by singlet oxygen through self-sensitized process. Therefore, the addition of singlet oxygen quenchers (for example, nickel dithiolato complexes and 4-diazobicyclo (2,2,2) octane) may help to increase photostability of photochromic compounds.

7.4.4   PHOTO SENSITIZATION

Photosensitization is the reaction to light that is mediated by the dyes or pigment molecules. The first law of photochemistry states that light must be

$$SPO \xrightarrow{hv} PM \xrightarrow{hv'} PM^1 \xrightarrow{ISC} PM^3 \xrightarrow{O_2} PM\left[PM^-.O_2^+\right] \rightarrow PM.O_2$$

SCHEME 7.2 Photooxidation of spirooxazines (SPO = spirooxazine dyes; PM = photo merocyanine form; ISC = inter system crossing).

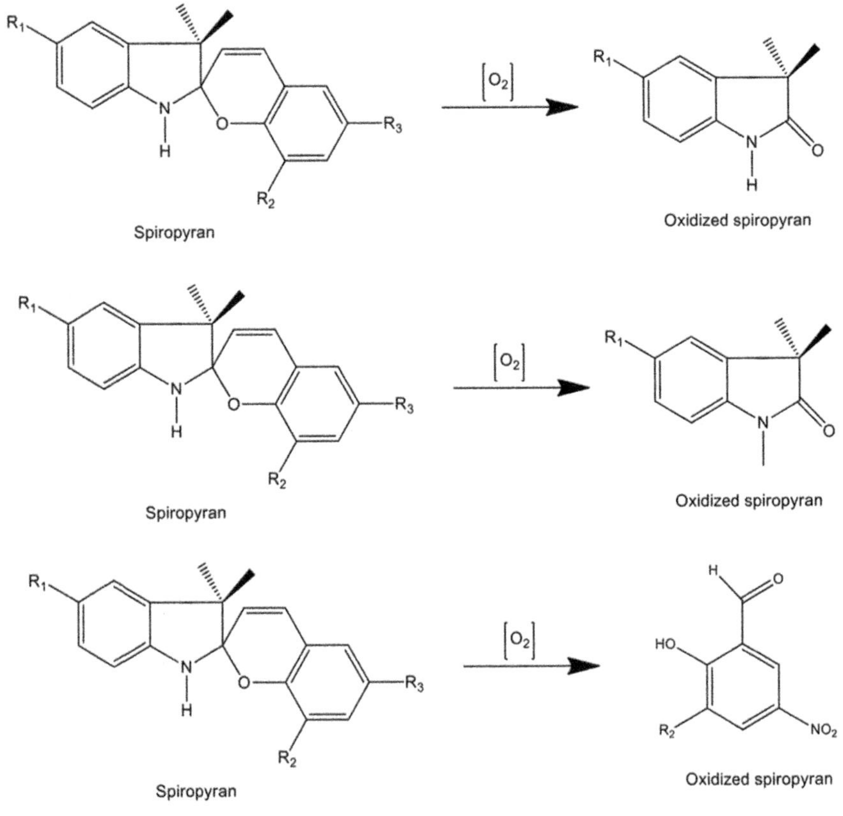

SCHEME 7.3   Possible photooxidation reactions of spiropyran (R – functional groups).

absorbed by a chemical substance (chromophore) for a photochemical reaction to occur. The excitational energy transfer from the triplet state to other states or molecules may initiate the photodynamic action, which can be classified in two types. The first type of photosensitization involves the direct interaction of adjacent molecules of photosensitizer (i.e., during excitation states), which allows the electron transfer to form the radical species. The second type of photosensitization are made highly reactive components that caused the direct transfer of the excitational energy from the triplet state to the ground state.

7.4.5 PHOTO TENDERING

In the presence of UV light, the material supplies hydrogen to the dyes, which causes reduction. As hydrogen is removed from the material, it undergoes oxidation. In this case, the material is the only responsible factor for the fading of dyes.

These factors are directly proportional to their photochemical degradation and result in color loss of the textiles. The last factor is nature of the interaction with antioxidants and UV stabilizers. These additives help to inhibit the photocatalytic activity, which helps to improve the photostability of photochromic dyes. On other hand, the UV absorbers will not affect the original shade of the substrate. The commonly used UV absorbers are 2-hydroxy-benzophenone, 2-(2H-benzotriazol-2-yl)-phenols, phenyl esters, substituted cinnamic acids, and nickel chelates (see Figure 7.11). However, the basic requirement of application of UV absorber is it must not change the color of dyed or printed substrates. There are two main reason that limits the use of UV absorbers: high cost and high concentration (0.5% to 2% on the weight of fiber) needed to achieve sufficient protection. Therefore, it is suggested to use the combinations of UV absorbers with other light fastness products [32]. Generally, UV absorbers must have the following characteristics:

- Must have high absorptivity of the UV spectrum, particularly 290 nm to 400 nm.
- Must be stable to sustain for long-time exposure under the above conditions.
- Must be chemically inert to other additives in the substrate.

A commercially well-known UV stabilizer is hindered amine light stabilizers (HALS); Scheme 7.4 shows their reaction during UV irradiation. These are derivatives of 2,2,6,6-tetramethylpiperidine. During UV irradiation, there is a transformation of HALS 1 to nitroxyl radical 2, which is an active species. The nonradical amino ethers may form NO-R with the reaction of NO and radical R. In the latter, the reaction of aminoether (NO-R)

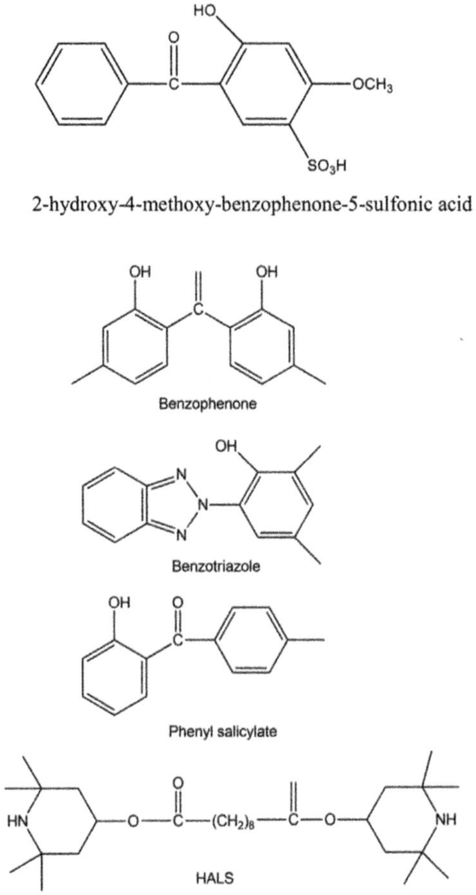
HO
O
C
OCH3
SO3H
2-hydroxy-4-methoxy-benzophenone-5-sulfonic acid
OH
OH
C
Benzophenone
OH
N
N
N
Benzotriazole
OH
O
C
Phenyl salicylate
HN
O
O
C
(CH2)8
C
O
NH
HALS

FIGURE 7.11 Examples of UV absorbers.

with peroxy radicals forms nitroxyl radicals, which is the mode of action of HALS during UV irradiation.

7.4.6 ANTIOXIDANTS

Antioxidants are organic substances that are added to minimize the photo-oxidation of textile materials. The commonly used antioxidants are gallic acid, caffeic acid, ascorbic acid, and hindered phenols and amines. These agents work either by trapping the radicals or by decomposing the peroxides.

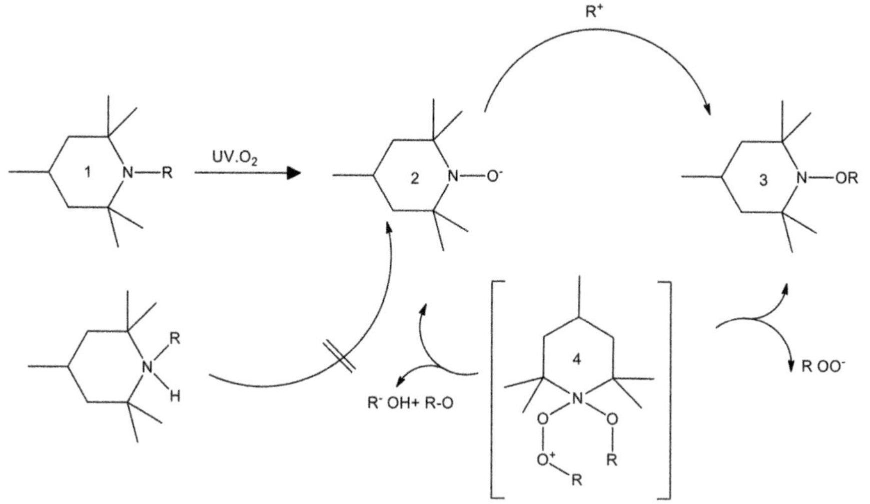

SCHEME 7.4 Mechanism of HALS during UV irradiation.

They absorb free singlet oxygen formed during the quenching of excited dyes (see Figure 7.12). Therefore, antioxidants terminate the chain reaction by removal of free radical intermediates. This is possible when they oxidize themselves, so that they could be act like reducing agents such as thiols, ascorbic acid, or polyphenol.

7.4.7 FACTORS AFFECTING LIGHT FASTNESS

There are many factors that affect light fastness:

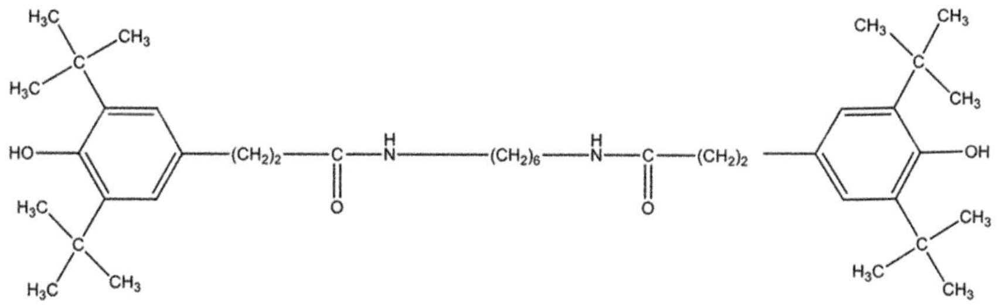

FIGURE 7.12 N,N hexamethylene bis (3,5-di-t-butyl-4-hydroxyhydrocinnamide).

- Spectral power distribution of the light sources and continuous irradiation
- Distance of light source from samples
- Relative humidity
- Duration of the test
- Ambient temperature
- Sample temperature

The light fastness properties of colored textile materials were assessed under accelerated test conditions by utilizing high-pressure mercury lamps or xenon arc lamp. For textile samples, flash-type lamps are not suitable, because the formation of merocyanine for every flash may vary. It is therefore advantageous to use an electronic shutter and a continuous discharge lamp for fatigue tests. These lamps are simulated to the natural daylight (6500 K). Therefore, it generates a huge amount of heat that could be transferred into the substrate. These temperature changes may show significant effects on the photo and thermochromic systems; in many cases, it increases the thermal bleaching reaction, followed by loss of chromic behavior. Few photochromic colors (spirooxazines) are more sensitive to temperature. Nevertheless, it changes the kinetics and thermodynamics of the ground state of photochromic materials with respect to their activation energies, thermodynamics, and equilibrium constants; also, the reduction of temperature may increase the photochemical reaction [32]. For example, spiropyrans are based on the nitro-substituted ones, which provide good colorability. However, spiropyran is not recommended for many potential applications because of their short life cycle. Under continuous irradiation, spiropyran rapidly reach the photostationary equilibrium; therefore, it decreases the absorption and initiates the monoexponential thermal bleaching kinetics after ceasing the irradiation. In our laboratory, we performed the continuous measurement of photochromic colored substrate, which is shown in Scheme 7.5 (continuous irradiation).

Moisture content and environmental temperature, oxygen, and UV content of the light source are the main environmental factors that affect the light fastness and stability of photochromic colored materials. An increase in any of these will result in an increase in the rate of photochemical degradation. For example, under the presence of UV light, the moisture reacts with atmospheric oxygen to form hydrogen peroxide and another active radical. These active agents thus formed will degrade the fiber (e.g., nylon,

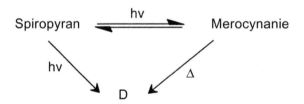

SCHEME 7.5 Thermal and photodegradation of spiropyran (D-degradation products) during continuous irradiation.

cellulose); also the absorbed moisture may initiate to swell the polymer chain, resulting in poor resistance to light under high moisture and heat. The maximum optical yields can be obtained on stationary irradiation, which also controls the thermal bleaching rate with respect to the photochromic reaction [32].

Feczko et al. [33] prepared photoresponsive polymer by micro-encapsulation of photochromic compounds. The nanocapsules of 5-chloro-1,3-dihydro-1,3,3-trimethylspiro[2H-indole-2,3'-(3H) naphtha [2,1-b](1,4) oxazine] were micro-encapsulated on poly (methyl methacrylate) and ethyl cellulose by emulsion solvent evaporation methods. However, both the polymers failed in being useful for practical textile application. Zhou et al. [34] prepared photochromic printed fabric by using melamine-formaldehyde microcapsules. The melamine-formaldehyde microcapsules were mixed along with the print paste and printed on the fabric. The results showed that the improved light fastness by photochromic microcapsules and the lifetime of photochromic microcapsules could extend from 6–7 h to 69–75 h under continuous UV, when stirring rate, emulsification time, and mass ratio of core materials/wall materials were 1000 rpm, 5 min, and 1:1, respectively.

To develop the textile-based UV sensor, PET fabric were printed with photochromic dyes via screen printing techniques. This experimental work was supplemented by a study of the influence of UV absorbers for the modification of sensor properties to alter their sensitivity to an external UV stimulus. A PET substrate was used according to standard ISO 105-F04:2001 for photochromic screen printing. Different concentrations (0.25, 0.5, 1.0, 1.5, 2.0, 2.5, and 3.0 g/30 g acrylate paste) of pigment was used for this study; 1.5–3% of the UV absorber was also used, and the printed samples are shown in Figure 7.10. This concentration was selected based on the visual assessment of the strongest color change. For printing, a standard printing composition acrylic-based binder was used, which is frequently used in standard textile pigment printing. In accordance with the manufacturer's recommendation, a drying temperature of 75°C and time of drying of 5 min were selected. The photochromic prints were fixed by ironing at a temperature of 190°C for a time of 20 s.

The intensity of color change usually depends on the concentration of photochromic pigment as well as the intensity of UV irradiation. The intensity of UV irradiation may be controlled by UV absorbers, but only in the case, when the spectral absorption spectra of the UV absorbers and photochromic pigments are similar. This means that the UV absorber and the photochromic pigment absorb in the same wavelength ranges. In this case, an influence on the depth of color photochromic change was expected. In this investigation, UV absorbers Cibatex APS and Cibafast HLF from CIBA Specialty Chemicals, Switzerland, were applied. Both UV absorbers were applied by screen printing. As a comparison, a control print was arranged with only the acrylate paste to eliminate the influence of the complex paste on the photochromic effect. The study of photo stability or durability of the printed fabric, incorporation of UV absorbers shows very good photochromic property even after 12 years (see Figure 7.13) of exposure by mostly artificial light sources in interior. On the other side, if was prepared textile print without UV absorber, then the sample degrade within a month as well as loss of photochromic property (See Figure 7.14).

Vik and Viková [15, 35–38] developed a new spectrophotometer under the Laboratory of Color and Appearance Measurement (LCAM), Technical University of Liberec. The FOTOCHROME is used to determine the development of color as well as degree of fatigue during continuous irradiation, which helps to measure the photochromic behavior of textiles or other substrates [39]. Due to controlled excitation of light sources by using the shutter, it is possible to measure photochromic properties of materials with respect to one or multiple cycles (Figure 7.4). Most of the photochromic materials (mostly P-type) are sensitive to the surrounding temperature. Hence, this instrument consists of temperature control during the measurement; this system also allows the measurement of the thermal sensitivity of photochromic samples. [36, 39]. This instrument can also allow be used to determine light fastness on photochromic materials.

As discussed in the previous chapters, thermochromic colorants can change their color with respect to the environmental temperature; these colorants restore their original form once the temperature is decreased.

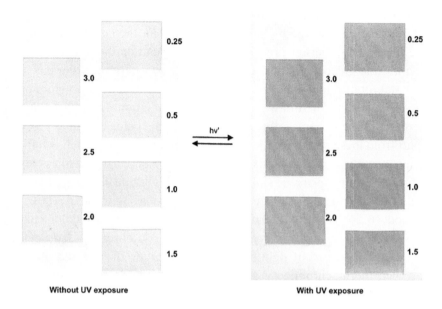

FIGURE 7.13 Screen-printed photochromic fabrics with UV absorbers (numerical values indicate the concentration of photochromic pigments).

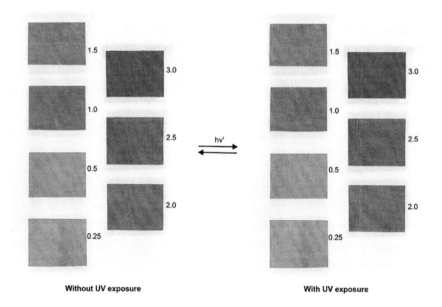

FIGURE 7.14 Screen-printed photochromic fabrics without UV absorbers (numerical values indicate the concentration of photochromic pigments).

However, the decoloration of thermochromic pigments may occur when they are exposed to the higher temperature than the coloration temperature. In general, thermochromic colorants depend on the temperature and their thermal history. In most of the cases, thermochromic colorants show poor stability to light exposure; this is due to the chemical structure of the ink formulation. Therefore, prolong light irradiation can affect the total thermochromic system, followed by the dynamic color changing property. Apart from these reactions, thermal oxidation may occur, which is initiated by the reaction of molecular oxygen from the system. Light fastness results of thermochromic printed samples are shown in Figure 7.15. Light fastness of thermochromic fabrics can be measured according to the AATCC Test method 16:2012. Up to 12 hours of exposure, there is no color change; after 24 hours, a huge difference in the color is observed. It is evident that the temperature of light sources may degrade the colored particles; however, thermochromic dyes have week stability during the continuous irradiation. The results can be graded to 5–6 with different samples.

7.5 WASH FASTNESS

These are accelerated laundering tests designed to evaluate color fastness to laundering of textiles and textile end products, which are expected to withstand frequent laundering. Stainless steel balls of 0.6 mm diameter are required to enhance the abrading effect of the laundering process. To evaluate the color fastness, change in color and staining of adjacent fabrics is assessed. To check staining of other fiber types, multifiber adjacent fabric is used. It is made of yarns of various generic kinds of fibers, each of which forms a strip of specified width providing even thickness of the fabric, i.e., acetate, cotton, nylon, polyester, acrylic, and wool. Assessment of colorfastness and staining

FIGURE 7.15 Light fastness test of printed fabrics with the thermochromic pigment.

is done using grey scales by visually comparing the difference in color or contrast between the untreated and treated specimens with the differences represented by the scale. In color fastness for dry cleaning, the method is the same as color fastness for washing, except that instead of different detergents, dry cleaning solution (percholoroethylene or tetrachloroethylene) is used. The drying condition of the sample depends upon the type of garment and care instruction, i.e., line dry, drip dry, flat dry, tumble dry, etc.

Wash fastness test can be carried out by using any suitable standard like ISO, AATCC, etc. [31, 32]. Using mild laundering conditions, it is not possible to conclude the result by using of traditional gray scale due to the dynamic color changing properties of photochromic textiles or fibers. Therefore, wash fastness can be carried out with respect to the color difference before and after washing. Often, the number of washes depends on the standards. It may be expressed in the ΔE between the background (before UV irradiation) and developed colors after 1 min exposure to UVA irradiation [14]. Anna and Christie [14, 19, 20] studied factors affecting the performance of photochromic materials. They applied two different photochromic dyes (naphthooxazine and naphthopyrans) onto the cotton and polyester fabric by screen printing techniques. To identify the washing performance

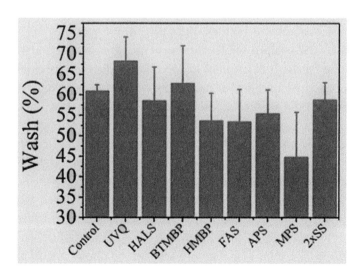

FIGURE 7.16 Durability to washing on photochromic wool fabric. (Reprinted from Marzieh Parhizkar et al. [13]. Reproduced courtesy of Journal of Engineered Fibers and Fabrics, P.O. Box 1288, Cary, North Carolina 27512-1288, USA. Tel: (919) 459-3700 Fax: (919) 459-3701 Internet: www.jeffjournal.org.)

of photochromic printed textiles, they carried out repeated wash fastness tests. The results showed that the degree of photocoloration increased with initial washing and then decreased with subsequent washings for the spiro-oxazine colorants. Similarly, for naphthopyrans, the degree of photocoloration decreased gradually with an increase in the number of washings. This may be due to loosening of polymeric binder matrix around the photochromic colorant molecules with initial washing and further conversion between colorless and colored forms of the spirooxazines, which are more rigid in structure than the naphthopyrans. They concluded that the wash fastness is strongly associated with the binder and not with the colorants.

Parhizkar et al. [13, 40] reported that incorporation of UV stabilizers improved durability to washing, while the other treatments showed reduced durability compared to the control sample (See Figure 7.16). However, silane combinations may influence the photochromic properties before and after washing. The photochromic properties are decreased after specific washing, and there is no significant difference with respect to the silane combinations (see Figure 7.17).

7.6   RUBBING FASTNESS

Generally, in the colored textiles, the unfixed dye particles are mechanically held on the surface while rubbing with skin or other possible contacts with the fabric. It determines how well the fabric will resist the colors against abrasion or rubbing. Therefore, it is necessary to determine rubbing fastness of the colored textile fabrics. The test is also important for the fabrics colored with functional dyes. There are two types of rubbing fastness test: dry and wet rubbing. In dry and wet rubbing, colored fabric may have ruptured with the finger containing white fabric, but in wet rubbing, a wet white fabric is used. During rubbing, the unfixed or loosely dye particles from the surface may get ruptured from the colored fabrics, and it is adsorbed by the white fabric. Rubbing fastness can be determined by the Crock meter; it has 16 mm finger and 9N load is applied, and the test track length is 104 mm. The standard test can be done with 10 cycles per 10 second. In general, the staining results are higher with the concentration of dyes, and it is directionally proportional. In printing, rubbing fastness purely depends on the type of binder used. The polyester/cotton (35/65%) fabric was printed with thermochromic pigments according to the pigment

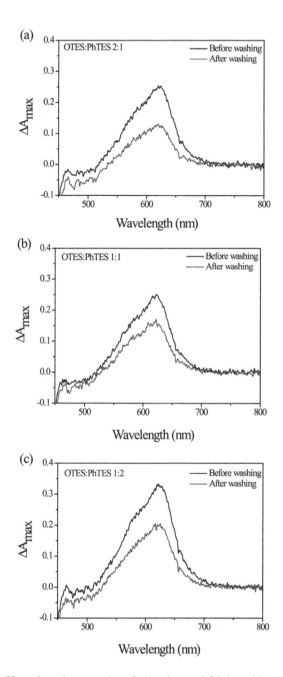

FIGURE 7.17 Photochromic properties of sol-gel-coated fabrics with respect to washing. (OTES – Octyltriethoxysilane; PhTES – Phenyltriethoxysilane) (Applied Reprinted permission from Marzieh [40]).

printing method. After printing, fabrics are dried at 100°C and cured at 120°C for 3 min. Rubbing fastness can be done for this fabric (see Figure 7.18), and results shows level 3 for wet rubbing and level 4 for dry rubbing of grey scale for staining (standard ISO 105-A03). Wet rubbing results show moderate fastness, it is due to the film formation of binder and fabric. Binders are polymers and it is transparent, during the curing it polymerize and it forms a thin film on the surface of fabric by holding the thermochromic pigment. Contrary to dyestuffs, which are able to penetrate into fiber structure, pigments are stick on the surface of supporting structure (fiber, yarn, etc.). However, rubbing fastness may be increased by using the proper or more effective binder (Figure 7.18).

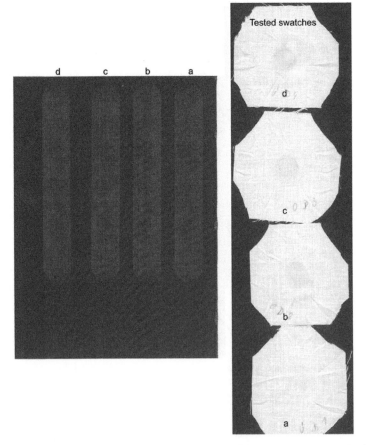

FIGURE 7.18   Rubbing fastness testing of printed fabric with a thermochromic pigment (a, b – dry rubbing; c, d – wet rubbing tests along with their tested swatches).

7.7 SOURCES OF FURTHER INFORMATION

Globally, there are several organizations preparing national and international level standards. These standards provide enormous information on the methods of testing, types of instruments, parameters for color fastness on textile materials, etc. The reader can find more information and advice on textile color fastness by contacting the national standard bodies given in Table 7.2.

TABLE 7.2 List of Selected Organizations

Organization name	Website
American National Standards Institute (ANSI)	www.ansi.org
Association Española de Normalización y Certification (AENOR)	www.aenor.es
Association Française de Normalization (AFNOR)	www.afnor.org
Badan Standard isasi Nasional (BSN)	www.bsn.go.id
British Standards Institution (BSI)	www.bsigroup.com
Bureau of Indian Standards (BIS)	www.bis.org.in
Bureau de Normalization (NBN)	www.nbn.be
Czech Office for Standards, Metrology and Testing (UNMZ)	www.unmz.cz
Deutsches Institutfür Normung (DIN)	www.din.de
Egyptian Organization for Standardization and Quality (EOS)	www.eos.org.eg
European Committee for Standardization (CEN)	www.cen.eu
Ente Nazionale Italiano di Unificazione (UNI)	www.uni.com
Finnish Standards Association SFS (SFS)	www.sfs.fi
Instituto Português da Qualidade (IPQ)	www.ipq.pt
Institute of Standards and Industrial Research of Iran (ISIRI)	www.isiri.org
International Organization for Standardization (ISO)	www.iso.org
Japanese Industrial Standards Committee (JISC)	www.jisc.go.jp
Kenya Bureau of Standards (KEBS)	www.kebs.org
Korean Agency for Technology and Standards (KATS)	www.kats.go.kr
Nederland'sNormalisatie-instituut (NEN)	www.nen.nl
Polish Committee for Standardization (PKN)	www.pkn.pl
SchweizerischeNormen-Vereinigung (SNV)	www.snv.ch
Standardization Administration of China (SAC)	www.sac.gov.cn
South African Bureau of Standards (SABS)	www.sabs.co.za
Swedish Standards Institute (SIS)	www.sis.se
Thai Industrial Standards Institute (TISI)	www.tisi.go.th

TABLE 7.3   List of Selected Professional Bodies

Organization name	Website
American Association of Textile Chemists and Colorists (AATCC) in USA	www.aatcc.org
Association pour la Détermination de la Solidité des Teintures et Impressionssur Textiles (ADSOL) in France	www.adsoletextile.fr
Comisión Española de Solideces (CES) in Spain	www.upc.edu/intexter
Deutsche Echtheits kommission (DEK) in Germany	www.dek-nmp511.de
Society of Dyers and Colorists (SDC) in UK	www.sdc.org.uk

Apart from these standards, several countries having specialized committees or professional bodies that provide the standards for color fastness (Table 7.3).

KEYWORDS

- **fatigue resistance**
- **light fastness**
- **rubbing fastness**
- **wash fastness**

REFERENCES

1. Choudhury, A. K. R., (2015). In: *Principles of Color Appearance Measurement*, Woodhead Publishing Limited, Cambridge, UK, pp. 1–25.
2. Choudhury, A. K. R., (2015). In: *Principles of Color Appearance Measurement*, Woodhead Publishing Limited, Cambridge, UK, pp. 117–173.
3. Choudhury, A. K. R., (2015). In: *Principles of Color Appearance Measurement*, Woodhead Publishing Limited, Cambridge, UK, pp. 26–54.
4. Hurren, C., (2008). In: *Fabric Testing*, Woodhead Publishing Limited, Cambridge, UK, pp. 255–274.
5. Richards, P. R., (2015). In: *Textile Fashion*, Woodhead Publishing Limited, Cambridge, UK, pp. 475–505.
6. Periyasamy, A. P., Dhurai, B., & Thangamani, K., (2011). Salt free dyeing: A new method of dyeing of Lyocell fabrics with reactive dyes, *Autex Res. J., 14*.
7. Bide, M., (2010). In: *Color Measurement*, Woodhead Publishing, Cambridge, UK, pp. 196–217.

8. Gangakhedkar, N. S., (2010). In: *Color Measurement*, Woodhead Publishing, Cambridge, UK, pp. 221–252.

9. Irie, M., & Uchida, K., (1998). Bull. Synthesis and properties of photochromic diarylethenes with heterocyclic aryl groups, *Chem. Soc. Jpn.*, *71*(5), 985.

10. Irie, M., Lifka, T., Uchida, K., Kobatake, S., & Shindo, Y., (1999). Fatigue resistant properties of photochromic dithienylethene: by-product formation, *Chem. Commun.*, *8*, 747.

11. Lucas, N. L., Van Esch, J. M., Kellogg, R. L., & Feringa, B., (1998). A new class of photochromic 1,2-diarylethenes, synthesis and switching properties of bis (3-thienyl) cyclopentenes, *Chem. Commun.*, *21*, 2313.

12. Parhizkar, M., Zhao, Y., & Lin, T., (2014). In: *Handbook of Smart Textiles*, Xiaoming Tao, ed., Springer International Publishing, pp. 155–182.

13. Parhizkar, M., Zhao, Y., Wang, X., & Lin, T., (2014). Photostability and durability properties of photochromic organosilica coating on fabric, *J. Eng. Fiber. Fabr.*, *9*(3), 65.

14. Little, A. F., & Christie, R. M., (2011). Textile applications of photochromic dyes. Part 3: Factors affecting the technical performance of textiles screen-printed with commercial photochromic dyes, *Color. Technol.*, *127*(5), 275.

15. Viková, M., (2011). Photochromic textiles, *PhD Dissertation,* Heriot-Watt University, Edinburgh, pp. 53–147.

16. Billah, S. M. R., Christie, R. M., & Shamey, R., (2012). Direct coloration of textiles with photochromic dyes. Part3: Dyeing of wool with photochromic acid dyes, *Color. Technol.*, *128*(6), 488.

17. Aldib, M., (2015). An investigation of an instrument-based method for assessing color fastness to light of photochromic textiles, *Color. Technol.*, *131*(4), 298.

18. Viková, M., (2004). In: *AIC 2004 Color and Paints*, Proceedings of the Interim Meeting of the International Color Association., José, L. C., & Hanns-Peter, S., ed., Associacao Brasileira da Cor: Porto Alegre, Brazil, pp. 129–133.

19. Little, A. F., & Christie, R. M., (2010). Textile applications of photochromic dyes. Part 2: Factors affecting the technical performance of textiles screen-printed with commercial photochromic dyes, *Color. Technol.*, *126*, 164.

20. Little, A. F., & Christie, R. M., (2010). Textile applications of photochromic dyes. Part 1: Establishment of a methodology for evaluation of photochromic textiles using traditional color measurement instrumentation, *Color. Technol.*, *126*(3), 157.

21. Cheng, T., (2008). Photochromic wool fabric by sol-gel coating, *PhD Dissertation*, Deakin University, pp. 10–153.

22. Chowdhury, M. A., Butola, B., & Joshi, M., (2015). Improving photostability of thermochromic colorants with ultraviolet absorbers and hindered amine light stabilizers, *J. Text. Apparel. Technol. Manag.*, *9*(2), pp. 1–9.

23. Viková, M., Periyasamy, A. P., Vik, M., & Ujhelyiova, A., (2017). Effect of drawing ratio on difference in optical density and mechanical properties of mass colored photochromic polypropylene filaments, *J. Text. Inst.*, *108*(8), 1365.

24. Periyasamy, A. P., Viková, M., & Vik, M., (2016). Optical properties of photochromic pigment incorporated polypropylene filaments: Influence of pigment concentrations and drawing ratios, *Vlakna a Text.*, *23*(3), 171.

25. Viková, M., & Vik, M., (2011). Alternative UV sensors based on color-changeable pigments, *Adv. Chem. Eng. Sci.*, *1*, 224.

26. Periyasamy, A. P., Viková, M., & Vik, M., (2017). A review of photochromism in textiles and its measurement *Text. Prog.*, *49*(2) 53.

27. Aldib, M., (2013). An investigation of the performance of photochromic dyes and their application to polyester and cotton fabrics, PhD Dissertation, Heriot-Watt University, Edinburgh, pp. 5–177.

28. Aldib, M., & Christie, R. M., (2013). Textile applications of photochromic dyes. Part 5: Application of commercial photochromic dyes to polyester fabric by a solvent-based dyeing method, *Color. Technol.*, *129*(2), 131.

29. Saravanan, D., (2007). UV protection textile materials, *Autex Res. J.*, *7*(1), 53.

30. Feczko, T., Varga, O., Kovacs, M., Vidoczy, T., & Voncina, B., (2011). Preparation and characterization of photochromic poly (methyl methacrylate) and ethyl cellulose nanocapsules containing a spirooxazine dye, *J. Photochem. Photobiol. A Chem.*, *222*(1), 293.

31. Understanding and improving the durability of textiles, 1st edn., Patricia, A., & Annis, ed., Woodhead Publishing Limited, 2012.

32. Chakraborty, J. N., (2014). *Fundamentals and Practices in Colouration of Textiles*, Second edition, Woodhead Publishing India, New Delhi, pp. 1–19.

33. Feczko, T., Samu, K., Wenzel, K., Neral, B., & Voncina, B., (2013). Textiles screen-printed with photochromic ethyl cellulose-spirooxazine composite nanoparticles, *Color. Technol. 129*, 18.

34. Zhou, Y., Yan, Y., Du, Y., Chen, J., Hou, X., & Meng, J., (2013). Preparation and application of melamine-formaldehyde photochromic microcapsules, *J. Sensors Actuators B Chem.*, *188*, 502.

35. Vik, M., & Viková, M., (2007). Equipment for monitoring of dynamism of irradiation and decay phase of photochromic substances. *Czech Patent PV 2007/858 PS3546CZ.*

36. Viková, M., Vik, M., & Christie, R. M., (2014). Unique deveice for measurement of photochromic textiles, *Res. J. Text. Appar.*, *18*(1), 6.

37. Viková, M., & Vik, M., (2015). The determination of absorbance and scattering coefficients for photochromic composition with the application of the black and white background method, *Text. Res. J.*, *85*(18), 1961.

38. Viková, M., & Vik, M., (2015). Description of photochromic textile properties in selected color spaces, *Text. Res. J.*, *85*(6), 609.

39. Christie, R. M., (2013). In: *Advances in the Dyeing and Finishing of Technical Textiles*, Gulrajani, M. L., ed., pp. 1–37.

40. Parhizkar, M., (2012). Photochromic fabrics with improved optical performance, PhD Dissertation, Deakin University, pp. 7–199.

INDEX

A

abrasion, 355, 373, 400
abridged fluorescence colorimetry, 237
absorbance, 11, 84, 85, 141, 242, 300, 322, 332, 333, 336–339, 341, 342, 346–349, 378, 379, 382, 387
absorption, 3, 9–11, 19, 36, 38, 39, 41, 43, 44, 49, 53, 55, 71, 77, 78, 79, 84, 87, 94, 109, 128, 142, 162, 243, 244, 284, 288, 301, 302, 305, 309, 311, 332, 349, 350, 357, 363, 381, 386–388, 394, 396
  coefficient, 10, 363
  filters, 10
  pattern, 94
accessory optic system (AOS), 24
accumulative capacity, 39, 40
acetate, 81, 82, 125, 127, 138, 141, 398
acid-base, 56, 57
acids, 64, 66, 71, 75, 355
acrylate paste, 395, 396
acrylic, 52, 60, 133, 137, 146, 346, 347, 395, 398
  fabric, 52
  soft binder, 382
Adams-Nickerson LAB (ANLAB), 184, 185, 188
adjacent fabrics, 373, 398
air-solid surface, 12
alcohol, 28, 64, 146
aliphatic chain fatty acid, 64
alkalis, 71, 355
alkyl side-chains, 58
alkylamine radical, 44
alumina, 224
alumino borosilicate, 37
Ambient temperature, 394
American Association of Textile Chemists and Colorists (AATCC), 398, 399, 404

amide, 64
amines, 392
aminoether (NO-R), 391
amphiphilic molecules, 58
analytes, 86, 87
angular displacement, 22
anilines, 37
anils, 49
anion, 389
anisotropic, 9, 228
anomalous trichromacy, 27
anthocyanidines, 84
anthracene, 41, 47, 48
antioxidants, 119, 377, 391, 392, 393
aperture, 21, 170, 226, 250–253, 261, 264, 271, 273, 292, 294, 298, 300, 355
apparatus, 251, 285
aqueous alkaline surfactant treatment, 52
architecture field, 59
aromatic
  amines, 64
  hydrocarbons, 41
ascorbic acid, 392, 393
Association
  Española de Normalización y Certification (AENOR), 403
  Française de Normalisation (AFNOR), 403
  pour la Détermination de la Solidité des Teintures et Impressionssur Textiles (ADSOL), 404
Atmospheric
  contaminants, 374
  ozone, 385
  triplet oxygen, 387
auxiliaries, 332, 374
AvaLight-DH-CAL, 296
AvaSpec spectrometers, 296
axial illumination, 284, 288

B

background effect, 258
bandpass, 233–235, 242, 243
bandwidth, 57, 63, 64, 235, 243, 244
BEDOT-arylene, 93
benzo, 52, 138
benzochromenes, 52
benzothiophene
    aryl groups, 376
    rings, 55
bidirectional reflectance distribution function (BRDF), 226
Binders, 402
bipolar, 21
bi-polaron, 92, 94, 95
bipolarons, 91
birefringence, 57, 344
bismethylene succinic anhydrides, 54
bispectral measurement, 238
black
    body, 4–7, 95, 157–159, 161, 170
        radiation, 4, 5, 95
        radiator, 4, 6, 7, 157–159, 170
    trap (BT), 333
blend electrospinning, 75
boiling, 55, 56
Boltzmann constant, 5
bonding configuration, 43
borosilicate, 37
Bouguer-Lambert-Beer law, 10
boundary surface, 223, 345
brain, 1, 21, 25–28, 87
Brightness, 16, 17, 198, 205

C

caffeic acid, 392
calcein, 84
calibration, 64, 99, 159, 238, 240, 262, 266, 278–280, 294, 312, 332, 347
    standards, 283
candela, 14, 15, 97
carbon, 28, 42, 51–53, 128, 385
    atoms, 46
    disulfide, 28
cationic agent, 382
cationization, 60

Caucasians, 27
cavity, 5, 254
cellulose acetate (CA), 142
central color defects, 28
chemical
    agents, 71, 79
    bonds, 376
    electrical current, 37
    energy, 37, 387
    reactions, 44, 56
    species, 109
    substances, 35
chemochromic
    colorants, 71
    materials, 71
    system, 87
chemochromism, 36, 100
chiral
    center, 59
    nematic liquid crystals, 59
chlorine radical fragment, 44
cholesteric liquid crystals, 57, 59
cholesterol, 59
Chroma, 16, 17, 169, 183, 189, 190, 193, 195, 198, 199, 201, 202, 204–206, 210, 211, 213, 268, 316, 317, 327, 328, 373
chromaticity, 2, 10, 61, 156, 160, 161, 167, 169, 178, 180–182, 186, 189, 193, 200, 209, 213, 216, 317, 319–321
chromene, 52, 298
chromic, 35, 36, 73, 89, 100, 148, 270, 348, 371, 389, 394
    compounds, 371
    dyestuff, 35
    materials, 35, 100, 111, 148, 315, 389
    phenomena, 35, 36
    polymers, 111
    textile materials, 371
chromism, 35, 36, 68
chromogenic materials, 71
chromophore, 8, 10, 42, 77, 78, 86, 111, 146, 390
ciliary muscles, 21
cinnamic acids, 391
circadian clock circuits, 37
classical dyes, 373
cleavage, 43, 44, 52

coaxial monofilament, 128
cobalt (II) chloride, 84
coefficient, 11, 118, 266, 332, 336–339,
   342, 363
color
   deficiency, 27
   difference formula, 194, 206, 207,
     211, 213, 217, 329
   fastness, 371–375, 398, 399, 403
   matching functions, 171, 173, 174,
     215–217, 231, 232, 245
   play, 57, 63
   quality, 2
   vision, 25, 28, 29, 64, 179, 197, 229
colorant, 18, 19, 35, 50, 52, 110, 116,
   122, 145, 268, 306, 310, 311, 331, 332,
   363, 364, 383, 384, 396, 398, 400
coloration, 37, 38, 42, 51, 55, 73, 75, 89,
   92, 122, 133, 137, 287, 315, 320, 326,
   327, 331, 348, 355, 374, 376, 378, 379,
   389, 398
   decoloration cycles, 376
   process, 374
   technology, 385, 386
Colorfulness (M), 198
colorimeter, 172, 173, 230–232, 280
colorimetric
   characteristics, 292
   humidity, 84
   parameters, 236, 284, 298, 326,
     329–331, 339
   problems, 309
   properties, 357
colorimetry, 1, 2, 10, 31, 156, 162, 165,
   173, 176, 182, 183, 195, 214, 217, 219,
   221, 225, 228, 229, 231, 233, 235, 237,
   268, 280, 326, 338, 365
complementary metal-oxide semiconductor (CMOS), 243, 246
cones, 21–24, 26, 27, 30, 44, 97, 196,
   197, 216
congenital color deficiencies, 28
Conical observation, 227
continuous irradiation, 284, 394, 398
copolymer structure, 96
copper chloride, 84
cornea, 20, 21

Correlated color temperature (CCT), 157,
   160, 168, 297
corroborative measurements, 11
cortex, 24, 28
cortical regions, 24
cotton (CO), 60, 116, 130, 132, 133, 135,
   137, 138, 332, 336, 342–345, 382, 383,
   398–400
crystalline (Cr), 57–59, 91, 359, 360
   phase, 58
   properties, 57
curvature, 21, 324
cuvette, 11, 284, 285, 288

D

dark current, 240, 247, 248
daylight, 10, 29, 97, 116, 156, 158–162,
   164, 165, 169, 195, 196, 207, 237, 329,
   394
decoloration, 142, 376, 378, 398
decomposing, 392
decomposition, 373, 388
degree of polymerization (DP), 124
delocalization, 40
delocalized polaron
   form, 92
   lattice, 92
delustering particle, 363
destabilizing effects, 377
deutan defect, 27
deuteranomaly, 27
deuteranopy, 27, 30
deuterium lamp, 287
Deutsche Echtheits kommission (DEK),
   404
Deutsches Institutfür Normung (DIN),
   210, 403
   DIN99, 209, 210
diameter, 21, 22, 61, 123, 139, 141, 146,
   175, 251–253, 257, 273, 289, 292, 294,
   296, 309, 363, 398
diarylethenes, 54, 55, 375, 376
dibenzanthracene, 41
dichromatic
   deficiency, 29
   reflection, 9
dichromats, 29

dielectric properties, 224
differential scanning calorimetry (DSC),
    125, 126
diffuse
    reflection, 221–223
    transmission, 285
digital printing, 110, 116, 144, 148
dipolar, 42
Directional, 227
    observation, 227
    unit point, 14
disaggregation, 52
discharge, 156, 164–166, 284, 287, 303,
    304, 318, 377, 394
    lamp, 284, 303, 318, 394
discotic mesogens, 58
Disperse Red 1, 75
dispersions, 56
disulfoxides, 37
dominant effect, 13
drawing ratio (DR), 123, 354, 356,
    358–363, 365
    DR1, 75, 76, 356
drift test, 304
dye, 50–55, 57, 59, 64, 65, 67, 68, 70–72,
    75, 76, 78–81, 84, 86, 87, 109–111,
    116, 118–120, 124, 126, 130–133,
    135–138, 141, 144–146, 156, 233, 268,
    331, 355, 363–365, 372–377, 379, 381,
    383–391, 393, 398, 400
    dyed fabric, 133, 138, 379
    dyeing, 52, 122, 131–139, 146, 286,
    374, 375, 382, 384, 386
    encapsulating pores, 377
    fiber bods, 373
    functionalized copolymer, 75
    leaching, 75
    transfer, 375
dyestuff, 18, 35, 131, 138, 221, 224,
    315–317, 324, 327, 375, 402

E

Einstein equation, 3
Electrical
    conductivity, 49
    current, 36

electrochemical redox reaction, 87
electrochromic
    colorants, 87
    glasses, 147
    properties, 96, 115, 135
electrochromism, 36, 87, 91, 92, 100, 135
electrolytes, 83
electromagnetic
    energy, 386
    radiation, 2, 10, 13, 19, 25, 215, 229,
    354
    spectrum, 382
electromagnetism, 6
Electron
    donor, 42
    transfer, 42, 90, 389, 391
        reactions, 42, 90
electronic
    absorption spectra, 38
    gain, 247, 248
    shutter, 303, 394
    transitions, 89, 91, 292
electrons, 6, 44, 98, 129, 236, 240, 386,
    387
electrospinning, 76, 110, 140–143
emission, 3, 5, 36, 47, 95, 97, 99, 100,
    142, 164, 294
energy bands, 36
energy flux, 305
epithelium, 21
equilibrium, 50, 56, 57, 70, 97, 294, 313,
    314, 350, 394
ethylgallate, 64
excimers, 46
excitation
    energy, 166, 387
    purity, 182, 322
excitational energy transfer, 390
exergonic process, 39
exhaust dyeing, 110, 131, 138
exhaustion, 134, 135, 375, 383
exothermal reaction, 40
external
    light, 297, 318
    stimuli, 36, 109–111, 269, 331, 348
extraction, 57, 128

F

fabrics, 67, 115–118, 133, 135, 138, 139, 144, 146, 147, 251, 261, 310, 379, 384, 386, 397, 398, 400–402
fading, 50, 51, 138, 301, 302, 350, 355, 372, 373, 378, 383, 385–388, 391
fatigue, 48, 50, 55, 138, 300, 303, 315, 318, 348, 371, 372, 375–377, 380, 383, 384, 388, 394, 396, 404
  resistance, 138, 315, 371, 372, 375–377, 380, 383, 384, 404
  resistant
    characteristics, 375
    photochromic reactions, 55, 376
femtosecond laser, 285
field of view (FOV), 243
filament, 128, 129, 159, 356, 359, 360, 363, 364
first-order differential equation, 10
flavylium salts, 84
fluorescence, 47, 49, 78, 144, 237, 238, 387
fluorescent, 99, 162, 164, 165, 169, 236–238, 294, 310
fluorine, 142, 292
fovea, 21, 22
fovea centralis, 22
frequency, 2, 3, 6, 14, 27, 351
Fresnel's equations, 12, 255
FTIR spectroscopy, 233
fulgides, 54, 55, 65
fulgimide, 54, 55
Full width at half maximum (FWHM), 234
Functional
  colors, 376
  dyes, 373, 400

G

gallic acid, 392
gamma, 385
gamut, 19, 53, 174
ganglion
  cells, 21
  neurons, 25
gases, 56, 73, 145, 386

Gasochromic materials, 71
gasochromism, 91
gelatin filters, 10
geometric conditions, 9, 225
geometrical optics, 305
geometry, 195, 228, 230, 231, 249, 255, 267, 272, 273, 285, 286, 290, 292, 294, 363
germicide lamp, 287
glass filters, 10, 156, 162
glasser cube-root, 186
graphics software packages, 188
gratings, 230, 232, 233
gray scale, 243, 246, 247, 399
ground state, 39, 40, 42, 46, 47, 77, 78, 236, 386, 387, 391, 394
G-type nerve agents, 81–83
Guide to the Expression of Uncertainty in Measurement (GUM), 276

H

halochromic, 75, 76, 136–138
  materials, 71, 75, 76
  nanofibrous material, 75
  nonwoven fabric, 75
  system, 75
  textiles, 75
Hemispherical, 227
  illumination, 300
  observation, 227
Hering opponent color theory, 26, 27
heterolytic cleavage, 42, 43, 50, 51
hexaphenyl bisimidazole, 44
hierarchical cluster analysis, 87
hindered
  amine light stabilizers (HALS), 119, 120, 127, 332, 347, 355, 378, 381–384, 391–393
  phenols, 392
HOMO (highest occupied molecular orbital), 386
homogeneous, 120, 262, 305, 310
homogenization, 288, 355
homolytic cleavage, 43
horseshoe-shaped, 180, 182
H-type, 81–83

hue, 3, 16, 18, 29–31, 157, 169, 183, 188–195, 198, 199, 201, 204–206, 211, 213, 316, 317, 319, 324, 326–331, 372, 373
human color vision, 19, 31
humid conditions, 84
humidity, 83–85, 87, 268, 269, 357, 377, 394
hunter lab, 185
hydrazones, 37
hydrogen
  atom, 45, 46
  tautomerism, 46
  transfer, 46
hydrolysis, 75, 121
hydrophobic, 52, 64, 113, 142
hydroxyethylcellulose, 84, 85
Hygrochromic materials, 71, 83, 120

I

illuminance, 14, 214, 294, 296
illuminant, 4, 157–160, 162–165, 169, 170, 185, 186, 198, 200, 214, 231, 232, 237, 245, 246, 330, 338, 376
incident
  flux, 294
  light, 8–10, 13, 18, 21, 59, 224, 226, 232, 237, 241, 243, 257
indole, 55, 121, 123, 145, 354, 356, 377, 395
indolinonaphthoxazine, 51
indolinospirodipyrans, 51
infrared, 4, 5, 7, 21, 38, 39
Innumerable factors, 372
inorganic, 12, 37, 55, 56, 87, 89, 121, 136, 146, 164, 354, 387
  compounds, 55, 56, 164
  photochromic, 37
integration, 10, 59, 177, 240, 247, 314
intensities, 9, 10, 12, 22, 23, 38, 168, 287
interference filters, 232, 247, 267, 302, 344
Interferometry, 233
International Commission on Illumination (CIE), 2, 14, 155, 157–170, 172–181, 183, 185, 188, 190, 192–195, 198–201, 207, 208, 210, 214–217, 219, 228–232,

235, 238, 239, 245, 253, 315, 319–321, 329, 333
  CIE 2015 cone-fundamental-based cie colorimetry, 214
  CIE CAM, 195, 329
  CIE CAM02, 200, 217, 315
  CIE CAM97S, 199
  CIE chromacity diagram, 319
  CIE colorimetry, 2
  CIE LAB, 199, 217
  CIE LCH space, 189
  CIE standard observer, 170, 217, 231
  CIE standard-observer, 2
  CIE XYZ, 175, 179, 188, 195, 199, 315
    tristimulus values, 195
  CIE1994, 208
  CIE2000, 210, 213
  CIELAB, 61, 187, 188, 190–192, 194, 195, 199, 203, 206–209, 211, 213, 231, 269, 270, 324, 326, 329
  CIELAB chromaticity plane, 61
  CIELUV, 192–194, 199, 207
International Organization for Standardization (ISO), 162, 318, 332, 357, 372–374, 377, 379, 395, 399, 402, 403
intrinsic, 9, 58, 71, 72, 83, 156, 292
  attenuation, 292
  property, 9
ionic
  characteristic, 43
  species, 36, 42
  structure, 42
ionochromism, 36
irradiance, 14, 157, 162, 164, 226, 242, 294, 296, 297, 299
irradiation, 38, 39, 42, 46–50, 52, 53, 117, 238, 284–286, 288, 289, 292, 300, 312, 315, 317, 322, 325, 327, 328, 330, 354, 355, 377, 379, 389, 391, 394–396, 398, 399
isomer, 42, 48, 49, 112, 113, 389
isomeration, 44
isomerizes, 49
isomers, 42, 45, 48, 53, 55, 111, 375
isothermal lines, 56
isotropic (I), 57–59, 224, 226–228, 342

liquid, 57, 58

J

Japanese Industrial Standards Committee (JISC), 403
Judd polynomial function, 183–185, 239

K

kelvins (K), 5, 7
Kenya Bureau of Standards (KEBS), 403
keto-enol, 56, 57
keto-form, 46
kinetics, 53, 298, 303, 305, 313, 314, 318, 324, 331, 349, 394
  description, 349
  measurement, 283
  model, 313, 314, 348, 350
Korean Agency for Technology and Standards (KATS), 403
Kubelka–Munk
  function, 312, 347, 349, 357, 361, 364, 365
  theoretical model, 310
  theory, 310, 331, 333, 347, 363

L

Laboratory of Color and Appearance Measurement (LCAM), 300, 303, 317, 318, 320, 332, 357, 396
lactim-lactam tautomerism, 56
Lambert's cosine law, 222, 225
Lambert-beer law, 10
lamps, 15, 156, 159, 162, 164, 165, 167, 169, 170, 244, 287, 292, 294, 303, 377
large possible aperture (LAV), 250, 273
lateral geniculate nucleus (LGN), 24
lead, 28, 38, 47, 84, 182, 270, 363
lenses, 51, 230, 285
leuco dye, 57, 59, 64, 65, 68, 70, 118, 120, 384
Lewis acids, 64
light
  emitting diodes (LEDs), 72, 167, 168, 284, 344, 357

fastness, 65, 119, 120, 146, 326, 355, 377, 379, 385, 388, 391, 393–396, 404
  propagation, 13
  ray, 12
  scatterers, 13
  stabilizers, 377
lightness, 16–18, 169, 183–190, 195, 198–202, 206, 211, 213, 274, 324–326, 331, 373
linear
  relation, 188
  sensor, 244, 286
  structures, 12
  transformation, 178, 179, 216
liquid, 56, 145, 267, 285, 332
  crystal, 10, 55, 57–59, 63, 64, 111, 376
    phases, 57, 58
    tunable filters, 10
    type, 57, 111
  crystalline polymers, 58
  filters, 10
locus of,
  monochromatic colors, 180
  spectral colors, 180, 181
low
  aromatic stabilization energies, 55
  density polyethylene, 125, 127
  fabrication cost, 92
  fidelity peripheral vision, 22
  surface energy silane, 377
lower critical solution temperature (LCST), 113
lowest unoccupied molecular orbital (LUMO), 386
luminance, 14, 15, 17, 18, 28, 97, 100, 169, 175, 177, 178, 180, 181, 183, 188, 192, 198, 200, 202, 206, 244
luminescence, 84, 95, 98, 100, 237
luminosity, 26, 27, 173, 178, 244
luminous, 14, 15, 173, 175, 188, 215, 231, 286
  efficacy, 15
  flux, 14, 15
  intensity, 14
Lux, 14
lyotropic liquid crystal phase, 58

M

MacAdam measurement, 183
macromolecular systems, 68
magnetic
    fields, 12
    vectors, 2
mass coloration, 110, 122, 148, 355, 374
matrix, 62, 70, 72, 75, 121, 145, 178, 201, 245, 247, 248, 269, 346, 381, 400
Matte surfaces, 224
Maxwell's
    equations, 12
    laws, 6
Maxwellian mode of observation, 170
Measurement
    accreditation, 279
    procedure, 271
mechanical properties, 354–359
melamine-formaldehyde microcapsules, 146, 395
melt
    blown technology, 138
    flow index (MFI), 356
membrane, 20, 44, 111
memorabilia, 388
mercury
    lamps, 394
    metal halide lamp, 378
merocyanine, 42, 50–52, 389, 390, 394
    colorants, 42
mesogenic properties, 58
mesomorphic, 57
mesophase, 57
Mesopic vision, 22
metal halides, 166
metal-free phthalocyanine, 46
metallo-porphyrines, 84
metameric, 168, 170, 171, 176, 177, 245
metamerism, 171, 176, 197, 217, 232, 243, 246
methyl methacrylate, 146, 395
methylene, 84, 85
micro roughness, 224
microcapsules, 61, 111, 116, 119, 145–147, 395
microencapsulation (ME), 60, 61, 67, 116, 145, 147

Microflash, 292, 293
micron-diameter capsules, 61
microphysical structure, 9
microscopic scale, 12
microstructures, 224
Mie's
    scattering, 13, 309
    theory, 309
modern diode array spectrophotometers, 285
moisture, 36, 83, 99, 120, 124, 386, 394, 395
molecular
    backbone, 55
    orbitals, 36
    rearrangement, 50, 57, 111
    recognition, 87
    species, 56
    structure, 50, 56, 374
monochromatic
    illumination, 284
    light, 2, 4, 10, 178, 180, 215, 234, 238, 285
    radiation, 15
    stimuli, 172, 173
monochromator, 232, 237, 238, 242, 243, 267, 285, 302, 303, 318
monoexponential thermal bleaching kinetics, 394
M-type cone, 27
Multifiber
    adjacent fabric, 398
    irradiation design, 292
Munsell, 29, 169, 183, 185, 239, 326
    colors, 183
    renotation system, 183

N

nanofibers, 75, 76, 109, 141–144, 148
naphthooxazine, 399
naphthopyrans, 52, 53, 133, 301, 350, 383, 399, 400
narrowband, 64, 165, 248
nematic (N), 58, 59
nerve
    cells, 24
    fibers, 21

impulses, 21
neurons, 24
neutral
    colors, 53, 213, 309
    density filter (F), 285
    EDOT, 93
    form, 42
nickel
    chelates, 391
    dithiolato complexes, 389
nitrobenzenes, 42
nitrogen oxides, 386
nitroxyl radicals, 392
N-methylomorpholine (NMMO), 123,
    124
nonlinear, 112, 183, 188, 200, 209, 365
nonlinearity, 187, 188
non-photochromic products, 375
nonradiative transitions, 40
nonspectral
    colors, 16
    purples, 180
nonspherical particles, 13
nonuniform temperature, 56
nonwoven, 75, 139, 140, 379, 380
    fabric, 138, 139
Norbornadiene, 40, 41
nucleophilic attack, 389
nylon, 52, 133, 134, 394, 398

O

odorants, 87
olfactory
    bulb, 87
    receptors, 87
        epithelia cells, 87
o-nitrobenzyl derivatives, 37
opaque, 9, 84, 135, 223, 249, 253, 305,
    309, 343
open-ring isomers, 55
ophthalmic plastic lenses, 53
opponent
    cells, 26
    chromatic scales, 184
optic
    chiasm, 24
    disc, 24

nerve, 21, 24, 28
optical
    absorption, 44
    density, 337, 347, 354, 357, 358,
        361–365
    fibers, 225, 290, 292
    filter system, 377, 379
    inhomogeneity, 310
    optical effect, 60
    power, 14, 248
    properties of solids, 31
    transitions, 94
    waveguide, 225
organic, 12, 25, 37, 38, 46, 54–57, 83, 84,
    87, 88, 121, 138, 147, 233, 354, 375,
    376, 392
    compounds, 38, 46, 56, 57, 88
    light emitting diodes, 92
    materials, 376
    neurovisual deficits, 25
    photochromic
        materials, 37
        systems, 375
    photo-conductors, 376
    pigments, 12
    polymers, 83
oxazine, 51, 121, 123, 146, 300, 301, 319,
    350, 354, 356, 377, 385, 386, 395
    ring, 51
oxazones, 37
oxygen, 42, 52, 53, 377, 388, 389, 393,
    394, 398

P

paper chemical agent detectors, 80, 81, 83
PEDOT, 94, 115, 136
penthacene, 41
percholoroethylene, 399
percolation, 97
perimidine, 49
permeability, 44, 99, 111
peroxides, 392
peroxy radicals, 392
phenols, 64, 391
Phenyltriethoxysilane, 401
phosphorescence, 42, 98, 99, 387
phosphorescent, 98, 99, 237

photo
  bleaching, 301, 350, 371, 375, 387
  catalytic activity, 391
  chemical, 40, 55, 355, 375, 377, 387,
    389–391, 394
    reaction, 40, 355, 377, 390, 394
    stabilities, 55
  chromic organo silica-coated fabric,
    378
  excited molecule, 46
  merocyanine, 50
  sensitization, 389
  switches, 48
  tautomerism, 46
  tendering, 391
PHOTOCHROM, 317, 320, 333, 347,
  348
PHOTOCHROM3, 303, 304, 308, 357
Photochromic
  behavior, 55, 116, 348, 381, 396
photochromic
  colorant, 37, 400, 145, 400
  compounds, 38, 49, 50, 54, 110, 111,
    116, 145, 288, 348, 354, 375, 376,
    389, 395
  dyes, 51, 52, 111, 116, 121, 132–134,
    138, 142, 144, 313, 314, 326, 365,
    375, 377, 379, 381, 383, 385, 386,
    387, 389, 391, 395, 399
  effect, 42, 50, 52, 139, 140, 287, 290,
    294, 297, 302, 326, 331, 377, 379,
    381, 396
  filament, 365
  fulgides, 54
  group, 42
  inks, 50, 52, 352
  materials, 37, 41, 47–49, 51, 53, 54,
    142, 145, 286, 288, 318, 355, 371,
    375–377, 379, 388, 394, 396, 399
  mechanism, 50, 54
  media, 41
  microcapsules, 146, 395
  organosilica-coated fabric, 377, 378
  pigment, 116, 117, 122, 123, 139, 312,
    327, 332, 337, 347, 354, 356, 359,
    377, 381, 396
  process, 41, 42

properties, 55, 111, 331, 354, 377,
  396, 400
  reaction, 43, 45–47, 53, 55, 375, 376
  response, 314, 361
  sensors, 315, 318, 329, 348, 354
  species, 41
  spiropyrans, 50
  studies, 38
  systems, 302, 303, 318, 348
  tautomerism, 46
photochromism, 36, 38, 42, 44, 45, 52,
  53, 100, 294, 303, 319, 324, 327, 354,
  355, 365
photocoloration, 138, 379, 381, 383, 384,
  400
photocyclization, 50, 51, 54
photodecomposition, 388, 389
photodegradation, 51, 371, 375, 379, 384,
  389, 395
photodimerization, 46, 47
photofading, 387
photogenerated species, 53
photoirradiation, 49, 109, 111, 112
photomultiplier tube (PMT), 285
photons, 3, 8, 44, 224, 305, 385, 386, 388
photooxidation, 371, 375, 389, 390, 392
photopic, 14, 15, 22, 99, 173, 178, 215
  efficiency function, 14, 15
photoreaction, 47, 55
Photosensitization, 389–391
photosensitizer, 387, 390
photostability, 50, 116, 372, 376–379,
  381–384, 389, 391
photostationary
  equilibrium, 394
photostationary state (PS), 112, 289
Phototropism, 37, 38
pH-sensitive,
  dye, 64
  nanofibrous nonwovens, 75
phthalocyanine, 47
p-hydroxybenzoic acid methyl ester, 64
physicochemical interactions, 386
planar
  molecular structure, 69
  structures, 53

Planck's
  constant, 3
  law, 6, 157, 158, 160
Planckian radiator, 162–164, 169
planetary atmospheres, 309
polarizability, 49
polarized filters, 290
polaronic charge carrier, 94
poly(3-alkoxythiophenes), 68
poly(3-alkylthiophenes), 68
poly(3,4-(ethylene-dioxy)thiophene), 92, 94
poly(4-methacryl aminoazobenzene), 46
poly(HEA-co-DR1-A), 75
poly(methyl acrylate) (PMA), 113, 114
poly(methyl meth-acrylate) (PMMA), 113, 141–143
poly[3-oligo(oxyethylene)-4-methylthio-phene], 69
polyamide, 116, 133
  polyamide-6 (PA6), 75, 76
polyaniline, 92, 135
polycatenar mesogens, 58
polychromatic, 231, 233, 236, 237, 284
polydimethylsiloxane (PDMS), 135
polyemeraldine, 92
Polyester (PET), 52, 72, 116, 132, 133, 138, 144, 145, 318, 379, 383, 385, 386, 395, 398–400
polymer, 51, 56, 60, 62, 68, 70, 91–95, 111–116, 124–128, 135–142, 145, 146, 286, 355, 363, 377, 389, 395, 402
  dispersion (PD), 60, 62, 63
polyphenol, 393
polypropylene (PP), 81, 122–125, 127, 128, 139, 354–356
polypyrrole, 92
polystyrene (PS), 142
Polythiophene side chain melting, 70
Porous
  ceramics, 83
  organosilica sol-gel, 70
Positive
  photochromism, 41
  solvatochromism, 77
Praying-Mantis
  accessory, 289

optical accessory, 290
printing, 60, 67, 68, 86, 115–120, 144, 146, 249, 318, 332, 336, 338, 374, 395, 396, 400, 402
prisms, 230, 232, 233
protanopy, 28, 30
pseudo-isochromatic plates, 29
P-type, 47–49, 53–55, 396
  photochromic, 53
  photochromism, 53
purity, 16, 17, 164, 181, 182, 255, 317, 320, 322–324, 331
Purkinje effect, 97
pyran, 42, 43, 50–52, 141, 289, 298, 319, 385, 386
  ring, 42, 52
pyridinium N-phenoxide betaine dye, 70

Q

quadrature, 204, 279
Quadricyclane, 40, 41
quality control, 207, 214, 255, 348, 372
quantum
  dots, 387
  number, 6
  yield, 42, 376
quasi-parallel rays, 227

R

Radiance factor, 227
Radiant
  energy, 2, 242
  flux, 14, 15, 157
  power, 15
radicals, 392
radio waves, 385
radiometry, 13, 229
raw TLC, 60, 62
Rayleigh-Jeans law, 5, 6
Rayleigh scattering, 13, 292, 309
receptors, 21, 23, 24, 26, 27, 196, 215
rectilinear path, 305
red-green
  deficiency, 27
  discrimination, 27
redox process, 387

reflected
light, 8, 9, 14, 17, 61, 62, 235, 237, 240
radiance terminology, 227
reflection, 11, 12, 36, 221–226, 228, 239–241, 248, 249, 255, 257, 284, 285, 305, 342, 357, 363, 365
factor, 227
spectroscopy, 11
refraction, 12, 36, 267, 305, 342
refractive
index, 9, 12, 13, 21, 49, 255, 309, 310, 342–346
indices, 21, 223, 344, 345
relative spectral power distribution, 159, 161
reproducibility, 84, 232, 290, 298, 304, 375
resin, 13, 111, 146
resistance, 48, 61, 83, 129, 138, 142, 355, 371, 372, 385, 386, 389, 395
resolution, 64, 86, 233, 234, 242, 245, 285, 329, 338, 344
Resonance, 233
energy, 40
retina, 21, 24–26, 28, 97, 175, 215
retino
collicular, 24, 25
cortical visual pathway, 24
hypothalamic, 24, 25
occipital, 24
pretectal, 24, 25
tectal visual pathway, 24, 25
Retro reflective surfaces, 225
Reversible
coloration process, 38, 73
interconversion, 45
transformation, 109, 111
reversion, 41, 49, 298, 300, 303, 314, 318, 348, 350
rhodopsin, 22, 44, 45
rods, 21–23, 26, 27, 58, 97, 100, 197, 196
rubbing, 331, 400, 402, 404

S

Saturation, 16–8, 23, 180, 188, 189, 193, 198, 201, 206, 215, 372, 373

scattering, 9, 11–13, 117, 118, 182, 285, 300, 306–311, 317, 332, 336–338, 341, 347, 363
coefficients, 309
particles, 12
Schweizerische-Normen-Vereinigung (SNV), 403
screen printing, 110, 116, 118, 119, 144, 148, 318, 332, 395, 396, 399
sensitivity, 21–24, 27–29, 71, 75, 80, 97, 100, 196, 197, 215, 216, 233, 240, 244, 285, 287, 296, 301–303, 318, 329, 331, 348, 354, 359, 363, 395
sensors, 37, 50, 55, 71–73, 75, 76, 83, 84, 86, 87, 92, 143, 144, 229, 230, 243, 244, 248, 315, 326, 329–331, 348
sensory
cells, 22
elements, 21
shade intensity, 312, 315, 326, 363
Short-term drift test, 308
short-wavelength cutoff filters, 237
silane, 121, 400
silica
coating layer, 377
fibers, 292
nanoparticles, 130, 131
Silver halide, 37
simulator, 162, 237, 294
singlet oxygen, 387, 389
smart textiles, 109, 147, 354
smectic A (SmA), 58, 59
smectic C (SmC), 58, 59
sodium ions, 44
solar
cells, 92
light, 37
protection glasses, 53
solarization, 292, 294
sol-gel, 56, 70, 110, 111, 121, 122, 148, 378, 381, 401
Solvatochromic materials, 71, 77, 79
solvatochromism, 77–79
Spectraflash, 292, 293, 355
Spectral
content, 15, 237
data, 255, 284, 315

linearity, 9
luminous function, 14
parameters, 284, 318, 332, 348
power, 2, 19, 156, 159, 161–169, 177,
  214, 236, 287, 294, 295, 329, 377, 394
  distribution (SPD), 2, 19, 156,
    159–163, 165, 167–169, 177, 214,
    236, 287, 294, 329, 377
  range, 162, 215, 267, 286, 296
reflectance, 9, 214, 236, 240, 241, 255,
  261, 270, 333
regions, 18
resolution, 233
transmittance, 9, 10, 245, 247
spectrometer, 232, 233, 235, 242, 267,
  289, 294
spectrophotometric
  description, 312
  measurements, 290
  system, 64
spectrophotometry, 10, 332, 339, 342,
  347, 365
spectroradiometer, 230, 242, 243, 246,
  280, 294
spectroscopic techniques, 233
spectrum, 2, 4, 7, 18, 19, 21, 36, 38, 41,
  49, 53, 55, 57, 77, 156, 157, 162, 168,
  172, 180, 215, 225, 226, 229, 230,
  233, 234, 237, 238, 242–246, 248, 267,
  285–290, 302, 305, 306, 336, 345, 347,
  385
specular, 9, 13, 221–224, 228, 230, 292,
  355
spike potentials, 26, 27
spiro
  compounds, 37
  cyclohexadienones, 49
  dihydroindolizines, 49
  indolinonaphthoxazines, 52, 133
    ring, 50
  lactone, 64
  linkage, 51
  manner, 50
  naphthooxazine, 133, 383
  oxazines, 134, 348, 389, 390, 394, 400
  pyran, 50–52, 65, 113–115, 142, 389,
    390, 394, 395

stabilizers, 119, 376, 381, 391, 400
stereoisomers, 56, 57
stoichiometric parameter, 91
stray light, 233, 242
stretched-exponential decay law, 97
substrate, 72, 87, 99, 118, 120, 136, 144,
  253, 258, 294, 301, 318, 326, 327, 332,
  335–338, 345, 347, 352, 377, 391, 394,
  395
superficial layer, 379
superior colliculus, 24
superoxide radical, 387–389
supramolecular system, 68
surface wettability, 111, 377
Swedish Standards Institute (SIS), 403
Synthetic
  chiral molecules, 59
  fibers, 133, 387

T

tautomeric sites, 47
Tautomerism, 45–47
telecom multimode, 296
telescopic table, 296
tele-spectroradiometer (TSR), 242
tenacity, 125, 348, 357–361, 363
tetrachloroethylene, 399
textile
  applications, 50, 56, 98, 146
  fabric, 12
  materials, 109, 110, 119, 133, 301,
    331, 337, 354, 355, 372, 374, 392,
    394, 403
  sensors, 35, 37, 312, 354
thalamus, 24
thermal
  bleaching, 42, 50, 348, 394, 395
  energy, 44
  equilibrium, 95
  exposure, 69
  performance, 60
  reactions, 375
  sensitivity, 303, 318, 348, 350, 354,
    365, 396
  stability, 48, 55, 57, 142, 375, 384
thermochromic, 55–59, 62, 64–68, 70,
  109–113, 118, 120, 123–131, 142–144,

147, 270, 381, 382, 384, 394, 396, 398, 400, 402
  colorants, 55, 398
  dyes, 57, 111, 130
  liquid crystal (TLC), 57–63, 67
  paint, 147
  phenomena, 56
  pigments, 381
  polymers, 68
  synthetic filaments, 125
  system, 398
thermochromism, 36, 38, 55–57, 64, 68, 69, 100, 111, 127, 270, 271
thermocouple, 285
thermodynamics, 40, 394
thermographic recording materials, 59
thermography, 59, 71, 111
thermostats, 283
thiazine, 84
thiophene, 55, 94
tobacco, 28
total reflected intensity, 223
toxic amblyopia, 28
toxicity, 57
transducer cells, 21
transitional metal oxides, 91
translucent, 8, 9, 223, 224, 253, 258, 271, 285, 306, 309, 343, 347, 349
transmission, 8, 11, 223, 225, 226, 228, 249, 261, 285–288, 292, 309, 349
transmittance, 11, 12, 255, 284, 285, 333, 335, 338, 342–345, 347
transparent, 9–11, 13, 20, 62, 68, 135, 221, 223, 224, 284, 285, 305, 402
triplet
  excitations, 42
  state, 41, 42, 390, 391
  triplet
    absorption, 42
    excitation, 41
    photochromism, 41
  type, 42
tristimulus values, 2, 173, 177, 178, 180, 188, 193, 195, 199–201, 214, 230, 232, 246
tritan color vision deficiencies, 28
tritanomaly, 28

tritanopy, 28, 30
T-type, 48–50, 54
tungsten halogen lamps, 162

U

Ultraviolet (UV),
  absorbers, 119, 120, 127, 331, 381–384, 391, 392, 395–397
  B solar radiation, 294
  blocking capabilities, 388
  content, 159, 394
  exposure, 140, 357, 379
  index, 37, 315
  irradiation, 37, 42, 46, 50, 51, 112, 116, 141, 146, 288, 292, 294, 312, 315, 318, 330, 331, 348, 377–379, 384, 389, 391–393, 396, 399
  lamp, 287, 302
  light, 35, 37, 38, 48, 50, 140, 164, 288, 292, 348, 350, 355, 377, 381, 383, 387, 389, 391, 394
  photons, 387
  protection, 387
  range, 37, 162, 242
  rays, 385
  region, 4, 162, 288, 336
  sensor, 37, 112, 331, 365, 395
  spectrum, 391
  stabilizer, 381, 391
  stimulus, 395
uniform
  color spaces, 182
  distribution, 278, 279

V

velocity, 138, 139, 297
visible
  light, 19, 48, 49, 95, 142, 387
  region, 37, 38, 53, 162, 290, 336
  spectrum, 2, 4, 53, 59, 157, 207, 230, 232, 306, 333, 347
visual
  assessment, 157, 357, 395
  cortex, 24
  field, 24, 28
  pathways, 24, 25

sensation, 2, 15
signal, 24
volatile organic compounds, 87
V-type, 81, 82, 83
Vulcanization, 374

W

wash fastness, 372, 379, 398–400, 404
wavelength, 2–4, 7, 12–14, 16, 17, 23,
24, 38, 39, 41, 49, 51, 59, 60, 73, 84,
99, 100, 156–159, 161, 162, 164, 167,
169, 172, 173, 180–182, 201, 215, 224,
226, 230, 232–243, 266, 267, 285, 292,
302, 305–309, 311, 319, 320, 336–338,
344–346, 357, 377, 379, 385, 386, 396
band, 157, 230
ratio, 13
wet rubbing, 400, 402
white standard (WS), 240, 241, 247, 333
Wien's displacement law, 7
wool, 120, 121, 133, 378, 379, 381, 398,
399

Wright-Guild observer, 178

X

xenon, 162, 164, 287, 292, 301–304, 318,
377, 394
arc lamp, 377, 379, 394
discharge lamp, 164, 287, 303, 318
Xenotest, 119, 120, 377, 379, 382
X-rays, 95

Y

Y tristimulus, 178
Yellowness-blueness, 198
Young-Helmholtz theory, 26, 27

Z

zinc chloride, 84
zwitterionic, 42, 78

Printed and bound by CPI Group (UK) Ltd, Croydon, CR0 4YY

23/10/2024

01777704-0011